C0-AJZ-313

M. Pehnt, M. Cames, C. Fischer, B. Praetorius,
L. Schneider, K. Schumacher, J.-P. Voß

Micro Cogeneration

Towards Decentralized Energy Systems

Martin Pehnt, Martin Cames,
Corinna Fischer, Barbara Praetorius,
Lambert Schneider, Katja Schumacher,
Jan-Peter Voß

Micro Cogeneration

Towards Decentralized Energy Systems

With Contributions by Michael Colijn, Jeremy Harrison,
Yasushi Santo, Jon Slowe, and Sylvia Westermann

With 59 Figures

Dr. Martin Pehnt
IFEU Institut für Energie- und Umweltforschung GmbH
Wilckensstr. 3

69120 Heidelberg

Dr. Corinna Fischer
Freie Universität Berlin
Environmental Policy research Centre (FFU)
Ihnestr. 22

14195 Berlin

Dr. Barbara Praetorius
Katja Schumacher
Deutsches Institut für Wirtschaftsforschung
Königin-Luise-Str. 5

14195 Berlin

Lambert Schneider
Martin Cames
Jan-Peter Voß
Öko-Institut e.V.
Novalisstr. 10

10115 Berlin

ISBN 10 3-540-25582-6 **Springer Berlin Heidelberg New York**
ISBN 13 978-3-540-25582-6 **Springer Berlin Heidelberg New York**

Library of Congress Control Number: 2005931799

This work is subject to copyright. All rights are reserved, whether the whole or part of the material is concerned, specifically the rights of translation, reprinting, reuse of illustrations, recitation, broadcasting, reproduction on microfilm or in any other way, and storage in data banks. Duplication of this publication or parts thereof is permitted only under the provisions of the German Copyright Law of September 9, 1965, in its current version, and permission for use must always be obtained from Springer-Verlag. Violations are liable to prosecution under the German Copyright Law.

Springer is a part of Springer Science+Business Media
springeronline.com
© Springer-Verlag Berlin Heidelberg 2006
Printed in The Netherlands

The use of general descriptive names, registered names, trademarks, etc. in this publication does not imply, even in the absence of a specific statement, that such names are exempt from the relevant protective laws and regulations and therefore free for general use.

Cover design: E. Kirchner, Heidelberg
Production: A. Oelschläger
Typesetting: Camera-ready by the Authors

Printed on acid-free paper 30/2132/AO 543210

Contents

Introduction

The electricity systems of many countries are currently undergoing a process of transformation. Market liberalization has induced major mergers and acquisitions in the electricity sector, but has also forced companies to seek out new business areas. Environmental regulations, like the Kyoto process and the European Emissions Trading Scheme, are exposing the sector to external pressure. New technologies – such as renewable energy, combined heat and power (CHP), or "clean coal" technologies – are emerging. Recent worldwide experiences with blackouts have once more put security of supply on the agenda. In Germany, the nuclear phase-out and decommissioning of outdated coal plants will lead to a need for replacement of more than one third of the current generation capacity by 2020.

The need for replacement is an extremely important driving force for the current transformation, forcing conventional and new technologies to compete for a role in the future energy supply. The overall transformation of electricity systems is neither driven nor shaped by technical or societal modifications alone, but rather by a rich diversity of processes in the realms of technology, politics, society and economy.

Achieving sustainable development in the energy sector entails specific qualities characterizing the changes which need to be undertaken. Climate change and limited fossil resources call for a reduction of non-renewable primary energy input and greenhouse gas (GHG) emissions by 50 to 80 % by 2050 (Enquête 2002). The resulting structural transformation will require innovation in many different realms, including the development of new technologies, new forms of corporate organization, new user routines, new institutional arrangements for governance, new conceptions regarding how problems should be understood, and new means of measuring electricity system performance.

One possible developmental path is decentralization of the electricity system. Distributed power generation in small, decentralized units is expected to help in reducing emissions and saving grid capacity, while also providing opportunities for renewable energy. It could thus form a constituent part of a more sustainable future. Broad implementation of distributed generation, however, would imply thoroughgoing structural change as well as a surge in innovation.

Recently, decentralization and developing means for autonomous or individual energy supply also appear to be *en vogue*. A trend towards smaller technical systems has, since the 1970s, been advocated by many writers as a return to life on a human scale. The economist and writer E. F. Schumacher, for instance, wrote that technological development should be given "a direction that shall lead it back to the real needs of man, and that also means: *to the actual size of man*. Man is small, and, therefore, small is beautiful. To go for giantism is to go for self-destruction" (Schumacher 1973). This vision of decentralized, and often autonomous, technological systems has been often replicated and has also been applied to energy systems. For instance, in its 2002 memorandum the Club of Rome demanded that, whenever possible, a decentralized energy supply should be established (Club of Rome 2002). Visionary thinkers like Jeremy Rifkin have stated that, in the new age of decentralized energy production, everybody could, in principle, generate and consume his own energy (Rifkin 2002).

The present book focuses on one such element of distributed generation options which could play a role within the development of sustainable energy systems for the future, actually a micro-aspect within the overall transformations that are already going on and will be going on over the coming years. This is the combined production of electricity and heat in small units that are directly embedded in the buildings where the heat and electricity are to be used. This configuration is referred to as *micro cogeneration*.

Compared to the currently dominant pattern which combines electricity production in central plants, supplying 100,000 buildings at once, with separate on-site heating systems, micro cogeneration would make a fundamental difference in electricity systems if it actually became widely implemented. It not only integrates technological as well as cultural and institutional components, but also entails the potential for reducing the ecological impacts of electricity production. However, as many chapters in this book will seek to illustrate, the current context for micro cogeneration in many countries is not a very bright one.

Micro cogeneration thus offers a rewarding opportunity for studying the conditions facing radical innovations in potentially unfavorable regime contexts. At the same time, when market and economic factors become favorable, micro cogeneration may have the potential for reaching a considerable market size, thereby helping to advance other downstream or system innovations, such as the "virtual power plant" or new household energy-management systems, combined with altered consumer awareness.

More recently, the interest in micro cogeneration has also been fuelled by an enthusiastic interest in fuel cells, which could, amongst other applications, also be used in individual buildings as CHP devices. But micro cogeneration goes beyond fuel cell technology and involves various other conversion technologies.

This book aims at assessing the potential contribution of micro cogeneration towards a sustainable transformation of electricity systems. We examine the role it should or may play within a sustainable energy strategy, assess related implementation conditions and discuss possibilities to improve the context for introducing micro cogeneration on a larger scale. The issue demands a multifaceted answer that considers the various factors involved in real world applications of micro cogeneration. This book, therefore, combines the perspectives of engineering and life cycle studies, economics, sociology, applied psychology, and political science. Not only various academic disciplines, but also different national perspectives need to be taken into account, because the success chances of micro cogeneration largely depend on both the "hardware" and "software" of a country: on the one hand, the existing infrastructural context which micro cogeneration has to fit into (e.g., building stock, dominant fuels, district heating infrastructure) and, on the other hand, the political and economic framework, including support schemes, innovation policy, energy prices, and micro cogeneration legislation. Therefore, authors from several countries where significant micro cogeneration-related developments are now taking place were invited to contribute to this book.

Structure of the book. The core of this book is based on research carried out within a project called "Transformation and Innovation in Power Systems" (www.tips-project.de), funded by the German Ministry for Education and Research (*Bundesministerium für Bildung und Forschung, BMBF*), under the auspices of the Socio-Ecological Research Framework (*Sozial-ökologische Forschung*), launched in 2001.

Chap. 1 defines the book's terrain: what is micro cogeneration? What are adequate conversion technologies? Which further technological components are required for establishing a functioning micro cogeneration system?

Micro cogeneration is part of a larger transformation process. In **Chap. 2**, we investigate the relevance of this process for the future performance of micro cogeneration and, vice versa, the role different kinds of energy scenarios attribute to micro cogeneration. In **Chap. 3**, we discuss important parameters which determine market perspectives for micro cogeneration, and try to assess the potential for it in the German market under different conditions.

How far this market potential is actually exploited depends primarily on the economic performance of micro cogeneration (**Chap. 4**). Here, we calculate its economic viability from different perspectives. Another prerequisite for successful market development is the environmental superiority of this innovation. Therefore, an environmental life cycle perspective on micro cogeneration is presented in **Chap. 5**.

Because micro cogeneration can be considered to be one of the most extreme examples of decentralization, the consumer gains in importance. Under certain circumstances, the boundary between consumer and producer is even blurred or eliminated. In **Chap. 6**, the users, particularly early pioneers who are necessary for spreading information about and realizing the systems, are the object of detailed description. Not only the consumers, but also energy companies are major actors involved in developing, or retarding, micro cogeneration development. **Chap. 7** looks at the setting of energy markets and entrepreneurial actors relevant to implementing micro cogeneration in Germany; it inquires about their interests, motivations, and strategies to foster, or to hold back, this innovative technology. The institutional framework of micro cogeneration in Germany involves not only directly CHP-related legislation, but also general energy legislation, innovation policy, and funding for research and development. **Chap. 8** tries to precisely locate the field on which micro cogeneration has to prove itself.

Successful development of micro cogeneration requires compatibility with existing and future energy systems. This concerns not only security of supply and availability of fuels, but also technical compatibility with electricity networks. This embedding of micro cogeneration is investigated in **Chap. 9**.

Whereas the TIPS study funded by the German ministry focused on micro cogeneration from an analytic point of view, the experiences of a micro cogeneration practitioner are described in **Chap. 10**, dealing with the various types of micro cogeneration operators and the range of the unforeseen problems occurring in the everyday operation of micro cogeneration.

Micro cogeneration is being developed, and in fact has been more successfully implemented, in other regions worldwide. We therefore invited micro cogeneration experts from the four most important micro cogeneration countries outside Germany – Great Britain, the Netherlands, Japan, and the United States of America – to report on micro cogeneration hard- and software, and the respective peculiarities in these countries (**Chap. 11 to 14**). Following our conclusions in **Chap. 15**, the reader is referred to a substantial body of literature and World Wide Web resources (**Chap. 16**).

The authors would like to thank the guest authors, namely, Sylvia Westermann, Michael Colijn, Jeremy Harrison, Yasushi Santo, and Jon Slowe, for their contributions. We also acknowledge valuable contributions from Raphael Sauter (FU Berlin); Katherina Grashof, Sabine Poetzsch, and Jens Gröger (Öko-Institut); Regina Schmidt and Bernd Franke (IFEU); as well as Lars Winkelmann (Berliner Energieagentur GmbH).

The authors gratefully acknowledge the funding of the TIPS research project by the German Ministry for Education and Research.

Martin Pehnt, Martin Cames, Corinna Fischer, Barbara Praetorius, Lambert Schneider, Katja Schumacher, Jan-Peter Voß

Heidelberg, Berlin, July 2005

Supported by the German
Ministry for Education and Research,

SÖF : Sozial-
ökologische
Forschung

1 Micro Cogeneration Technology

Martin Pehnt

1.1 Defining Micro Cogeneration

The principle of cogeneration has long been known. As early as the first decade of the 20th century, a number of cogeneration units were already supplying heat and electricity to houses and companies. Cogeneration, or combined heat and power production (CHP), is "the process of producing both electricity and usable thermal energy (heat and/or cooling) at high efficiency and near the point of use" (WADE 2003). It thus incorporates three defining elements: 1) the simultaneous production of electricity and heat; 2) a performance criterion of high total efficiency; and 3) a locational criterion concerning the proximity of the energy conversion unit to a customer.

While the discussion on micro cogeneration, or micro CHP, has only recently gained momentum, the technological roots of micro cogeneration go back to the early development of steam and Stirling engines in the 18th and 19th century, respectively. Today, several technologies exist that are capable of providing cogeneration services, such as reciprocating engines, gas turbines, Stirling engines, and fuel cells. But, in principle, the exhaust heat from any thermal power plant, such as gas combined-cycle power plants or coal power plants, can be used for cogeneration applications.

Advances in the technology, as well as a general trend towards smaller unit sizes of power plants, have led to an increased interest in small CHP units, with the hope of ultimately developing units that can provide electricity and heat for individual buildings. This is what we call micro cogeneration which we define as the

> simultaneous generation of heat, or cooling, energy and power in an individual building, based on small energy conversion units below 15 kW_{el}.

Whereas the heat produced is used for space and water heating inside the building, electricity produced is used within the building or fed into the public grid.

Opposed to, for example, the EU cogeneration directive, according to which micro cogeneration is defined as "a cogeneration unit with a maximum capacity below 50 kW_{el}" (EU 2004), we restrict the definition of micro cogeneration to systems below 15 kW_{el} for the following reasons. Firstly, these systems are clearly systems for use in single-family dwellings, apartment houses, small business enterprises, hotels, etc., which can be distinguished from systems supplying heat to a district or neighborhood (i.e. district heating systems). Secondly, systems in this small power regime substantially differ from larger ones with respect to electricity distribution, ownership models, restructuring of supply relationships, and consumer behavior. Also, compared to conventional CHP, based on district heating, no additional heat distribution grid is required. Systems below 15 kW_{el} can be directly connected to the three-phase grid. Moreover, the barriers all CHP systems have to face are more pronounced in the case of such small systems.

As shown in Sect. 2.2, taking an integrated perspective, the innovation of micro cogeneration combines a set of novelties in a number of dimensions. At first sight, though, micro cogeneration often appears as a purely technical innovation. The obvious aspect of novelty is the introduction of a new machine, with new functionalities and different connections to other technical components of the electricity and heat supply system. In this chapter, we will look at the different technical components of a micro cogeneration plant as well as the technologies available for micro cogeneration purposes.

The technological core of micro cogeneration is an energy conversion unit that allows the simultaneous production of electricity and heat in very small units. In addition to this core, further technology components are

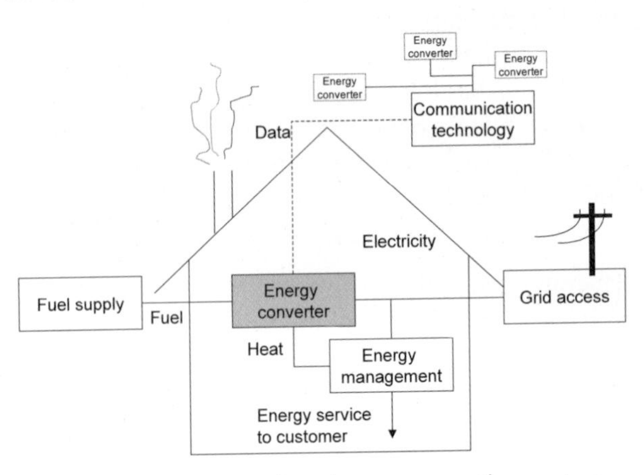

Fig. 1.1. Technological components of a micro cogeneration system

involved in a micro cogeneration system (Fig. 1.1), such as well-developed grid access, including possible metering and control devices. In the remainder of this chapter, the various technological components of such a system will be described in detail.

1.2 Conversion Technologies

A conversion technology serves to convert chemical energy that is stored within a fuel into "useful" forms of energy, i.e. electricity and heat. A number of different conversion technologies have been developed which have domestic CHP applications (Fig. 1.2). The conversion process can be based on combustion and subsequent conversion of heat into mechanical energy, which then drives a generator for electricity production (e.g. reciprocating engines, Stirling engines, gas turbines, steam engines). Alternatively, it can be based on direct electrochemical conversion from chemical energy to electrical energy (i.e. fuel cell). Other processes include photovoltaic conversion of radiation (e.g. thermo photovoltaic devices) or thermoelectric systems.

In principle, most conventional cogeneration systems can be down-scaled for micro cogeneration applications. However, some of them have yet to be successfully implemented for very small applications. Micro gas turbines, for instances, have only been developed with capacities above 25 kW_{el} and are thus not categorized as micro cogeneration technologies according to our definition.

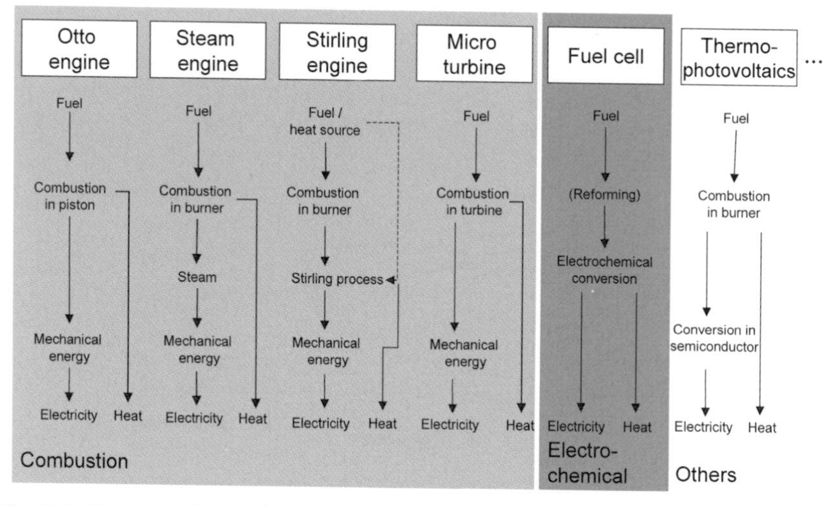

Fig. 1.2. Cogeneration technologies and conversion steps

1.2.1 Reciprocating Engines

Reciprocating engines are based on conventional piston-driven internal combustion engines. For micro cogeneration applications, typically, spark ignition (Otto-cycle) engines are used, comparable to those used in automobiles. In an Otto engine, a fuel, for instance natural gas, is mixed with air and compressed in a cylinder. This mixture is than ignited by an externally supplied spark. The now hot, expanding gas moves a piston, thereby causing the crankshaft to rotate. The mechanical energy produced by this combustion is then used to drive a generator. The exhaust heat as well as the heat from the lubricating air cooler and the jacket water cooler of the engine are recovered using heat exchangers, and then supplied to the heating system.

Reciprocating engines operate with less excess air compared to gas turbines. This leads to higher combustion temperatures, causing thermal NO_x production, due to the oxidation of the nitrogen contained in the air. There are two possibilities to reduce the amount of NO_x released. The engine can be operated in a lean mode, i. e. with excess air, so that reaction temperatures of the reaction are lowered. The second option is to operate the system almost stochiometrically (i. e. with an air/fuel ratio $\lambda = 1$) and to use a three-way catalyst.

The electrical efficiency of reciprocating engines, defined as net electrical energy output divided by natural gas input, depends strongly on the electrical capacity of the system (Fig. 1.3). At sizes below 15 kW_{el}, efficiency generally does not exceed 26 %. Thermal efficiency depends on the system and its level of heat integration (e.g., whether condensing heat

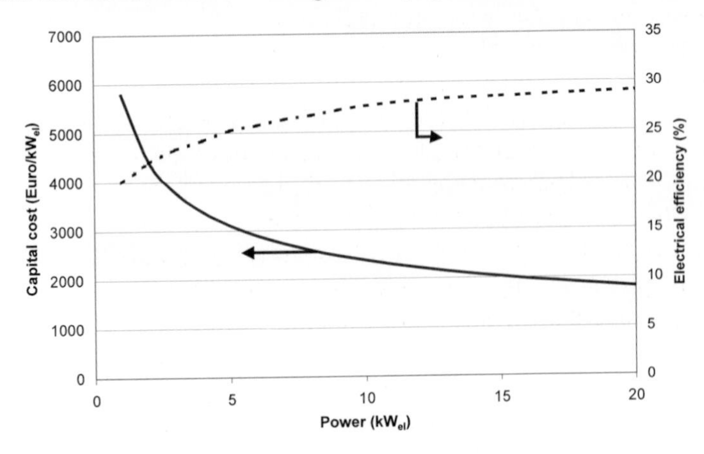

Fig. 1.3. Capital cost and electrical efficiency of reciprocating engines as a function of the size of the system (Source: ASUE 2001)

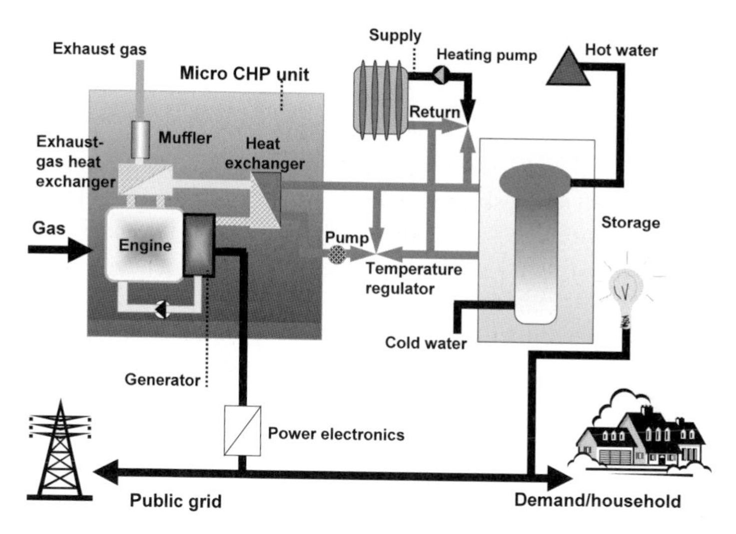

Fig. 1.4. Principle components of a reciprocating engine system (adapted from Ecopower)

is used). Combined electrical and thermal efficiency (total efficiency) varies between 80 and well above 90 %. Similar to electrical efficiency, capital costs per kW_{el} depend on the electrical capacity of the system. A significant decline of capital costs (scale effect) can be observed particularly as systems reach the 10 kW_{el} range (see Chap. 4).

Current Developments

Reciprocating engines are commercially available and produced in large numbers by a variety of companies worldwide. The market leader is the Germany-based company Senertec. The Senertec model – called Dachs ("badger") – generates 5.5 kW_{el} and a thermal power of 14 kW (Fig. 1.5). It achieves 25 % seasonal electrical efficiency and thermal efficiencies above 80 % (depending on the building, over 90 % when using a condensor). As of fall 2004, Senertec had sold 10,000 of these models.

Other companies providing micro cogeneration units include Power Plus (recently purchased by the boiler company Vaillant), with its 4.7 kW_{el} Ecopower module, capable of modulating its capacity, and the US-based Vector CoGen (5 and 15 kW_{el}), using a Kawasaki combustion engine. The latter is currently optimized for series production. According to the product specifications, Vector CoGen units achieve electrical efficiencies of around 28 to 34 % and total efficiencies between 70 and 79 %. In Japan, the companies YANMAR, Sanyo and AISIN also develop reciprocating engine based power stations.

Fig. 1.5. Senertec Dachs (left) and PowerPlus Ecopower (right): examples of reciprocating engine micro cogeneration for apartment houses and small commercial enterprises (Sources: Senertec and Ecopower)

An interesting development for single-family house applications is Honda's small 1 kW_{el} system named Ecowill (Fig. 1.6), which was developed jointly with Osaka Gas and other companies. Honda's cogeneration unit combines the GE160V – the world's smallest natural gas engine – with a lightweight generation system. The system is based on a $\lambda = 1$ Otto engine, with a three-way catalyst and oxygen feedback control to reduce the quantity of NO_x emissions. Osaka Gas distributes this system and has already sold more than 10,000 units in 2004 (see Chap. 13).

Fig. 1.6. Honda Ecowill micro cogeneration engine (Source: ASUE)

1.2.2 Stirling Engines

Unlike spark-ignition engines, for which combustion takes place inside the engine, Stirling engines generate heat externally, in a separate combustion chamber. In the Stirling engine, developed in 1816 by Robert Stirling, a working gas (for instance helium or nitrogen) is, by means of a displacer piston, moved between a chamber with high temperature and a cooling chamber with very low temperature. On the way from the hot to the cold chamber, the gas moves through a regenerator, consisting of wire, ceramic mesh or porous metal, which captures the heat of the hot gas and returns it to the gas as the cold gas moves back to the hot chamber.

Stirling engines can be designed in different configurations, distinguished by the position and number of pistons and cylinders and by the drive methods (cinematic and free-piston) (Educogen 2001). The mechanical energy of the Stirling engine is used to drive a generator.

Due to the fact that fuel combustion is carried out in a separate burner, Stirling engines offer high fuel flexibility, in particular with respect to biofuels, and, because of the continuous combustion, lower emissions. In principle, other heat sources, such as concentrated solar irradiation, can be used. Companies such as Solo and Sunmachine have developed parabolic mirrors for that purpose.

Stirling engines have the potential to reach high total (electrical plus thermal) efficiencies. Their electrical efficiencies, however, are only moderate. So far, 20 % seasonal average efficiency has been achieved in larger systems, with a predicted > 24 % for future models. Small Stirling engines are designed for low cost; consequently, they achieve lower electrical efficiencies than larger units, typically around 10 to 12 %.

Current Developments

Stirling engines are in between the pilot and demonstration phases and marketing. There are still field trials being carried out; but initial commercial products are already defined and on the verge of series production. The New Zealand-based company WhisperTech is developing a Stirling engine called WhisperGen, with a capacity of up to 1.2 kW_{el} and 8 kW of heat (Fig. 1.7). In the WhisperGen, four sets of piston cylinders are put in an axial arrangement. The British utility Powergen, part of Germany's E.ON, has ordered 80,000 WhisperGen power stations, due to be delivered by mid 2005. A prerequisite for this is the establishment of series production facilities. As Stirling engines require very precisely produced components, the scale-up from small-scale to series production presents a considerable challenge.

Fig. 1.7. Example of small (≈ 1 kW_{el}) Stirling engine micro cogeneration: WhispherTech

MicroGen markets a linear free-piston Stirling engine with 1.1 kW_{el}. MicroGen anticipates that, by 2007, series production will be launched. EnAtEc micro-cogen B.V., from the Netherlands, is in the stage of field-testing its 1 kW_{el} unit, with an electrical efficiency of 10 %. The Swiss company powerbloc GmbH is focusing its efforts on a 1.1 kW_{el} Stirling engine.

With respect to systems above 1 kW electrical capacity, the German companies Solo, Mayer&Cie. and Sunmachine have been developing Stirling machines. The Solo engine has sold 30 units up until the middle of 2004 (Fig. 1.8). Solo and Sunmachine are also experimenting with wood pellet burners and solar concentrators.

Fig. 1.8. Examples of larger (≥ 3 kW_{el}) Stirling engine micro cogeneration: Sunmachine (left) and Solo (right)

1.2.3 Fuel Cells

A fuel cell converts the chemical energy of a fuel and oxygen continuously into electrical energy. Typically, the fuel is hydrogen. The energy incorporated in the reaction of hydrogen and oxygen to water will be partially transformed into electrical energy (Pehnt 2002).

The "secret" of fuel cells is the electrolyte, which separates the two reactants, H_2 and O_2, in order to avoid an uncontrolled, explosive reaction. Basically, the fuel cell consists of a sandwich of layers that are placed around a central electrolyte: an anode at which the fuel is oxidized; a cathode, at which the oxygen is reduced; and bipolar plates, which feed the gases, collect the electrons, and conduct the reaction heat (Fig. 1.9). To achieve higher capacities, a number of single fuel cells can be connected in series. This is called a fuel cell stack.

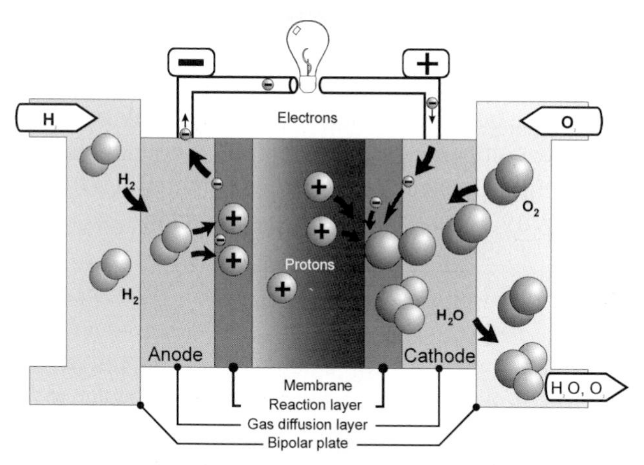

Fig. 1.9. Basic construction of a fuel cell-example Polymer Electrolyte Fuel Cell

Fuel cell micro cogeneration units are either based on Polymer Electrolyte Fuel Cells (PEFC; also Proton Exchange Membrane Fuel Cell, PEMFC), using a thin membrane as an electrolyte and operating at about 80° C, or Solid Oxide Fuel Cells (SOFC), which are high-temperature fuel cells working at 800° C. Recent efforts have been working toward the development of high-temperature molten carbonate fuel cells for this low-power segment.

Typically, natural gas is the available fuel for micro cogeneration applications. It mainly consists of the hydrogen-containing methane (CH_4), which is converted into hydrogen in a so-called reforming reaction. This takes place either in a separate device, the reformer, or, as in the case of high-temperature fuel cells, inside the stack (internal reforming).

Taking natural gas as the dominant fuel for fuel cells: In the short- and medium-term perspective, low temperature fuel cells (Proton Exchange Membrane Fuel Cells, PEMFC) in the low-power range may reach seasonal electrical efficiencies on the order of 28 to 33 %; in the long-term it is possible to achieve up to 36 % for domestic systems. However, it is so far unclear whether fuel cell systems can achieve the same thermal efficiencies as promised by the competing technologies. This is due to the fact that the heat cannot be extracted at well-defined points in the system, but rather at many dispersed heat sources, leading to greater measures being required for insulation and heat exchange.

Current Developments

In the last decade, considerable efforts have been made to further develop this technology, particularly in terms of mobile applications. However, fuel cells are not yet commercially available for micro cogeneration applications. Here, both R&D activities and initial field trials are being carried out. Some companies recently had to announce that further R&D is required before their systems can become commercially attractive, with acceptable longevity, technical performance, and system cost.

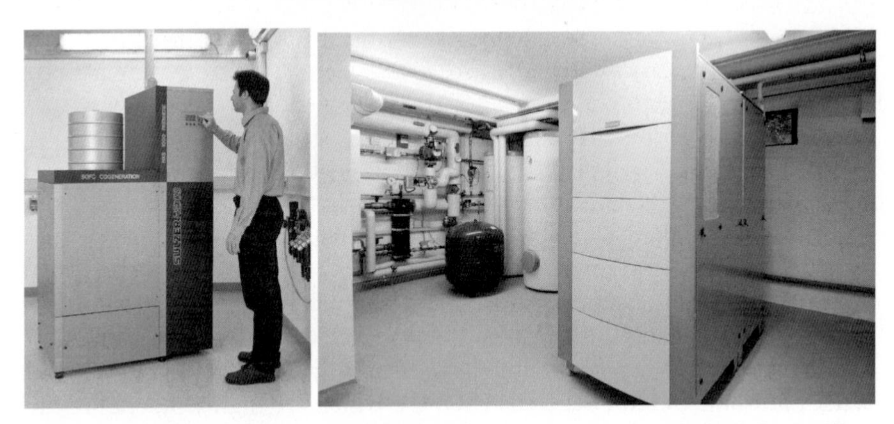

Fig. 1.10. Examples of fuel cell micro cogeneration: Sulzer Hexis (left), Vaillant (right) (Sources: Sulzer Hexis, EWE AG)

Among the most advanced companies in this sector are Sulzer Hexis, marketing a 1 kW_{el} SOFC system with electrical efficiencies between 25 and 30 %, which is currently being field-tested with more than 100 test units and further developed for series production, and Vaillant, which integrates a Plug Power PEMFC stack into a heating system with capacities of 4.6 kW_{el} and 7 kW_{th}. The latter system is currently being field-tested. IdaTech, together with RWE Fuel Cells and Buderus, is developing a 4.7 kW_{el} PEMFC system, and European Fuel Cell GmbH is planning the field-testing of a 1.5 kW_{el} PEMFC system. The target electrical efficiency of these systems is above 30 %; some fuel cell manufacturers even anticipate 35 %. As seasonal efficiencies, 32 % seems more likely for micro cogeneration systems. From 2004, the time perspective until readiness for marketing is about 8 to 10 years.

It is interesting to see that most fuel cell developers have been co-operating with major boiler manufacturers to ensure a broad market in the heating sector (see Chap. 7).

1.2.4 Other Technologies

A number of other technologies for micro cogeneration energy conversion are currently under development. Among the more advanced concepts, machines based on Rankine steam cycles are being developed. The Rankine cycle is the ideal prototype for steam engines in use today. Various companies, such as the Australian Cogen Micro with a 2.5 kW_{el} system, Energetix' Inergen system with a 1 kW_{el} generator, Climate Energy LCC, or the Baxi Group are pursuing this path, using different types of expanders for the steam, such as free-piston engines, scroll expanders, or reciprocating engines, and different types of working fluids, such as steam, organic substances, or two-phase mixtures of steam and water. However, none of these products is commercially available yet.

One advanced example of a Rankine type engine is the SteamCell, developed by the German company Enginion. In the SteamCell, feed water is compressed, heated with a compact burner, and transformed into steam, which then drives an innovative piston engine (Fig. 1.11). Following the expansion, the steam condenses and flows back into the tank. The SteamCell has a rated electrical power of 4.6 kW_{el} and a target electrical efficiency of 17 %. Field tests are planned for 2005, with forecasted market introduction by 2006.

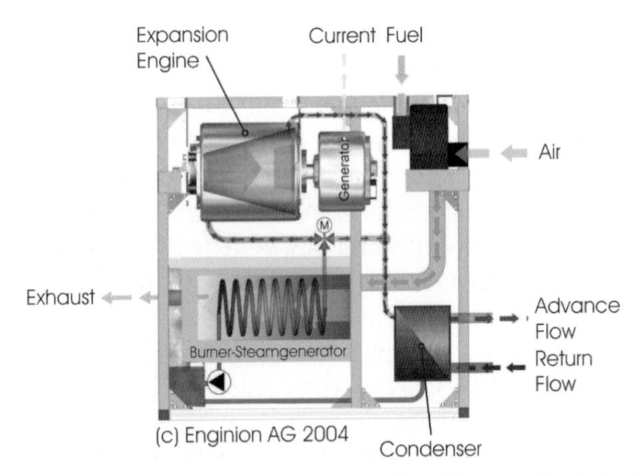

Fig. 1.11. Principles of the steam expansion engine (Source: Enginion)

Due to the continuous combustion in the burner, the hydrocarbon and carbon monoxide emissions remain very low, whereas nitrogen oxide emissions are determined by the maximum combustion temperature, which is carefully controlled in the system. The external heat supply also offers the possibility of using a variety of fuels.

A similar steam cogeneration system is currently being field-tested by the Germany-based company OTAG. In this system, called "Lion" (3 kW_{el}), steam is produced, injected into a piston from alternating sides, and expanded. The moving linear piston then produces electricity.

Generally, the electrical efficiencies of all Rankine cycle machines are low, on the order of 12 to a maximum of 20 %.

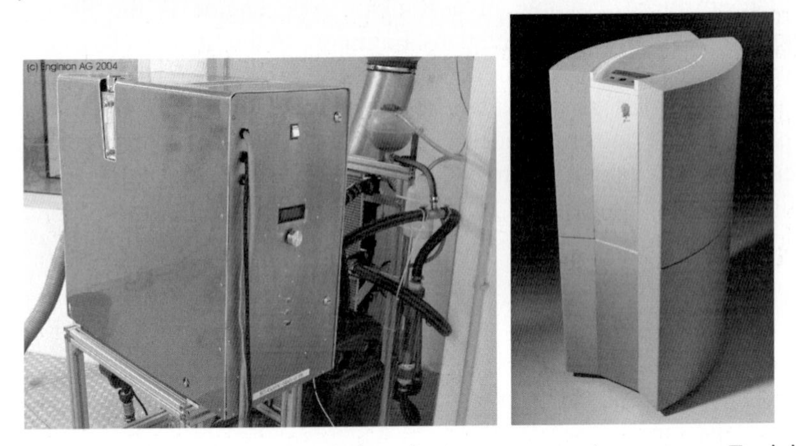

Fig. 1.12. Examples of steam cycle based micro cogeneration systems: Enginion SteamCell (left) and OTAG Lion (right)

There are other technologies, which are not based on electrochemical processes (fuel cell) or combustion engines (reciprocating, Stirling, or steam engine), but rather on semiconductors that convert the waste radiant energy of a heat source directly into electricity by means of a modified photovoltaic cell. This technology, called *Thermophotovoltaics* (TPV), applies low band-gap photovoltaic materials. The system consists of a burner and a ceramic emitter (optionally with a filter), which is heated by the burner and emits light, which is then turned into electricity using a photovoltaic cell.

Several developments are required before TPV systems can be successfully introduced into the market. Particularly, the emitter material must be optimized to match the wavelength range to the cells. Several rare earth materials are currently being investigated for that purpose. Also, solar cell design and fabrication, as well as system engineering, are required to optimize performance. Several research institutions, such as the Colorado based National Renewable Energy Laboratory (NREL), and companies, such as Sarnoff Corporation and Edtek, are developing TPV systems.

Thermoelectric devices also directly convert electromagnetic radiation into electricity. However, they do not apply photovoltaic materials, but instead make use of the fact that pairs of dissimilar conductors, e.g. differently doped semiconductor materials such as Bismuth Telluride, produce a current when there is a temperature gradient between them. California based Hi-Z Technology, Inc. developed a 1 kW_{el} thermoelectric generator to generate electricity from the waste heat of engines (e.g. Diesel) as a substitute for the truck engine alternator. In principle, thermoelectric technology can also be employed for micro cogeneration purposes. However, both thermophotovoltaic and thermoelectric devices are not expected to enter the micro cogeneration market in the short or medium term.

1.3 Grid Integration, Communication Technology, and Virtual Power Plants

For operation in the context of a larger system, for instance as a "virtual power plant", effective devices are required for communication between micro cogeneration units and system operator. These devices should support the optimal operation of the cogeneration system by effectively matching the operation of the individual systems with the demands of the user and of the grid operator.

Networking of several micro cogeneration devices is possible on several levels:

- Microgrids that physically connect the micro cogeneration units to several customers without further transferring information between the units, thus forming a more or less independent grid,
- information technologies connecting the micro cogeneration units to a data server, and
- “virtual power plants”, which combine the information technologies with a central management system.

Communication interfaces are currently being developed by several manufacturers, allowing additional features such as web-based control of the power plant, as well as alarm devices, automated data collection and alerts to maintenance companies in case of failures, etc.

With respect to households, similar communication devices could be developed to form energy management systems for a household. Such *home load management* could actively influence loads, depending on external signals (for instance time-dependent electricity rates), defer and prioritize loads, and, ultimately, act as a “home energy broker”, automatically selling to or buying electricity from other customers.

Taking this a step further, the micro cogeneration unit could be externally controlled by a central operator to exploit additional benefits and services. In several projects, such as in the Vaillant field test program, suitable communication pathways for such communication strategies are being investigated. For instance, an easy and cheap method of communication could be unidirectional ripple control technologies, which allow the utility to turn the power plant on or off in periods of peak or low demand. Data management through the internet, Powerline technology, SMS or other forms of bi-directional data flows would be even more advanced.

The concept of communicative networking ultimately leads to the *virtual power plant*. A virtual power plant consists of a number of geographically distributed power generation units – generally decentralized and low electrical capacity – which are integrated into one larger operational unit by means of a joint control and operator interface (Feldmann 2002; Jänig 2002; Stephanblome and Bühner 2002; Arndt and Wagner 2003). The term “virtual” does not refer to the energy flows, but rather to the plant itself, which is not at one location, but is rather dispersed among a number of generators.

Micro cogeneration units can be elements of such a virtual power plant. Often, larger CHP units and renewable electricity generating systems, such as wind power and photovoltaics, are mentioned in the context of virtual power plants. Generally, the coordinated connection of individual power plants allows for the balancing of fluctuating rates of generation caused by renewable energy systems (wind, solar irradiation) or by fluctuating demand (CHP systems).

Virtual power plants rely on advances in several technological areas to successfully meet customer demand, technical and safety standards, cost pressures, and environmental performance. Particularly, information and communication technologies and a management system are required for successful integration of the respective energy systems. The management system often includes a forecasting tool to anticipate future generation and demand and to better integrate the individual systems. Communication links to external service providers, such as weather forecasts, electricity stock exchanges, and so forth are required if more sophisticated forecasting and optimization are to be realized. Also, optimization and simulation tools may form part of the management system of a virtual power plant.

Possible communication pathways include telephone communication, particularly ISDN and DSL, internet, ripple control, UMTS, and Powerline technology (where electricity lines are used as a communication medium). Low communication costs are often regarded as crucial for the economic viability of virtual power plants (Lewald 2001; Arndt and Wagner 2003). Lewald calculates that, for certain concepts, 10 % of the monthly total cost would be for communication, which does not mirror the additional benefits of virtual power plants.

From an energy-economic point of view, virtual power plants could offer various ways of reducing costs and increasing revenues (Roon 2003):

- *Cost reduction*: On the hand, there are effects related to "clustered interests", for instance, buying a number of CHP units or service contracts could lead to volume discounts, as well as discounts for the fuel needed. Also, lower interest rates could be offered to larger operators than to individual power plant operators. On the other hand, the integration of several systems into one operational unit could lead to lower O&M costs, due to automatic early-warning systems, or to minimized fuel use (and thus, lower fuel costs), due to optimized operational strategies (e.g. optimized operating points, merit order).
- *Increase of revenues*: CHP products, particularly electricity, can be marketed differently when many systems are pooled, because the specific transaction costs can be lowered and certain regulatory requirements can be fulfilled (e.g., in order to participate in the control

power market, a minimum capacity is required). These options include selling the electricity on spot or regulating energy markets.

With these positive economic aspects of virtual power plants in mind, one should note that the amount of power that can be devoted to commercialization is usually only marginal because, in many cases, the producer's own consumption of the electricity generated is more lucrative than feeding it into the grid, even considering new marketing possibilities. In addition, power generation is combined with heat generation, which has to be used locally. Furthermore, considerable institutional barriers impair the realization of alternative ways of commercialization. Other than with virtual power plants based on larger individual generation units, under present-day conditions, the potentially higher proceeds of connected micro cogeneration plants do not justify the high expense for installation and management of a virtual power plant (Roon 2003).

1.4 Conclusions

In Table 1.1, different conversion technologies are compared on the basis of selected criteria. In addition, Fig. 1.13 depicts the status of market development of the various technologies.

Table 1.1. Characteristics of micro cogeneration technologies

Conversion technology	η_{el}	η_{th}	Noise level	Pollutant emissions	Fuel flexibility	Market availability	Economic viability
Reciprocating engine	20-25	> 85	Medium	Rather high, depending on catalyst/engine technology and maintenance	Medium	Commercially available	Given for certain applications
Stirling engine	10-24 [a]	> 85	Low	Very low to medium [b]	High	Near to market	Cost reduction necessary
Fuel cell	28-35	80-85	Low	Zero (H_2) to almost zero (hydrocarbons)	Medium	Pilot plants, R&D	High cost reduction necessary
Steam expansion engine	n. a.	n. a.	Low	not yet measured, principally similar to Stirling	High	R&D	High cost reduction necessary

η: efficiency, el electric, th thermal. [a] depending on the Stirling concept
[b] depending on the burner type

Whereas *reciprocating engines* are commercially available, produced in large numbers and achieving high electrical and total efficiencies, they suffer from higher exhaust emissions compared to the competing micro cogeneration systems (see Sect. 5.1.2).

Because fuel combustion is carried out in a separate burner, *Stirling engines* offer lower emissions as well as high fuel flexibility, allowing, in particular, for the use of bio-fuels and solar irradiation. Stirling engines have the potential to achieve high total efficiency, though with only moderate electrical efficiency. Smaller Stirling engines have an even lower electrical efficiency and are primarily designed to be low cost.

Fuel cells are still in the R & D phase, with a number of pilot plants currently being tested. They offer the potential benefit of the highest electrical efficiency and almost zero local emissions. Additionally, the high ratio of electrical to thermal efficiency (power to heat ratio) might increase the CHP electricity generation potential, because with given (and limiting) heat demand, more electricity can be produced (see Chap. 4). In the last decade, considerable efforts have been made to further develop this technology. However, it remains unclear whether fuel cell systems can ever achieve a total efficiency equaling those promised by competing technologies. Also, the high capital cost of early product generations remains a major challenge.

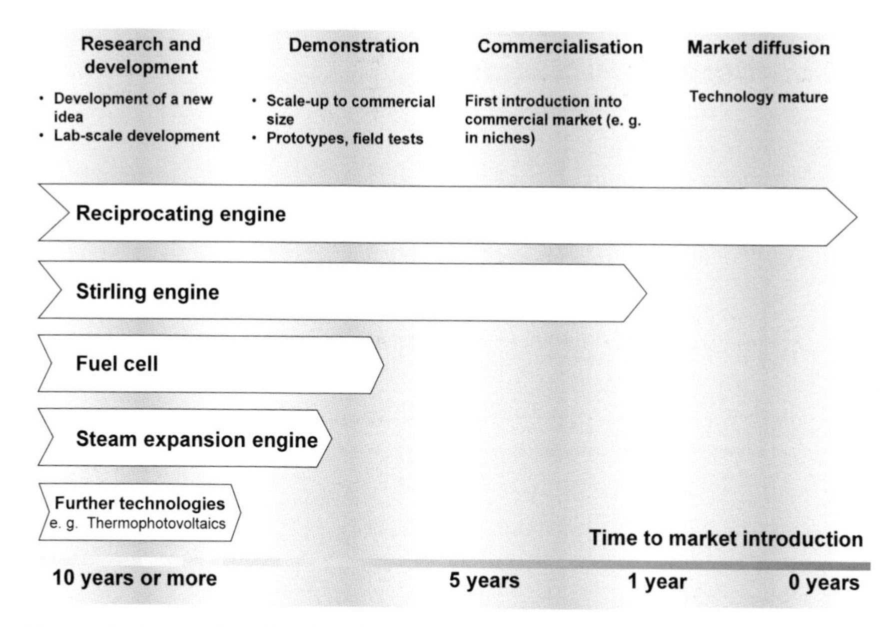

Fig. 1.13. Status of market development of micro cogeneration technologies

Steam expansion machines are still in the R&D phase. Potentially, they offer low emissions and high fuel flexibility, due to their external burner. In addition, some steam engine concepts envision a flexible flow of steam, allowing electricity and heat ratios from approximately 0.3 to pure heat supply. However, the technical demonstration of feasible CHP systems is still awaited and electrical efficiencies are expected to be low. *Other technologies*, such as thermophotovoltaic and thermoelectric systems, are still in the research stage.

2 Dynamics of Socio-Technical Change: Micro Cogeneration in Energy System Transformation Scenarios

Jan-Peter Voß, Corinna Fischer

2.1 Introduction

Energy is in flux. There is much talk nowadays about restructuring of markets, reorientation of business strategies, emergence of new technologies, reform of regulatory institutions, reinvention of the customer, or reconfiguration of the socio-technical architecture of energy production and consumption. Fundamental changes are said to be going on or are expected in the near future (Patterson 1999; Eising 2000; CSTM 2001; Smil 2003). This transformation of energy comprises electricity, natural gas, and the heating of buildings alike.[1]

The current transformation is closely related to the liberalization of formerly monopolized markets. These changes in the regulatory framework triggered further changes which go well beyond the stronger cost orientation and improved customer services that were the initial goals of the reform. The liberalization of electricity markets actually works towards a successive restructuring of the whole sector in its technical, organizational, and even cultural dimensions. The new market environment alters strategic requirements for businesses in dealing with risk and uncertainty entailing the adaptation of corporate organization and

[1] This has not always been the case. A couple of years ago, energy systems were mainly referred to with respect to their inertia and resistance towards change – be it for more economic efficiency or ecological soundness. Together with other public utilities such as water, railroads or telecommunications, electricity and natural gas even formed the subject of a special strand of research on 'Large Technical Systems', which aims to investigate the special role of these systems in public life and political governance (Hughes 1983; Mayntz and Hughes 1988; Coutard 1999; La Porte 1991; Summerton 1992).

technological portfolios. In fluid European markets, the strategic capabilities of companies (including size) become crucial for success. Short lead times and pay back periods, as well as flexible application of technologies, gain in importance.

This process links up with specific circumstances in local contexts. In Germany, for example, the changes through liberalization are combined with the politically decided upon phase-out of nuclear energy and the investment cycles of the coal-based generation infrastructure, creating a time window of a few years around 2015 in which one third of total electricity generation capacity needs to be substituted for by new plants. In this respect, there is indeed a fundamental transformation going on in electricity systems.

But there is also another dimension to the talk about transformation which is more programmatic than factual: the question of sustainable development. For energy systems, sustainability requirements are effectively represented by the need to achieve substantial reductions in the emission of greenhouse gases – up to 80% lower emissions in 2050 than in 1990. The already long-running discussion about ecological restructuring of energy systems is now linking up with the ongoing changes outlined above. The current situation of flux thus represents a window of opportunity which may be used to introduce more sustainable structures and means of energy provision. It is argued that now while the system is 'heated up', its future course of development is moldable, whereas in a few years' time, when new social and technical structures have emerged and become stabilized, the window will be closed again.

Against this background concerns are being raised, particularly from the perspective of sustainable development, about new path dependencies which could result from current decisions and practices. The emerging structures of tomorrow are being shaped in the political, technical, and economic processes of today. They therefore require special attention with respect to the systemic interactions that they set off and the long-term consequences resulting from them. The current phase of restructuration may lead to an ensuing phase of decades-long stability. If such a lock-in happens it should not be one on the wrong path. It is for this reason that prospective sustainability assessment of various innovations – in technology as well as institutions and cultural practices – and the system alternatives to which they are connected gains in importance.

This book focuses on a small aspect of this overwhelmingly complex problem set. It concentrates on micro cogeneration, which, compared to our conventional means of electricity and heat generation and supply, could make a fundamental difference if it actually becomes widely implemented. It has the potential to be pushed by and fed into a process of radical decentralization and, thus, become an architectural innovation which can turn the structure of energy systems upside-down. Former end-points of electricity flows through a system could become sites of power generation, consumers could become producers, transmission networks could shift function from transport to regulation, distribution networks could be upgraded by information and control technologies to connect various local generation and load-management facilities into virtual power plants. Added to the restructuring effect for electricity systems is the further consequence that micro cogeneration would primarily be based on gas instead of oil, which today is the dominant energy source for residential heating, and coal, which is today dominant for power generation. This implies a substantially increased role for gas supply and networks for electricity and heating. Such effects make micro cogeneration a potentially path-breaking technological innovation, if it can link up with complementary changes in institutional and cultural contexts, such as market structures, network regulation, and user practices. Some of the effects and requirements of developments along such a path are explored throughout the book.

In this chapter, we will not go into the details of the factors that constitute a working micro cogeneration configuration. These obviously include more than a smoothly running motor. Users would need to become acquainted with running a domestic power plant business, installers be prepared to take care of integrated electricity and heating systems, electricity and gas networks be reconfigured to take up changing load patterns, political regulation would need to adapt to a far more dispersed market structure in the sector. Many of these aspects will be taken up in the following chapters, while being investigated as drivers or impacts – or sometimes both – that will play a part in the evolution of new micro-patterns in energy provision and use. Here, we set out to explore the lay of the land in which micro cogeneration, as one innovation among many parallel efforts, finds itself, sketching out factors that may change this landscape over the coming decades. We will also tentatively explore how micro cogeneration itself, and possibly in interaction with other social and technical innovations, may play a role in reshaping the terrain.

This is quite an ambitious task. It deals with transformations on a very aggregate level, shaped by many underlying processes which are essentially unpredictable themselves and, yet, interact with each other in even more unpredictable ways. Moreover, we talk about long-term changes with a horizon of more than 20, and up to 50, years. Such an investigation, the analysis of future transformations in energy systems, cannot be done in any 'exact' manner. However, there have been and still are attempts at doing so: forecasting energy futures with the intention of predicting the likely mix of energy sources, employed technology, consumption levels, and other features of a system for tens of years ahead. They have not been very successful. And they will be so even less now, when stable framework conditions can no longer be assumed.

In this chapter, we will therefore approach the exploration of macro-scale dynamics in energy systems in different ways. The first is a discussion of transformation dynamics in a rather conceptual manner (Sect. 2.2). The intention is to create an understanding of the breadth and complexity of the energy transformation now taking place and draw attention to specific aspects which could be critical with respect to micro cogeneration development. For this, we draw on concepts and theories that have been derived from historical studies of socio-technical change in energy systems and other infrastructure sectors. This will allow us to develop a general understanding of the diversity of elements and processes on different levels which are coming together in what we conceive of as the parameters of the energy transformation.

The second approach we take in our exploration is to make use of the rich offering of scenarios, visions, and expectations about energy futures which are formulated by diverse actors from the sector itself (Sect. 2.3). Even though none of these future visions can be taken as a precise prediction of what is going to happen, a review of possible futures which are currently being discussed may serve to peg out the likely bandwidth of changes which may be linked to the ongoing energy transformation. And since the expectations embodied in these visions and scenarios orient actual investment decisions, political strategies, and use practices they may indeed be more or less translated into reality.

In concluding this chapter, we aim at pointing out some critical linkages between the general energy transformation dynamics and the possible path(s) of development of micro cogeneration (Sect. 2.4). This shall be the fruit to be harvested from the conceptual and bird's eye perspective which this chapter adopts. It shall provide a starting point and background foil for the more focused and detailed analyses of micro cogeneration innovation in the following chapters.

2.2 Driving and Embedded: Micro Cogeneration and the Dynamics of Socio-Technical Change

Studies of socio-technical change deal with the conditions of long-term change in large technical systems like electricity provision (Kemp 1994; Mayntz and Hughes 1988). Empirical studies of innovation processes show the close entanglement of social, technical, and ecological conditions and their manifold interactions in bringing about change and shaping production and consumption patterns (Bijker et al. 1987; Weyer et al. 1997). These studies have come up with a multi-level, multi-actor, multi-domain concept of socio-technical change which portrays the complex interactions that are involved in the development of novel forms of technology, and their embedding in and shaping of existing system structures (Geels 2002b; Rip and Kemp 1998). Applied to micro cogeneration, this means conceptualizing the innovation in a more comprehensive way than simply referring to a group of technical artifacts, i.e. reciprocating engines, Stirling engines, or fuel cells set up for small-scale heat and power generation. Instead, the perspective is shifted towards understanding the interplay of these material components with other components, such as the knowledge needed to design and operate them, user practices which fit technical functions, interests of market actors to produce them, maintenance networks to provide operational services, complementary technologies, regulatory provisions, and the like. It is the combination of such heterogeneous components that create a socio-technical "configuration that works" (Rip and Kemp 1998, p. 330). If any single component which is critical for the technology to work in practice is lacking or misadapted, the configuration is spoiled, i. e. innovation does not occur.

This perspective raises awareness about system design activities, strategic re-combinations, and alignments in the process of innovation which go far beyond the dimension of nuts and bolts. They involve the forming of innovation alliances and playing out market rivals, negotiations between producers and financers, testing of user acceptability, lobbying for favorable regulations, creating public expectations and images, and so on. Accordingly, the socio-technical configuration that finally comes into being is strongly shaped by factors such as the image attached to the artifacts by users, the interests and power of market players, institutional path dependencies, and public opinion. This may play a stronger role than technical functionality understood in a narrow sense. For micro cogeneration this means that the technical artifact which may finally become established can be quite different (in design, size, performance) depending on, for example, whether it is provided by electric utilities or heating system suppliers; whether maintenance is taken care of by users, local installers, or the suppliers themselves; whether it gains political support for ecological reasons or security of supply.

Another important aspect of recent conceptualizations of the dynamics of socio-technical change is a focus on the interaction of particular innovations with broader-context structures and ongoing changes within them. This multi-level perspective is inspired by evolutionary theory, which understands the emergence and change of complex systems as local variations which have to prove their fitness in the context of specific selection environments (Nelson and Winter 1982; Dosi 1988; Schneider and Werle 1998; Nelson 2000). If they survive, they become part of the selection environment for successive variations. This view is especially important for radical innovations, such as micro cogeneration, which face a selection environment that is geared towards central generation and long-distance transmission of electricity combined with separate heat production. The existing "regime" of energy provision (Kemp 1994; Rip 1995) may indeed represent a fundamental barrier for the widespread application of micro cogeneration technology, because it more or less subtly works towards the preservation of the existing structure: to which vested interests, actor networks, traditions, established mind sets, sunk costs, and more are attached. This resistance to micro cogeneration may take the form of adverse network regulations that do not grant indiscriminate conditions of access, technical standards which do not take account of the specificities of local generation, concerns raised about security of supply or the buying out of competitors on the market. Studies of socio-technical change, however, also state conditions under which radical innovations can prevail (Geels 2001; Geels 2002a; Berkhout et al. 2003):

- When they make use of niches within existing regimes to gain technological maturity and develop a network of actors and social institutions which supports them,
- when complementary innovations can link up with each other and become more independent from services provided within the regime,
- and, this is most important, when they can exploit pressures, interstiches, and ongoing changes on the regime level as windows of opportunity for widening their niche, broadening support networks, or promoting favorable standards and regulations.

Thus, even though existing structures of electricity provision in all their facets play an important role in shaping micro cogeneration innovations, they do not determine them. This is even less the case because they themselves undergo change, quite fundamentally and rapidly during recent years, and because this change is partly influenced by the dynamics of innovations themselves.

Against this conceptual background, it appears that the dynamics of micro cogeneration development does not have a single source or direction. It can better be understood as an evolutionary process from which a new socio-technical configuration emerges. This process is embedded in the structures of the existing energy regime. Taking into account that this 'selection environment' itself is undergoing evolutionary change, one can speak of co-evolution to grasp the interaction of heterogeneous processes of innovation (Norgaard 1994; Rip 2002).

In the current transformation of the energy system, co-evolutionary dynamics are becoming visible as the introduction of competition-oriented regulation shifts the criteria for fitness of technological innovation from aptness to central-system control towards flexibility with respect to volatile market conditions. Market-based regulation, on the other hand, could become successful, because development in gas turbine technology, distributed generation, renewable generation technologies, and so on have opened possibilities for competition in electricity generation markets. Another instance of co-evolutionary dynamics apparently is the difficulty for independent network regulation to be installed against the resistance of the highly organized interests of the electricity industry in Germany. Liberalization was made politically feasible only by letting industry negotiate conditions for network access on its own (Voß 2000). As soon as regional monopolies were opened to competition, however, the associational structures that supported the alliance of the electricity industry fell victim to internal quarrels, giving way to an incremental strengthening of network regulation – up to the establishment of an independent regulatory body eight years after liberalization.

Against the background of co-evolutionary dynamics and multi-level interactions, the relation of innovation dynamics in micro cogeneration and broader transformation processes in the electricity system cannot simply be reduced to a constellation in which one is mechanically influencing the other. Micro cogeneration innovation could be an important driver of regime change. As such, it needs to be assessed with respect to the sustainability of the broader changes it might help to bring about. This includes decentralization, reduced role of the transmission networks, changing actor constellations, and fuel-switch towards natural gas. Micro cogeneration could also be strongly influenced and shaped by broader developments on the regime level. For this reason, if one aims at assessing the actual potential of micro cogeneration to contribute towards sustainable development, it is necessary to investigate the factors which determine the fitness of micro cogeneration technology under a range of possible alternative paths of regime transformation. This concerns their strategic value to certain actors, their economic and environmental impacts, requirements with respect to institutional and technological infrastructures and how these all play out in different scenarios of transformation. In the chapters of this volume, both of these perspectives – micro cogeneration as a driver of transformation, as well as micro cogeneration as being embedded in transformation – are discussed for important dimensions of the relationship between innovation and transformation in energy systems.

To make the complexities of the dynamics of socio-technical change practically manageable for an integrated assessment of micro cogeneration, we propose to conceptualise it as an embedded innovation cluster (Fig. 2.1). The cluster comprises various interdependent innovation processes, social as well as technical. Giving credit to conventional talk about technology we can thus take the technical artefact, i. e., the energy conversion unit, as the focal innovation which is at the core of the cluster. Other technological, institutional, or cultural innovations whose development is linked to the focal innovation can in a first step be divided into three types, according to the dominant kind of relational influence they have with respect to the focal innovation (Fig. 2.1 for a schematic sketch of the micro cogeneration innovation cluster):

- Promoting innovations which provide favourable conditions for the focal innovation to prosper (e.g., Third Party Financing schemes, maintenance and service networks, remuneration of avoided transmission costs by network regulation, heat storage technologies),
- induced innovations for which the micro cogeneration technology is a prerequisite or driver (e. g., virtual power plants, integrated facility-management services, adaptive networks), and
- competing innovations which belong to alternative socio-technical configurations, and can thus impede micro cogeneration development or become inhibited themselves by it, depending on which innovation leads the race in occupying application potential or resources (e.g., district heating, thermal solar collectors, long-distance import of solar electricity).

From this brief excursion into the foundational aspects of conceptualizing technology and innovation, we can derive an understanding of how micro cogeneration as a new and emerging technology is embedded in broader processes of transformation. One lesson is that the relation between innovations in micro cogeneration and transformations in energy systems is not a unidirectional but rather a recursive one. This makes feedback likely to occur and creates the possibility for positive or negative feedback dynamics, for example, the mutual reinforcement of decentralization and micro cogeneration development or the curbing of micro cogeneration development by limited niches for distributed generation ("increasing returns", see Arthur 1997; Pierson 2000).

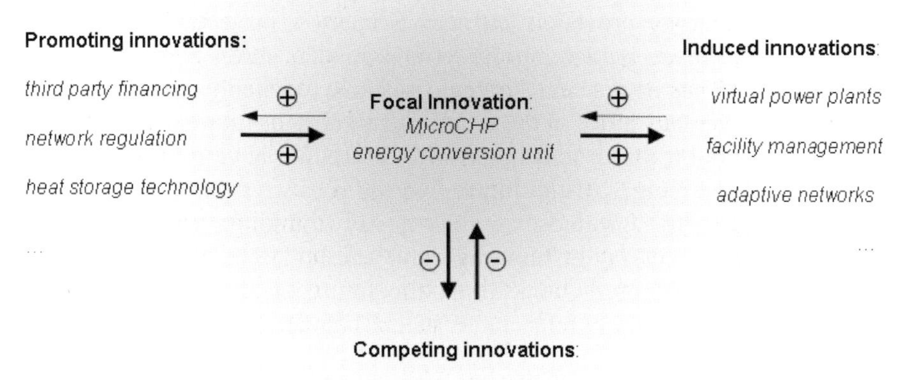

Fig. 2.1. Micro cogeneration innovation cluster

The actual manner in which they will both interact is therefore very difficult to predict. The complexity of this relationship is further demonstrated by a closer look at how micro cogeneration is interwoven with its social and technical context, and the innovation processes and changes taking place therein. In the following section, we will look at different visions of possible future contexts and discuss their implications for micro cogeneration.

2.3 Pluralism of Prophecies: Scenarios of Transformation in Electricity Systems

Transformation on the basis of co-evolutionary dynamics is not predictable. Nevertheless, expectations about the future are always there, and they shape ongoing change. These expectations also guide thinking about particular innovations, such as micro cogeneration, and they orient actual investment decisions and technological design activities. In order not to let these expectations implicitly crawl into the investigations and assessments which are undertaken in this book – and the evaluations of these by the reader – we try to make as transparent as possible the spectrum of future expectations with respect to the role of micro cogeneration in the transformation of energy systems. We do this by offering a review of selected future-oriented studies which differ remarkably with respect to objectives, methods, scope – and results. The variation of expectations reveals the great uncertainties which are connected to projecting technological and institutional changes in complex systems like energy provision and use. Scenarios, projections, and trend studies are all made by actors who have particular stakes in the future of energy. Their normative and strategic visions inevitably creep in when they assess the plausibility of the indefinite number of future developments which are possible (Grin and Grunwald 2000). We have to acknowledge that each of the images of the future and the pathways that lead there are constructed on the grounds of present-day thinking, which is itself embedded in present-day institutions, actor constellations and so on – even if some who make predictions about the future claim special visionary capabilities, which they refer to a special model or simply to expert authority.

However, even though none of the scenarios and diagnoses about the future of energy can be taken as an inevitably correct prediction, they can still tell us a lot about reality. They tell us what people who have a stake in the energy transformation, and who are actually the ones that will undertake it, are expecting of the future. Since these are the frameworks to which they refer in making decisions and undertaking actions, they may actually become transformed into agendas and programs which are indeed quite powerful in structuring "real" technological, political, economic, or cultural futures (Dierkes et al. 1992; van Lente 1993; van Lente and Rip 1998; Canzler and Dierkes 2001; Konrad 2004). Self-fulfilling or -defeating prophecies, as it were.

The prophecies which we have a look at in this chapter are based on different methods and refer to diverse application areas. Some are purely qualitative, building on narrative storylines, some are the result of numeric modeling. Some refer to energy – including electricity, gas, heating, mobility – others only to electricity. And some look at the world, or Europe, and others specifically at Germany. The overview does not aim at giving complete information on all the imagined futures, but attempts to sensitize us about the breadth of assumptions and expectations that various actors arrive at by using various methods. Choice of studies has therefore been led by three considerations: (1) to cover a wide scope of different approaches, (2) to capture the most influential prophecies (either because they are most discussed or because they are authored by important actors), and (3) to restrict the discussion to studies that are, with respect to their geographical and thematic focus, relevant for micro cogeneration. Besides explicit mention of *micro cogeneration*, this refers to the role of *CHP in general*, the role of *distributed generation* (DG), or the development of the *heat market.* Finally, we are sensitive to statements about the future development of *competing options,* from which limited application potential for micro cogeneration could be derived.

Our intention is that, by unfolding the diversity of future expectations, they can work (on us and the reader) as a "self-reflecting prophecy", i. e., a form of viewing the future that does not tempt us to follow too straightforward a path or argumentation on the basis of given assumptions, but rather to keep in mind the contingencies that are involved and strive for problem analyses and strategies which take a balanced stance with respect to alternative developments that may occur (Voß et al. 2005). In some respects, this may even be a reason to look for different kinds of strategies that put emphasis on experimentation, adaptability, cooperation, and robust strategies in order to be prepared for unexpected things to happen (Axelrod and Cohen 2000; Voß and Kemp 2005).

2.3.1 Types of Energy Scenarios

In looking at the diversity of future narratives about energy, it is helpful to build some broad types, which are characterized by differences in purpose, method, and scope. The following Table 2.1 gives an overview of different types of energy scenarios that we have analyzed. The paragraphs following the table give a brief introduction to each type of scenario-study and outline the findings with respect to relevant aspects.

2.3.2 Forecasting

Energy forecasting exercises aim at predictions of future energy demand and production structure, including primary energy consumption, types of end use, technology, and fuel mix. The methodological approach is based on economic modelling, including extrapolation of past trends via statistical data series, sometimes complemented by interviews with stakeholders. Energy forecasting usually builds on the assumption that political, social, and economic framework conditions are constant, that technology development moves on continuously, and that no other disruptive events will change the course of affairs. Within the framework assumptions, the models usually link input variables such as economic and demographic development, settlement structure, and primary energy reserves to output variables like primary or end-use energy consumption, fuel mix, prices, and emissions.

Table 2.1 Overview of different types of energy scenarios

	Forecasting	Technology foresight	Policy scenarios	Explorative scenarios
Studies included	Prognos 2000, Pfaffenberger and Hille 2004	Larsen and Sønderberg Petersen 2002 (Risø), Nakicenovic and Riahi 2002 (IIASA), Holst Jørgensen et al. 2004 (EurEnDel)	Fischedick et al. 1999/2001 (Wuppertal Institut), Matthes and Cames 2000 (Öko-Institut), Fischedick et al. 2002 (UBA), Diekmann et al. 2003 (DIW), Nitsch et al. 2004 (DLR), Velte 2004 (EurEnDel)	Patterson 1999, Shell International 2001, IEA 2003, Jäger et al. 2004 (IMV)
Purpose	Prediction of energy consumption and fuel mix	Anticipation of technological developments	Construction of desirable target states for energy system	Exploration of the breadth of possible energy system developments
Method	Quantitative (economic) modelling	Delphi, Expert study	Quantitative (economic) modelling	Identification of key factors, cross-impact analysis, narrative storylines
Treatment of technology	Efficiency improvements as external variable	Explicit treatment as variable	Explicit treatment as building block	Explicit treatment as variable
Time horizon	2020	2020-2100	2020-2050	2025-2050
Regional scope	Germany	Europe, World	Germany	Germany, World
Thematic scope	Primary energy, End-use Energy, electricity	End-use Energy	Primary energy, heating and electricity	End-use Energy, electricity, utilities

In our sample of future studies, the *Prognos Energiereport III* (Prognos AG 2000) and a study by Pfaffenberger and Hille (2004) represent forecasts for the German energy market in 2020. The Prognos report is widely referred to in energy science and policy in Germany. The study by Pfaffenberger and Hille is the most recent energy forecast available. It has been commissioned by the incumbent electricity-producer associations and deals in some detail with the prospect of CHP.

Both energy forecasts assume a certain market for CHP, but make the qualification that CHP development depends on its economic competitiveness. Because oil or gas are generally the cheaper alternatives, restrictions to competitiveness are seen especially on the heat side. Competition is exacerbated by a continuously decreasing heat demand, prompted by political regulation. Prognos assumes heat demand in households to decrease from 2064 to 1990 PJ from 1995 to 2020, and in small businesses from 776.8 to 500 PJ. Also, emission trading is projected to have a negative impact on the competitiveness of CHP, because competing systems, namely conventional individual heat systems, are not subject to emissions caps. Nevertheless, both forecasts assume total electricity production from CHP to increase. Pfaffenberger and Hille suggest an increase from 10 % of electricity production today to 16 % in 2020. Prognos predicts that many new gas power plants will use CHP technology. The growing share of CHP in electricity production is not due to a widening of district heating areas, but to increased densities of connection and a higher power-to-heat ratio. These developments for CHP in general, however, cannot simply be transferred to micro cogeneration. In fact, centralized CHP is in competition with micro cogeneration. Pfaffenberger and Hille explicitly examine several variants of small-scale co-generation, including fuel cells, Stirling motors, micro gas turbines down to 30 kW_{el}, and reciprocating engines down to 1 kW_{el}. The latter is attested to be of "minor importance", Stirling motors are not considered a viable option because of their low power-to-heat ratio, fuel cells and micro gas turbines are presented as technically interesting, but not yet marketable. Furthermore, it is assumed that fuel cells will become competitive more in the range of higher capacities.

Prognos, by contrast, does not specifically assess micro cogeneration development, but casually mentions that decentralized cogeneration in reciprocating engines may become an attractive option in case of increased transmission costs. With respect to distributed generation, Prognos states higher shares than Pfaffenberger and Hille. Whereas the former refers to reduced transmission losses through decentralized generation, the latter point out the economies of scale of central generation. With respect to competing options to micro cogeneration, both forecasting studies expect stable developments in conventional heating technologies (oil, gas, and district heating) and a slow, long-term increase in solar thermal energy use. Altogether, micro cogeneration is portrayed as being of minor relevance for future energy systems until 2020. This matches the generally conservative orientation of future outlooks which are based on the extrapolation of existing framework conditions and economic investment behavior. In this framework of future thinking, micro cogeneration may only become an option in case of excessive transmission costs and, even then, it is restricted by decreasing heat demand and strong competition with conventional heating technologies.

2.3.3 Technology Foresight

Technology foresight studies share with energy forecasting the aim of developing reliable expectations about the future by determining 'most probable' paths of future development. But they have a more specific focus than energy forecasting. They try to determine future advances in specific technologies, their costs and probable role in a future energy system. Methodologically, technology foresight studies are usually based on expert knowledge about application potentials, "technological maturity", and critical factors for further development of cost and performance, for example on the basis of learning curves. Three technology foresight studies are included in our analysis. The Risø Energy Report 1 (Larsen and Sønderberg Petersen 2002) focuses on Denmark, but also discusses European developments. It has been chosen because the specific technology assessments are embedded in a secondary analysis of important global energy scenarios and, therefore, integrate other influential viewpoints on the future of energy systems. The study covers global developments until 2050 and explicitly deals with a range of "clean power generation" and "cross cutting" technologies, CHP being one of the latter. A second study was carried out at the IIASA by Nakicenovic and Riahi (2002) as a secondary analysis of assumptions about technology development in a broad range of other studies. Its aim is to assess the role of change in various technologies (including generation technology, fuel production, carbon sequestration, transport, and energy consumption) across a number of scenarios, with a time horizon of 2050 to 2100. Ultimately, it aims at identifying core technologies that are robust across scenarios and can contribute to a sustainable energy system. The third technology foresight is the recent European Energy Delphi (EurEnDel) (Holst Jørgensen et al. 2004). EurEnDel sets out from a perspective that views technology development as a social process influenced by policy decisions and public acceptance. This leads to a cautious understanding of predictive capabilities using a combination of Expert Delphi methods with a qualitative assessment of the compatibility of specific technology developments with certain dominant cultural values. From this comparison, three alternative qualitative scenarios are generated. Here, we focus on the results of Expert Delphi which predict the timing of technology development in energy efficiency, transport, storage, and supply in a timeframe until 2020.

A first interesting aspect to note is that the studies by IIASA and Risø arrive at quite distinctly different conclusions, even though they largely refer to the same global energy scenarios. While IIASA sees nuclear energy as a "robust" option for the second half of the century, Risø puts it up to political will whether the technology will have a future. It also issues the statement that a *decentralization* of the electricity system is likely to occur - a topic on which IIASA does not comment at all, restricting itself to various generation technologies without discussing their embedding in a system of supply.

Risø also explicitly mentions *CHP* as a cornerstone of Danish energy policy and as a "promising technology" for achieving CO_2 reductions in the future, whereas it is not an explicit topic for IIASA and EurEnDel. Fuel cells, however, are a prominent technology in all reviewed studies. EurEnDel generally discusses fuel cells only for mobile purposes, but also mentions in a casual remark that "diffusion of residential CHP/fuel cells may contribute to increasing energy efficiency of buildings" (Holst Jørgensen et al. 2004, p. 83). IIASA identifies fuel cells (along with nuclear and photovoltaics) as one of a few "robust" technologies that will become relevant in the second half of the century. The Risø authors are more optimistic and expect specific types of fuel cells to be marketable in 2010 and play a major role in another 10-15 years. While IIASA does not make reference to fuel cell applications in CHP, Risø states that, "forty years from now, nations with access to natural gas may have completely replaced their conventional CHP installations with solid oxide fuel cells" and "micro cogeneration such as small-scale fuel cell systems based on natural gas may become commercially available in the near future. Benefits of CHP thus extend into sectors presently covered by e. g. domestic boilers" (Larsen and Sønderberg Petersen 2002, p. 17,19).

With respect to *heat market* developments, IIASA and Risø remain silent. EurEnDel quotes experts who expect 50 % of new buildings until 2026 will be low-energy houses, and emphasise the robustness of low-energy building technology across three scenarios which emphasize individual choice, ecology, and social equity as different value orientations for policy-making. Biomass for heating and district heating is expected to be "widely used" in 2017. While IIASA does not touch upon *decentralization* of energy systems, both Risø and EurEnDel issue the expectation that distributed generation will become more widespread due to benefits for the environment, wealth, quality of life, and security of supply. EurEnDel quotes Delphi statements about on average a 30 % share of power generation to be below 10 MW_{el} by 2021. They add, however, that R&D activities and appropriate regulation are prerequisites for this expectation to be realised.

As mentioned with respect to heat market developments, all three studies expect energy-efficient housing technology to become an important *competing option* to micro cogeneration. Also, reference was made to district heating, possibly based on biomass, as an ecologically attractive competitor. Interestingly, EurEnDel also discusses the construction of international electricity transmission lines to transport regionally produced renewable energy over long distances. This may be a potentially competing development to decentralized generation, which some experts expect to come into play between 2024 and 2042; 16 %, however, think that this will never occur. The Risø report, starting out from an analysis of current Danish energy policy, identifies limits to decentralized CHP in systems where it already has a high share, especially when combined with a high share of fluctuating input from renewable generating technologies.

Although the three technology foresight studies do not elaborate on micro cogeneration as a central technology of the future, we can summarise our review with the assessment that it is considered as a possibility and an interesting option by some. It also matches the expectations about a decentralization of electricity generation structures. However, the discussion of micro cogeneration is mainly a by-product of their consideration of fuel cells: a technology that attracts considerable attention and fuels expectations about fundamental change in energy systems. Other micro cogeneration technologies are not on the agenda. The foresight studies raise concern about restraints and competing options, such as decreasing heat demand, district heating, large-scale renewable energy import, or a large share of fluctuating electricity generation from renewable sources.

A general lesson put forward by the technology foresight studies is that there is a broad array of possible future technology options. As technology development depends on social factors, including policies and value preferences, it is difficult to make reliable forecasts with respect to future technologies. As a safeguard against strategic failure because of wrong expectations, technology foresight studies advise us to consider and support a broad portfolio of technologies and to identify "robust" strategies that may work in different contexts.

2.3.4 Policy Scenarios

Policy scenarios follow quite a different philosophy than the sorts of energy forecast and technology foresight studies which have been discussed above. Instead of trying to find out about the most probable course of energy system transformation, they take as a starting point the view that this course can be deliberately shaped by society. Visions for desired states of future energy systems and possible ways to implement policy goals are what they are concerned about. Policy scenarios are constructions of feasible and consistent paths of energy system development which can be achieved with certain activities under certain framework conditions. As such, they can give strategic guidance to the policy process, explicate consequences of action and non-action, and can be used to demonstrate the feasibility of general goals (e.g., decommissioning of nuclear power, reduction of dependence on fuel imports) in political discourse.

The usual structure of policy scenarios is a reference scenario (business-as-usual) which is contrasted with alternative scenarios that can be achieved if different political activities are undertaken. The reference scenario may be differentiated into a set of forecasts based on varying framework parameters. The scenarios which represent policy strategies can either be constructed starting from certain policy instruments, whose effect on the energy system is simulated with the help of (economic) models, or they can be constructed starting from policy goals (usually quantitative emissions reductions targets), from which necessary development steps and measures are derived via qualitative and quantitative backcasting (Robinson 1982; Robinson 2003). The variables from which policy scenarios are built are much the same as in energy forecasting. Economic and demographic development, housing structure, prices, and sometimes technology are common input parameters, energy use and emissions common output. Input parameters may be manipulated according to policy preferences, e.g. phase-out of nuclear power or growing shares of renewable generation. These kinds of scenarios make heavy use of economic modelling; but there are also exercises which produce qualitative scenarios through discussion of stakeholders' and experts' views. In our analysis, we include seven scenario studies for Germany that have become influential in political discussion there. The oldest one under review is the "nuclear phase out" study produced at the Wuppertal Institut (Fischedick et al. 1999/2001). Its aim was to construct possible nuclear-free futures for Germany, covering a timeframe up to 2030. Also, the "energy turnaround" study, worked out at the Öko-Institut by Matthes and Cames (2000), aims at outlining a sustainable future energy system with a time horizon up to

2020. The same goal is pursued by the "long-term scenario" study commissioned by the federal Environmental Agency of Germany (*Umweltbundesamt*, UBA), which extends the time horizon to 2050 and has also been conducted by the Wuppertal Institute (Fischedick et al. 2002). A politically relevant document is the report of a parliamentary commission on "sustainable energy provision under conditions of liberalization and globalization" until 2050 (Enquete-Kommission 2002); it, however, presents only the opinion of the ruling parties, the Social Democrats and Greens, because no consensus could be achieved. A study named "Policy scenarios" by the German Institute for Economic Research (*Deutsches Institut für Wirtschaftsforschung*, DIW) and other institutes (Diekmann et al. 2003) evaluates the impact of climate protection policies and suggests further possible policies following 2012, using various timeframes up to 2050. Another one commissioned by the Federal Ministry for the Environment, carried out by the German Aerospace Center (*Deutsches Zentrum für Luft- und Raumfahrt*, DLR), the Wuppertal Institute, and other institutes (Nitsch et al. 2004), concentrates more specifically on the "ecologically sound extension of renewable energy sources", and is the foundation for the CHP study of Krewitt et al. (2004), which is described in Sect. 3.3. And finally, the scenario part of the EurEnDel study (as opposed to the Delphi part) has also been analysed (Velte 2004).

Though we have assessed more policy scenario studies than other types of studies, the sample is the most homogeneous with respect to methods and results. The reference scenario is usually based on the forecasting results of Prognos. Policy scenarios are constructed via (economic) modelling. There is remarkable convergence about what is necessary to build a sustainable energy system.

Without exception, all climate protection scenarios assign CHP a crucial role and sketch a strong increase in electricity production from CHP. The Öko-Institut study (Matthes and Cames 2000), as well as the three to which the Wuppertal Institute has contributed (Fischedick et al. 1999/2001, Fischedick et al. 2002, Nitsch et al. 2004), assume at least a doubling of CHP electricity. While Fischedick et al. sketches this out until 2020, the Öko-Institut assumes a doubling as possible even by 2010. Diekmann et al. (2003) see new gas power plants being almost exclusively constructed as CHP plants. It is expected that the share of electricity produced by CHP will be 30 % in 2020 (Öko-Institut scenario "Policy") or up to 40 % in 2050 (Enquete-Commission scenario "Energy efficiency") If technical potential was fully exploited without concern for political and economic restrictions, even 40% in 2020 could be achieved according to the Öko-Institut (scenario "Potential"). Besides district heating, CHP is expected to be used in the industrial sector and in the supply of individual properties.

There is also convergence across climate protection scenarios with regard to natural gas and biomass becoming the standard fuels for CHP. It is especially decentralized, small CHP that will contribute to this development. For example, Nitsch et al (2004). expect that, of a total of 80GW newly installed capacity, seven GW will be in the form of gas-fired small heat and power plants until 2020 (scenario "Naturschutz Plus I"). Not all studies are explicit as to just *how small* CHP will become. Fischedick et al. (2002), Matthes and Cames (2000), and Nitsch et al. (2004) expect the supply of individual buildings to be an interesting option.[2] Improved ecological performance of micro cogeneration is expected with the arrival of fuel cells and micro turbines. Fischedick et al. elaborate in some detail possible micro cogeneration technologies. In their "sustainability scenario", CHP includes decentralised heat and power plants, micro gas turbines, Stirling motors, and fuel cells. Fuel cells are also expected to be a promising technology by both Diekmann et al. (2003) and Nitsch et al. (2004). Finally, Diekmann et al. emphasise that a massive expansion of CHP will not come on its own. Even if some expansion is portrayed by the business-as-usual scenario, strong policy intervention is necessary to compensate for price pressure stemming from the liberalization process and to lift CHP shares in electricity generation to the desired level.

All scenarios assume a decreasing *heat demand* due to improved insulation, energy efficient heating systems, and "smart" buildings. This development is even present in business-as-usual scenarios and is reinforced in the climate protection scenarios. To reconcile this development with the desired expansion of CHP power, climate protection scenarios postulate a massive switch from gas- and oil-based individual heating systems to district heating, both in private households and in the business sector. (For example, Fischedick et al. (2002) assume in their "sustainability scenario", that by 2050, 60 % of heat demand will be covered by district heating.) Besides switching to district heating systems, all but Fischedick et al. (1999/2001) (which focuses on the power sector and therefore does not discuss the issue) assume a switch towards individual heating systems based on solar thermal, biomass, or geothermal energy.

---

[2] Matthes and Cames remind us that micro cogeneration can be ecologically inferior to centralized CHP, and should therefore be considered as a "fall-back option" where centralized CHP is not available.

Besides convergence on CHP matters, there is also a relatively broad consensus in policy scenarios about decentralized and distributed electricity generation as a building block for a sustainable energy future. Decentralized generation is a prerequisite for employing many renewable energy sources, such as biomass, because of their low energy density, as pointed out by the Enquete-Commission. The same applies to CHP because of difficulties in transporting heat over great distances. Nitsch et al. (2004) see synergies between CHP and the use of renewable energy sources in so far as the former can help to shape distribution structures toward the needs of the latter. Reference is also made to public resistance against the locating of large power plants as another driver for decentralization. Velte (2004) expects in her scenario 1, "Change of Paradigm", that "more than one third of Europe's electricity production takes place in distributed generation facilities" (Velte 2004, p. 21).

For co-ordinated operation of distributed facilities and network management, Fischedick et al. (2002) and Diekmann et al. (2003) point to the importance of intelligent load management in "virtual power plants", which Diekmann and co-authors expect to be possible by 2030 . Matthes and Cames (2000) as well as Nitsch et al. (2004) raise concerns about whether decentralization is the only and optimal strategy for sustainable energy supply. They point to the possibility of a highly centralized system, with imports of huge quantities of electricity or hydrogen produced from renewable sources in the Earth's sun belt. Fischedick et al. think its possible to combine both strategies.

In the policy scenarios, three developments are emphasised which could compete with the introduction of micro cogeneration: decreasing heat demand, import of renewable energy, and some kinds of domestic renewable use. Decreasing heat demand means that a massive increase in the number of users of CHP (district heat or micro cogeneration) is necessary in order to provide sufficient heat sales to run CHP (see Sect. 3.3). Even so, Matthes and Cames (2000) assume in their scenario "Policy" that CHP heat sales can at best be stabilized by increased industrial use of CHP and improved power-to-heat ratios. Nitsch et al. (2004) are more optimistic, saying that "a further expansion of (decentralized) CHP is compatible with a strategy of substantially decreased heat demand and even offers considerable enlargement possibilities" (Nitsch et al. 2004, p. 28, authors' translation). Both studies also discuss the import of large quantities of renewable power or hydrogen from geographically remote areas as an option after 2030. With respect to possible competition between the use of domestic renewable energy sources and CHP the picture is more complex. There is compatibility when renewable fuels can be used for CHP. Biomass, biogas, and geothermal energy are suitable for this purpose. Fischedick et al. (2002) suggest, however, the use of renewable energy sources rather for district heating than for micro cogeneration, because of higher cost efficiency. In their "long-term scenarios", like Matthes and Cames (2000), they see fossil fuels in CHP as a „bridging technology" for later substitution by biomass or biogas. Furthermore, CHP, even if gas-fired, may structurally pave the way for an increased use of renewable energy. Distribution grids and load management systems that can manage distributed CHP generation can later be utilized for the uptake of power from windmills or photovoltaics. Against this background, Matthes and Cames (2000) as well as Fischedick et al. (2002) and Nitsch et al. (2004) consider CHP a building block for the expansion of renewable energy carriers. On the other hand, electricity produced by fossil-fuel-based CHP competes with electricity from windmills, hydroenergy, and photovoltaics. The studies differ in their assessment. Fischedick et al. (2002) suggest that CHP power, after a growth phase, will have to decrease again to make room for renewable energy use. Nitsch et al. (2004) expect that competition will only take place in the long run, whereas, up to 2030, there is enough conventional power to be substituted by both CHP and renewable energy. The Wuppertal Institute nuclear phase-out study explicitly states that there is no competition: "Climate protection goals can only be reached by combining all the strategic climate protection options available (…) e.g., extension of CHP, application of renewable energies in the power sector, fuel switch by extended gas combined cycle power plants." (Fischedick et al. 1999/2001, p. 33, authors' translation).

Finally, there is potential competition between heat produced on a CHP basis and renewable space heating, e.g., solar thermal applications, wood pellet heating, or geothermal technologies. Nitsch et al. (2004) reflect upon this topic, suggesting that biomass should be rather used in CHP than in heating alone, because CO_2 emissions reductions are greater. The complicated relationships suggest that (micro) cogeneration strategies need to be diligently adjusted to strategies for heating efficiency and renewable energy production.

To sum up, climate policy scenarios for the ongoing energy transformation assign a central role to CHP. Some explicitly include micro cogeneration in their considerations. From an environmental point of view, it is important, though, to fine-tune a micro cogeneration strategy to suit central CHP, renewable energy use, and strategies for reduced heat demand. The scenarios present a remarkable contrast to energy forecasting studies: While the latter do not see much growth potential for CHP, especially for district heating, which is even predicted to decrease, the policy scenarios construct possible developmental paths in which CHP comes out with a great potential and plays a major role in the energy transformation. This, however, is linked to respective policies and framework conditions.

2.3.5 Explorative Scenarios

Similar to policy scenarios, explorative scenarios do not aim at the prediction of the most probable future, but rather at the exploration of possible developments. In contrast to policy scenarios, however, they do not have such a strong focus on normative goals and policy output. Explorative scenarios are instead intended to raise awareness about the breadth of developments which may possibly take place, irrespective of their desirability. Accordingly, these scenarios comprise a much broader and heterogeneous set of variables in thinking about the future than do forecasts and policy scenarios. They do not restrict themselves to consumption quantities, prices, regulations, fuels, or emissions, but also include factors from various dimensions such as cultural values, social practices of energy use, public opinion, political institutions and power relations, market structures, and even completely new technological systems with different performance characteristics. Because many of these factors and their interrelations cannot be modelled in quantitative terms, such scenarios often work with qualitative or semi-quantitative methods. These range from the free creation of narrative storylines elaborated over the systematic unfolding of possibility spaces along the axes of a few main variables, to the analysis of system dynamics with the help of cross-impact matrices or conceptual modelling. Often they involve participatory processes making use of diverse sources of expertise. The result is usually a set of scenarios which represent the extremes of structural variation. In this review, we include four explorative scenario studies, which apply different methods and cover different aspects of the energy system.

The Shell scenarios (Shell International 2001) cover global energy system developments until 2050 and are based on a method developed for coping with uncertainty about the development of business environments. Two scenarios flesh out the storylines "dynamics as usual" and "spirit of a coming age", using the axes of resource constraints, technology development (mainly hydrogen and photovoltaics), and social and personal priorities as key variables. A quantification of key figures has been attached ex post. We include the Shell scenarios because they are among the most well-established explorative scenarios worldwide, with a tradition going back to the beginning of the 1970s.

The International Energy Agency (IEA) presents a study which first undertakes a general review of global and national energy scenario studies and then develops its own set of four scenarios out of them (IEA 2003). Three of these are explorative, one is normative; the three scenarios differing most from "business as usual" have been fleshed out in more detail. The explorative scenarios which are taken into account here are constructed around the main axes of speed of technology development and strength of concerns for the global environment. The IEA scenarios are included because they are the only explorative energy scenarios by an international governmental body.

In an explorative scenario exercise which has been carried out as part of an ongoing project on utility system development in Germany, "Integrated Microsystems of Supply" (*Integrierte Mikrosysteme der Versorgung*, IMV)" (Jäger et al. 2004), four scenarios have been constructed which describe the structure of electricity, gas, water, and telecommunications provision in Germany for the year 2025. Special attention is put on decentralization, service orientation, and interlinkages between the sectors. The scenarios rest on the variation of around 30 key factors; the four most extreme scenario frameworks have been fleshed out narratively and with some quantitative figures. We include these scenarios because they represent expectations of stakeholders who will practically shape the transformation in their daily business.

Methodologically, all three studies are based on participatory processes and stakeholder and expert consultations.

By contrast, Patterson (1999) is actually an essay about a fundamental transformation in electricity systems. It is concluded, however, with two elaborate scenarios of the shape of electricity systems in 2020, "traditia" and "innovatia", which are presented in form of fictive conversations by the people who inhabit them. The scenarios are freely created narratives, although rooted in analytic understanding of the dynamics of electricity system transformation as elaborated in the first part of the book. Because of its peculiarity we have made this study part of our sample.

Micro cogeneration is explicitly mentioned in all studies. It is treated in the most detailed manner in the IMV scenarios, where differentiated shares for micro cogeneration below 4.6 kW_{el} and mini cogeneration below 100 kW_{el} are given for each scenario. Depending on the political and social context, electricity generation shares reach numbers of 2 % (all mini cogeneration) in Scenario D, 3.5 % (0.5 % micro, 1 % mini) in scenario C, 4 % (1 % micro, 3 % mini), and up to 7.5 % in scenario A (2.5 % micro, 5 % mini). Generally, the shares of micro and mini cogeneration increase when scenarios assume a more participatory policy style, more ecological concern, a more competitive market environment with many actors, and a more rapid pace of technology development. Scenario A, characterized by all those features, is most supportive to micro cogeneration. They are assumed to function on the basis of reciprocating engines in the first years, and with larger shares of Stirling engines and fuel cells, partly fired with biogas, towards 2025. Micro cogeneration is expected to be embedded in "virtual micro-utility systems" which coordinate electricity, heat, and water production and consumption on the level of buildings or blocks with the help of Information and Communication Technology (ICT) for monitoring and control.

The other studies also consider micro cogeneration as part of the energy system transformation. The IEA study (IEA 2003) concludes after a comparison of three explorative scenarios, that CHP and micro cogeneration technology are the fifth and sixth most important technology areas out of 16 which appear across all scenarios. This leads them to conclude that the area is ripe for development (p. 108). The scenario "Clean but not sparkling" states that, "In nearly all OECD countries micro-generation and cogeneration in a distributed fashion would increase significantly" (p. 69), initially based on fuel cells, which would then lose out towards 2050 against micro turbines in competition over costs. Whereas Shell only takes short notice of micro cogeneration, as a side-effect accompanying fuel cell development (Shell International 2001, p. 48), Patterson even sees micro cogeneration as a driver for distributed generation options, stating that "as the cogeneration option becomes more widely recognized, its application is going to expand rapidly, onto sites with successively smaller loads (...). In due course the spread of cogeneration in turn may encourage on-site generation of electricity even when no heat is required." (Patterson 1999, p. 147)

Conventional CHP, however, is not given much attention in either of the studies. IEA just mentions in its "Clean but not sparkling" scenario that a strong efficiency increase with existing technologies represents a good environment for CHP. IMV again gives quantitative figures, which range from 17.5 % of total electricity in CHP (including mini and micro) in scenario A, over 15 % in B, 5 % in C down to none in scenario D. They also mention that in scenario A and B it is plausible that 2 % of total electricity are generated by biomass-fired CHP.

Developments with respect to distributed generation again attract great interest in explorative scenarios. Decentralization actually seems to be one of the main storylines which are explored in recent scenario work on the transformation of energy systems. In the course of the general analysis which precedes his scenarios, Patterson offers the following explanation:

> In a liberal competitive framework, investors are more likely to choose smaller generating units which can be built, commissioned, come into service and begin earning a return rapidly. Liberalization also clears away obstacles to on-site generation and cogeneration, and allows entrepreneurial participants to negotiate innovative contractual arrangements, including technical arrangements, among themselves with minimal interference from government. (Patterson 1999, p.148).

In his scenario "Innovatia" (Patterson 1999), he works this latent dynamic out into a story about electricity system regulation failing to assure the provision of system services (load-following), which results in frequent black-outs and rapidly increases the attractiveness of on-site generation for purposes of security of supply. On-site generation then later turns into "local generation, for clusters of users close together" (p. 163), such as housing estates acquiring their own local power station to supply complete electricity and heat services as an add-on to their customers. Similar to Patterson, Shell also comes up with decentralization as a general trend. As the third out of five features which the two scenarios have in common, they refer to "the shift towards distributed or decentralised heat and power supply for economic and social reasons" (p. 58). As with the "Innovatia" scenario by Patterson, it is their scenario "Spirit of a coming age" in which this trend fully unfolds. For Shell, distributed generation could plausibly be driven by fuel cell applications which "start with stationary applications to businesses willing to pay a premium to ensure highly reliable power without voltage fluctuations or outages. This demand helps drive fuel cell system costs below $500 per kW, providing a platform for transport uses and stimulating further cost reductions" (p. 48). They continue the storyline:

> Suppliers of home appliances become major manufacturers and distributors of stationary fuel cells to compensate for saturated OECD markets for their existing products. Commercial and residential buildings take advantage of these low-cost fuel cells, and established natural gas grids, to produce and trade surplus peak-time electricity through internet markets. Hot water is provided by surplus fuel cell heat.

In contrast to Patterson and Shell, IEA and IMV make decentralization a distinctive factor between different possible transformation paths. IEA says about its scenario "Dynamic but careless" that it privileges centralized generation for economies of scale and lower costs (IEA 2003, p. 81).[3] The

[3] However, even in this case decentralized generation is not ruled out completely, if it can compete on the basis of costs. This opens opportunities in areas which

scenario "Bright skies", on the other hand, features a strong increase in decentralized generation as it would allow a decrease in transmission losses (IEA 2003, p. 94). IMV differentiates between technological and organizational dimensions of decentralization in its scenarios about future developments in the utility system. In scenarios A and C, technological decentralisation is high, including large shares of distributed generation. In scenario A, this development is part of technology development driven by societal needs, and part of integrated customized utility services. In scenario C, distributed generation is embedded in fuel-cell-based strategies of international oligopolists.

With respect to developments in the heat market, IMV as well as IEA scenarios are interesting in so far as both speak of factors leading to decreasing heat demand as well as countervailing factors which could secure application potential for CHP. IMV scenario B mentions increased rates of population living in suburbs and increased living area per person as factors which can offset increased heating efficiency. IEA scenario "Bright skies" states that, in the residential sector, CHP would be more broadly used at the same time as solar systems become integrated into buildings for water heating and for power production, more efficient heat pumps would be used for space conditioning in buildings, and passive heating and cooling systems and architectures would also be developed. Building management systems using ICT to monitor and control the energy needs of entire buildings would become increasingly common (IEA 2003, p. 95).

The latter reference to the IEA scenario already shows that competing options such as solar thermal energy use, district heating, passive heating, or fluctuating renewable energy production are often mentioned to increase in parallel with micro cogeneration. This lack of making possible competition explicit is either a sign of imprecise reasoning in scenario construction or expresses the tacit assumption that competition will not become strong enough to block either option completely.

In overview of the explorative scenarios it can be said that micro cogeneration, together with large CHP and distributed generation, especially in fuel cells, is actually one of the central socio-technical configurations whose future is explored by scenario studies which search for possibly path-breaking developments that of the energy transformation may lead to.

---

are sensitive to supply shortages – areas widened after 2015 because of security concerns about terrorism.

2.4 Conclusions

Looking at the four different scenario types we have reviewed, namely forecasts, technology foresight, policy scenarios, and explorative scenarios, it is clearly evident that micro cogeneration is treated in very different ways. This concerns especially whether micro cogeneration is given explicit consideration as an element in the transformation of energy systems and, if so, the role assigned to it. First of all, it is important to notice that even scenario studies of the same type come up with diverging expectations with regard to the role of micro cogeneration. This supports the suspicion that projections about future developments are influenced by the specific actor perspectives and contexts from which they are undertaken.[4] A second aspect is that, beyond the idiosyncrasies of every scenario study, there are also commonalities among studies of the same type. Whereas in forecasts, micro cogeneration does not appear as an important feature of future energy systems or, if so, then only as back-up for premium security of supply, in technology foresight studies it is recognized as a possible application for fuel cells which are considered a potential breakthrough technology. In policy scenarios, it is put forward as part of CHP in general, as one of the most essential elements to achieve climate protection goals; and in explorative scenarios, micro cogeneration plays an important role in driving and being driven by possible developments towards technological decentralization.

The review has thus explicated different interpretations and expectations of actors and revealed varying strategies towards micro cogeneration. Energy policy based on forecasting exercises will be assured that general lines of policy do not have to change, because no changes are to be expected in the structure of the electricity system. And, if there was a market price for the extra security service of distributed generation, micro cogeneration could smoothly enter the market. Innovation policy oriented towards upcoming technologies and informed by technology foresight studies would take care to build up domestic R&D, in order to be part of and possibly lead a breakthrough in the fuel cell market. It is fine, if fuel cells are applied in micro cogeneration but what counts is a novel technological principle, not the performance of actual socio-technical configurations. Climate policy, looking out for blueprints of sustainable energy systems, will take policy scenarios which tell them that CHP is a necessary building block and that overall potential can be extended – if micro applications are added to the portfolio. Finally, companies, and other

4 Other insightful comparative reviews of several energy scenario studies can be found in Smil (2003) and Enquete-Kommission (2002).

actors who draw on explorative scenarios to cope with uncertainty and increase strategic capacities to act under different paths of energy system transformation, will include micro cogeneration in their portfolio as an option for the decentralization path, should it occur.

Since all of these – energy policy, innovation policy, climate policy, companies, and other actors – strongly influence the actual dynamics of energy transformation in general, and micro cogeneration innovation in particular, the different roles of micro cogeneration articulated in the scenarios are all "real" in the sense that they may have some kind of an effect on what the future will actually be. This will largely depend on interactions between the actors who hold the different views supported by the scenarios. Against this background, it is worthwhile to undertake a critical evaluation of the realities (re-)produced by the scenarios.

As a general conclusion from all reviewed scenarios, it appears that micro cogeneration sits somehow in a neglected corner of perspectives on energy transformation. It is either overshadowed by distributed generation, the fuel cell, general CHP, or decentralization. As such, it appears as a niche which develops as a by-product of several differently focused actor strategies. That it is still mentioned by all of these different perspectives, however, also shows that it is at the interface of many different concerns and developments in electricity, gas, and heating. There are also important synergies between micro cogeneration and (infrastructural) developments that are linked to more focal concerns like innovation in fuel cells, emission reductions from conventional CHP, structural change towards decentralization, and distributed generation for security of supply. Such synergies include electricity network management, virtual power plant technology, extension of gas grids and so on. It is doubtful, however, if the niche can be extended without the specific qualities and framework requirements of micro cogeneration being explicitly considered in energy transformation scenarios, so that they can become translated into robust strategies. One reason for this is that, besides synergies, there are competitive constellations (e.g., with the promotion of energy efficiency in buildings and renewable energy generation as central elements of policy scenarios for climate protection). These can play out to the disadvantage of micro cogeneration, especially if its role in the energy transformation does not become recognized and properly assessed.

In order to keep track of such interlinkages, it would be necessary to produce integrated scenarios which combine the various focal concerns and explore how they may interact or compete. This would be an adequate basis for coordination and careful adjustment of actor strategies such as policies, investment strategies, and consumer choice (Grin and Grunwald 2000; Voß et al. 2005).

Such scenarios would also need to shed light on some other blind spots of the scenarios we have reviewed (IMV is an exeption here); for example, surprisingly little information is given with respect to the grid. This, in spite of network access and back-up services playing a crucial role for the transformation of the energy system, electricity in particular. Another aspect is even more crucial. This refers to the actors and social institutions which are behind market price developments, regulatory measures, and technology development in the energy system. In the vast majority of scenarios, they are simply neglected. Different states of output variables are assumed without concern for the factors and processes that bring them about. Some of the policy outputs or technological developments assumed or promoted in the scenarios would seem very implausible or even impossible, if the scenarios took account of the development of actor coalitions or institutions blocking or promoting them.

Similarly, some variables such as market structure, consumer attitudes, or political institutions are mostly considered as "framework conditions" which are simply assumed to be constant or following trend extrapolations. This stands against the actual dynamics of socio-technical change as elaborated above. According to empirical studies of technological innovation and transformation, factors such as utility organization (unbundling, multi-utility, transnationalisation), user attitudes towards self-provision, collective action capacity of technology promotion networks, or institutions of energy policy-making would be important factors which restrict or enable certain paths of energy transformation. A more encompassing view on energy transformation which follows the purpose of grasping relevant dynamics and starting points for shaping strategies would require systematically exploring such factors "behind" the economic and technical parameters of energy provision. In any case, however, this much has become clear from our review of conceptual work and scenario studies: The future of energy systems is not something which can be known. It is being created.

3 The Future Heating Market and the Potential for Micro Cogeneration

Martin Pehnt, Lambert Schneider

As a cogeneration technology, micro cogeneration delivers two useful products: heat and electricity. Consequently, the successful implementation of micro cogeneration depends not only on the development of the electricity market, but also eminently on the trajectory of development of the heat market. Given the reduction in specific energy demand for space heating, the question arises whether the heat sector actually offers an attractive segment for a new technology line.

Based on a description of the current situation of micro cogeneration in Germany, it is therefore our aim to characterize driving factors for the development of the heat market and, based on this, derive a "technical market potential" for micro cogeneration in the residential sector in Germany. We have chosen to focus on the residential sector, because micro cogeneration appears to have a particularly promising future there, due to the existence of a large quantity of single-family houses with relatively low heat demand. In determining this technical market potential, certain technical and non-technical restrictions are considered, such as the existing fuel supply infrastructure or the number of suitable objects, which may present prohibitive barriers for micro cogeneration applications. Please note that in our estimation of the technical market potential, we do not consider further aspects that are important drivers for or barriers of the diffusion of micro cogeneration, such as its economic feasibility compared to other supply options (see Chap. 4), or information on and acceptability of the respective technology (see Chap. 6), administrative and legal constraints (see Chap. 8), etc. Consequently, the technical potential we derive in the following chapter for the residential sector in Germany should be considered as an upper limit for the development of micro cogeneration under the given circumstances.

3.1 Current Situation of Micro Cogeneration in Germany

Micro cogeneration is currently still rather insignificant in the German power generation portfolio. In Fig. 3.1, the installation of new micro cogeneration plants (< 15 kW_{el}) in Germany is estimated for the years 1990 to 1998 and 2002 to 2004.

Fig. 3.1 shows that only very few micro cogeneration plants were installed during the nineties, whereas, in recent years many more micro cogeneration plants have been installed: mainly the "Dachs", a successful product from the company Senertec, which was brought onto the market in 1996, and has since then sold more than 10,000 units. Recently, there has been a steady increase, with 50 % more plants being installed in 2004 compared to 2002. This trend is confirmed by the quantities of electricity fed into the grid under the new CHP law, from 1 April 2002 onwards, as reported by the German association of grid operators VDN. Under this legislation, from April to December 2002, new CHP plants smaller than 50 kW_{el} fed only 15 GWh into the public grid, while the quantity increased to 60 GWh in 2003 and 78 GWh in 2004.

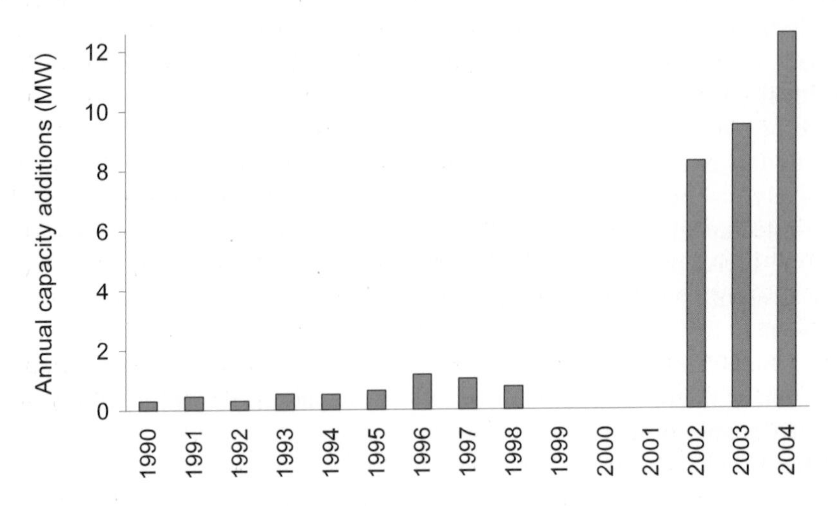

Fig. 3.1. Total annual capacity additions of micro cogeneration plants (< 15 kW_{el}) in Germany from 1990 to 1998 and 2002 to 2004[1]

---

[1] Data from 1990 to 1998 is based on a database maintained until 1999 by ASUE. Data from 2002 to 2004 is based on a recent manufacturer's survey by Öko-Institut (2005). The database by ASUE cannot be considered as complete; however, with more than 5000 reciprocating engines, it probably covers a large share of actual installations. For the survey by the Öko-Institut (2005), only 23

The steady increase was made possible by favorable legislation, including tax exemptions for natural gas and electricity and CHP bonus payments (see Sect. 8.1). In addition, increasing electricity prices plus bonus payments for biomass CHP (adopted in 2004) have worked to increase the volume of micro cogeneration plants sold.

Based on this information, and taking into account that some micro cogeneration plants have been decommissioned following the liberalization of electricity markets after 1997, we estimate that as of the end of 2004, there is a stock of about 60 MW micro cogeneration capacity in Germany, generating about 240 GWh electricity annually. This is still a very small share of overall electricity generation (about 0.04 %); and, though sales volumes are increasing, micro cogeneration is still a very small niche market, far away from broad market introduction.

3.2 Demand Drivers in the Heat Market

Micro cogeneration is particularly suitable in cases where existing gas heating systems need to be modernized, where new houses are built, or where consumers switch from oil or other fuels to natural gas. Thus, the prospects of micro cogeneration will depend on the development of the heat market.

Experience in the past has shown that the diffusion of innovative heat systems depends on favorable general conditions and requires time. The diffusion of condensing boilers is a good example that illustrates this point. Condensing boilers gain higher efficiencies because they use latent heat from the flue gas. With the Netherlands and France being pioneering countries, in Germany condensing boilers were introduced in large numbers in 1991. In that year, every major boiler manufacturer began offering condensing boilers. Today, the total number of installed condensing boilers amounts to 1.7 million, with 270,000 systems being sold per year as of 2002 (Fig. 3.2). Since 2003, gas condensing boilers achieved a market share of 39 % of all, or 53 % of gas boilers.

---

out of 70 manufacturers provided data. However, as the most important producers of micro cogeneration plants responded, we estimate that more than 90% of installed micro cogeneration plants are covered. The 2004 data is based on manufacturer's estimates as of September 2004.

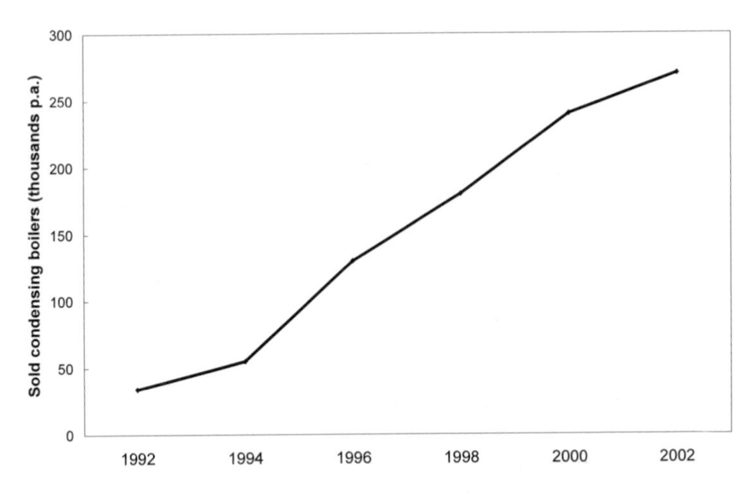

Fig. 3.2. Diffusion of condensing boilers into the German heat market (Source: IEU 2004)

The successful diffusion in Germany can be explained by a number of favorable market conditions that coincided in the early 1990s. Condensing boilers offered comparatively simple installation and reduced electricity consumption. The German Union of Sewage Water *(Deutsche Abwasservereinigung)* allowed the discharge of condensing water into the sewage system. Increased marketing efforts pointed toward the environmental advantages (higher efficiency, lower emissions) and lower fuel costs of the technology.

At the same time, in East Germany the modernization of many buildings and the restructuring of the economy after reunification stimulated the demand for new heating systems significantly. A number of support schemes by local utilities, the German states *(Laender)*, and from the federal level promoted the use of condensing boilers. In this regard, technical, environmental, economic, market, and strategic aspects simultaneously supported this innovation in the heating sector. Despite this coincidence of many positive factors, the diffusion took a long time to go from pilot testing to market success.

Turning to micro cogeneration, it is important to assess whether there is a similar coincidence of driving factors. If one wants to determine the micro cogeneration potential beyond the extrapolation of current trends, considering possible structural changes in the society or in the energy system, the picture becomes very complex, because one has to consider various aspects in determining the structure and level of future heat demand (Fig. 3.3). Each of these aspects will be discussed in detail below.

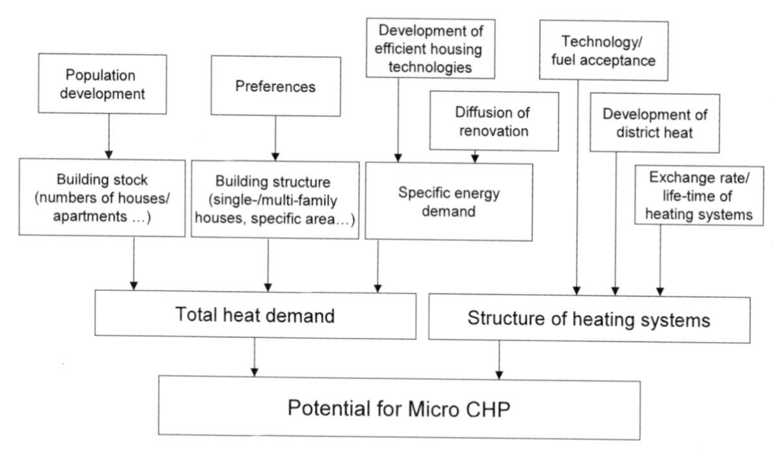

Fig. 3.3. Influential factors on the development of heat demand and micro cogeneration potential in the residential sector

Population Development

Future development of total residential living area depends, first of all, strongly on population development. Estimates of future German population diverge substantially. The Enquête Commission assumes a reduction from today's 82.2 to 67.8 million inhabitants by 2050, with significant decreases particularly after 2020 (Enquête 2002). The Federal Statistical Office of Germany (*Statistisches Bundesamt*) forecasts a reduction to only 75 million in 2050, in the business as usual forecast, with possible ranges of between 67 and 82 million (Destatis 2003).

Living Area

It is likely, however, that, parallel to the downward development of the population, the trend towards larger dwellings and a higher per-capita demand for living area will continue. From 1986 to 2001, the average living area per capita increased from 34.4 to 39.8 m^2 (Statistisches Bundesamt 2003a). This trend toward higher specific area demand will further continue.

Building Stock and Structures

Building stock and structures in Germany are dominated by detached houses with 16.98 million residential buildings existing there in 2001

(Statistisches Bundesamt 2003a). From the existing building stock, 62 % are single-family houses and 20 % are houses with two apartments with the rest having three apartments or more. Most newly built residential buildings are single-family houses, whose share increased from 65 % to more than 80 % in 2002 (Fig. 3.4). This trend towards a higher share of single-family houses will continue. Yet, the absolute number of all newly built residential buildings in 2002 was under the level of 1995. Nevertheless, for micro cogeneration, this trend means that small-sized systems suited for single-family houses will gain in importance.

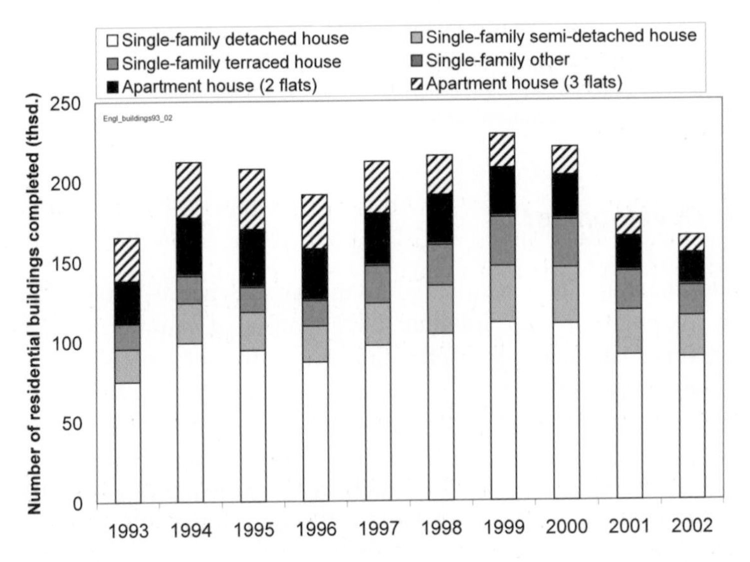

Fig. 3.4. Types of new residential buildings completed in Germany from 1993 to 2002 (Source: personal communication, Federal Office of Statistics and State Office for Statistics and Data Preparation of Bavaria)

Specific Energy Demand

To derive final heat demand, not only does the heated living area need to be determined, but also the specific energy demand. With high levels of insulation becoming state-of-the-art, and legislation requiring more stringent insulation standards (German Energy Savings Decree *Energieeinsparverordnung*, see Sect. 8.1.5), the specific heat demand of new buildings, as well as of the existing building stock, will decrease significantly. Krewitt et al. (2004) assume that, for instance, the specific space-heat demand of a single-family house (year of construction before 1986) will move from 262 kWh/m^2a to 61 kWh/m^2a in 2050, for large multi-family residences (year of construction before 1986) from 160 to 54 kWh/m^2a. This reduction evidently decreases overall residential space-heat demand along with the potential for CHP in the residential sector, because micro cogeneration units have to be designed even smaller than they are, in order to achieve more favorable economic conditions.

However, the annual distribution of heat demand will change in a favorable manner for micro cogeneration. As the seasonally varying space-heating demand decreases due to better insulation, the constant demand for service water will gain in importance. Micro cogeneration plants with small power-to-heat ratios could operate more continuously and longer, improving their economic performance.

Rate of Renovation

How quickly space heating demand will decrease depends on the rate of renovation of the building stock. Today, about 0.5 % of the buildings in Germany are renovated, and their energy needs optimised, each year (Fischedick and Nitsch 2002), leading to a rather slow decrease in energy demand. However, increased rates of renovation are politically targeted.

Structure of the Heating Supply

The structure of the national heating supply – the share between district heating, central heating systems for each building, and heating systems supplying one floor – is of relevance, because micro cogeneration is best suited for central heating systems and is in competition with district heating, where larger cogeneration units are applied. Whereas currently 13.7 % of the households in Germany are heated by district heating systems, almost 70 % use block or central heating, and only 7.9 % use heating systems supplying one floor (Statistisches Bundesamt 2003b). The remaining 9.1 % are individual heating systems. Decentralized warm water supply, e. g. from electric warm water boilers, also reduces the micro cogeneration potential.

The development of district heating networks determines which of the buildings could potentially be supplied with district heat. From an environmental point of view, district heating is not inferior to micro cogeneration, and typically exhibits better economic performance in areas with higher heating (see Chap. 4). A revitalisation and densification of urban areas would significantly increase the development options for district heating. In rural areas, the population density is much lower, with concomitant higher development costs for district heating. However, even in many rural communities, district heat could be realized. Taking the example of a German rural community, Nast (2004) calculates that 86 % of total heat demand is in locations where the total heat losses of the district heating system are below 20 %.

The structure of the national heating supply is also determined by the availability of appropriate fuels. For the German building stock in 2002, natural gas had a share of 47.7 % with an increasing trend (Statistisches Bundesamt 2003b). At that time, 31.8 % of flats were heated by fuel oil, 13.7 % by district heating, and 4.1 % by electrical heating systems. Only 1.6 % of flats had free-standing coal-fuelled heating systems. In new buildings, natural gas had a share of above 75 %, fuel oil 13 %, and district heat 8 % (BGW 2003), with rather large regional variations. Natural gas is the preferred fuel for micro cogeneration applications (see Sect. 9.1), and the large, decentralized natural gas infrastructure in Germany is certainly favorable for micro cogeneration diffusion.

In the future, renewable fuels, such as wood pellets, geothermal heat, and solar collectors, will play an increasing role (Sect. 3.3). Today (2003), only 4.1 % of the heat in Germany is supplied by renewable fuels. Micro cogeneration plants could in the future also use biomass as a fuel. Several engines on the market are already suited for biogas. The introduction of Stirling engines could facilitate the use of wood pellets. However, a combination of solar collectors and micro cogeneration technologies will usually not be economically feasible, as solar heat generation would reduce the potential operating time of a supplementary micro cogeneration plant significantly. In this regard, the diffusion of micro cogeneration would then also depend on the diffusion of solar collectors. In countries with sufficient solar radiation, solar collectors are likely to be economically and environmentally more attractive than micro cogeneration.

Lifespan of Heating Systems

Besides these changes in the building stock, the speed of micro cogeneration diffusion also depends on the exchange rate and lifespan of the heating systems. According to the German Energy Savings Decree, particularly old heating systems have to be replaced before 2007.[2] This, in

---

[2] According to EnEV §9, heating systems with gaseous and liquid fuels with a year of construction prior to 1978 have to put their heating systems out of operation by 31 Dec 2006. If the burner was replaced after 1 Nov 1996, or the

combination with the necessary replacement of a comparatively large number of systems installed prior to 1995 in East Germany, will lead to a significant demand for new heating systems. The National Union of German Gas and Water Utilities (*Bundesverband der deutschen Gas- und Wasserwirtschaft)* estimates that, in addition, approximately 50,000 heating systems will be converted from coal, oil, and electricity to natural-gas-fuelled systems.

In the next section, we develop scenarios for the future technical micro cogeneration potential. In doing so, we take the different drivers described above into account. Already at this stage, however, we can summarize some consequences of these trends:

- The overall heat demand will decrease significantly, despite an increase in specific residential living area per capita; however, the speed of this decrease will mainly depend on the renovation rate of buildings, which today is still very low.
- The structure of heating supply will further shift towards gas-based heating systems at the expense of oil. In addition, renewable fuels are expected to play an increasing role in German heating supply. Whether district heating will succeed in gaining a larger market share remains an open question. From an economic and environmental point of view, an extension of district heating, even in areas with single-family housing, seems possible. The probability of success, however, also depends on various other aspects.
- Given the "gas boom" in the early 1990s and the installation of many gas heating systems after reunification, it can be expected that there will be an increased demand for new systems between 2008 and 2015, when these systems have to be replaced (Krammer 2001). Krammer calculates that the largest share of the total gas heating demand in Germany, particularly in the medium term, will come from the modernization of existing gas heating systems.

3.3 The Technical Micro Cogeneration Potential: Scenarios for Germany

To get a better understanding of possible future development in the heating sector and the role micro cogeneration could play, we quantitatively

---

system modernized such that the legally required level of exhaust gas losses is met, the system does not need to be put out of operation until 31 Dec 2008.

investigate two scenarios of future development until 2050. Such a long-term perspective appears important, because investment decisions for power plants and heating systems lock-up capital for a long period. Today's decisions, for instance in favor of large condensation power plants or with regard to insulation standards for building renovations, will impact the diffusion of CHP technologies and alternative options in the long-term.

The following descriptions are based on a detailed energy model and set of scenarios in Krewitt et al. (2004), developed for the German Environmental Ministry (*Bundesministerium für Umwelt, Naturschutz und Reaktorsicherheit*), and using the building stock model of the Wuppertal Institut, the renewable energy model of the German Aerospace Center (*Deutsches Zentrum für Luft- und Raumfahrt*), and the system technology model of the Institute for Energy and Environmental Research (*Institut für Energie- und Umweltforschung*). Further detailed assumptions, such as potential evolution of the future building market, are described in detail in Krewitt et al. (2004).

3.3.1 Future Heat Demand in the Residential Sector

The wide range of possible development paths for future energy demand in Germany can be characterized by using two different scenarios, namely a "reference scenario" – a business-as-usual scenario as described by the Enquête Commission for Sustainable Energy Supply of the German Parliament (2002) – and a "sustainability scenario". The underlying assumptions of both scenarios are summarized in Table 3.1. In particular, both scenarios assume for Germany an increase in living area, from 3,156 to 4,066 million m^2, and a population decline from 82.2 to 67.8 million people.

Table 3.1. Future development in Germany: Boundary conditions for the reference and sustainability scenarios

Basic data	2000	2010	2020	2030	2040	2050
Population (mil)	82.2	82.1	80.8	77.9	73.3	67.8
Working population (mil)	37.4	37.6	37.2	34.9	32.3	29.6
Households (mil)	38.1	38.5	38.8	38.1	36.1	33.7
Apartments (mil)	37.4	40.1	42.0	41.0	39.5	38.0
Living area (mil/ m^2)	3156	3694	4016	4242	4150	4066
Heated useful area (mil/ m^2)	1458	1514	1539	1564	1530	1509
GDP (bill. Euro 2000)	2075	2499	2953	3367	3747	4089

The main rationale underlying the two scenarios is described as follows:

- In the *reference scenario*, neither stakeholders from industry and policy nor end-users are expected to change behavior. Energy policy measures which have been agreed upon so far will be implemented, and command and control measures will be adopted to assure implementation of state-of-the-art technology. Open markets for electricity and gas are assumed. The German eco-tax is frozen on its current level, according to existing legislation, and there are no specific targets with respect to the reduction of greenhouse gases.
  We assume that in the reference scenario the rate of energy-related renovation of buildings remains nearly constant over time, resulting in a heating demand reduction in the residential sector of only 20 % by 2050.
- The *sustainability scenario,* on the other hand, describes a development path which aims at achieving key sustainability targets, including a reduction of greenhouse gas emissions of 80 % by 2050 (compared to 1990) and a phase-out of nuclear energy. The exploitation of large efficiency potentials will be promoted by appropriate policy measures on both the demand and supply sides. All types of renewable energy sources are exploited to a maximum, as far as the integration of renewable energy sources into the changing energy supply system is feasible under the given structural, economic, and ecological constraints.
  The development of residential heating demand in Germany under the sustainability scenario reflects the fact that there is still a large potential for measures to increase energy efficiency in buildings (Krewitt et al. 2004). It is assumed that the building stock is renovated to a large extent by 2050, including the implementation of additional energy saving measures. As a result of such measures, the heating demand is reduced by nearly 50 % by 2050, particularly due to the decrease of demand for space heating.

The resulting final energy demand for space- and service-water heating is shown in Fig. 3.5. The most obvious difference between the two scenarios is the drastically lower long-term energy demand in the sustainability scenario, which is lower by almost a factor of two. Particularly from 2030 onwards, buildings with very low specific-energy demand diffuse into the market.

Reference Scenario

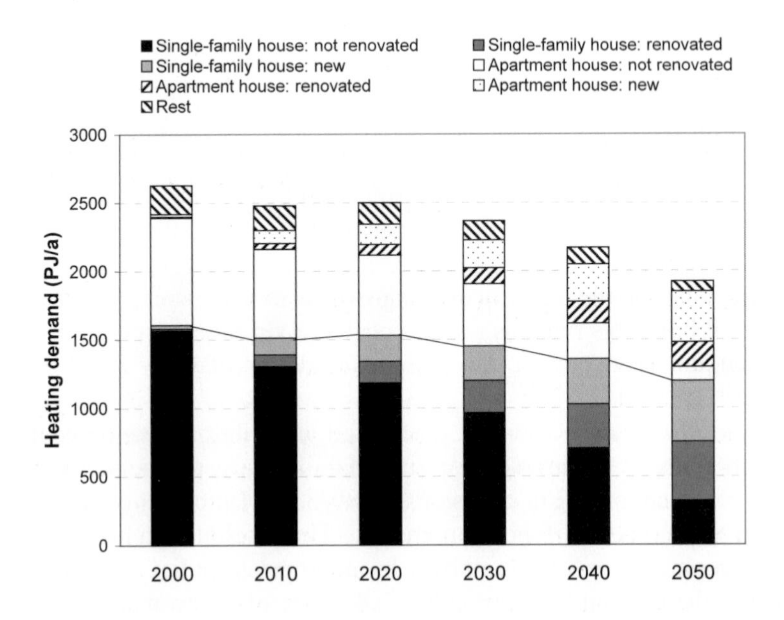

Sustainability Scenario

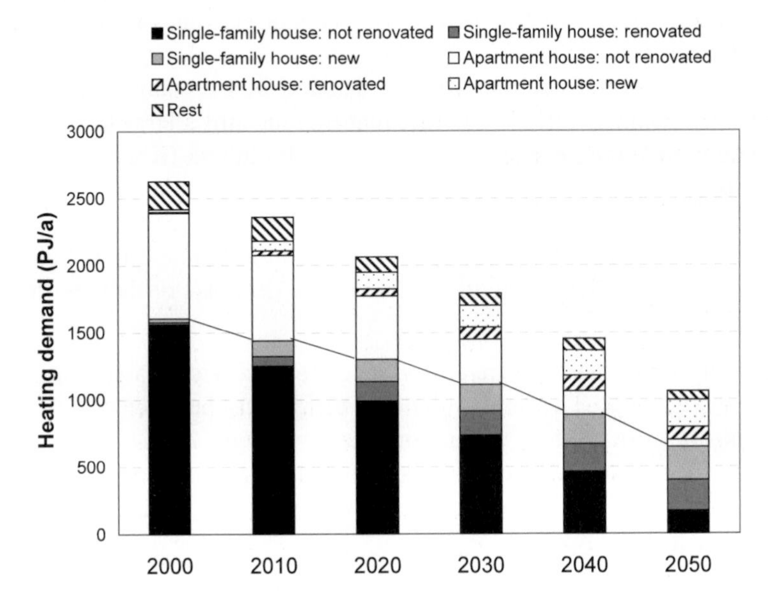

Fig. 3.5. Future development of space- and service-water heating demand for private households according to the reference and the sustainability scenario

In both scenarios, the energy demand for single-family houses dominates. By 2050, this demand becomes equally distributed between on not renovated, renovated, and new houses. Another notably increasing segment is the smaller multi-family houses, whereas larger houses, or even multi-story buildings (see “Rest” in Fig. 3.5), require only 8 % of total 2000 heating demand, and even less in subsequent years. This points to the importance of developing micro cogeneration systems suited for single-family houses if larger market shares are to be harvested.

3.3.2 Future Heating Supply in the Residential Sector

The next question arising is how this heating demand will be supplied. Based on the heating demand and the scenarios outlined above, Krewitt et al. (2004) specify a set of 25 reference applications, including a wide range of residential buildings (single-family houses and apartment buildings with various levels of energy demand, e. g. passive houses, low-energy houses, buildings with average heating demand, etc.); non-residential buildings like schools, hospitals and hotels; district heating networks with different consumer structures; and industry applications in the automotive and chemical industry. For each of these reference applications, detailed heat and electricity demand patterns are specified by using a CHP-planning software (BHKW-Plan 2003). Krewitt et al. (2004) then calculate the possible contribution of CHP to satisfy the heating demand (Fig. 3.6), with the constraint that in the sustainability scenario a CO_2 reduction of 80 % should be achieved. This implies a large contribution of renewable energy carriers to total energy demand.

- In the *reference scenario*, heating supply structure does not change radically. Conventional boilers still dominate the heat market in 2050, providing 69 % of space- and industrial-process heating. Heating from cogeneration even decreases slightly over the next decades, until it grows again after 2020, reaching a share of 13.5 % in 2050. As there is no ambitious greenhouse gas reduction target, renewable energy carriers only marginally contribute to the heating supply (<10 %).
  Corresponding to the more centralized structure of the reference scenario, most new power plants are large condensation power plants (almost 60 GW_{el}) and large cogeneration power stations.
- The *sustainability scenario*, on the other hand, is characterized by a significantly reduced energy demand and a rapidly increasing share of renewable energy sources (particularly biomass) and cogeneration. In the year 2050, the contribution of CHP systems to heating supply has

increased from 12.3 % in 2000 to 25.5 %. In this scenario, a large part of the cogeneration heating supply is based on district heating systems. It was assumed that, until 2050, 50 % of the existing residential buildings, and even 60 % of the new buildings, are connected to a district heating network. This large share is based on analyses of the structural potential of district heating in Germany and the successful development of district heating in other countries, such as Denmark, the Netherlands and Austria. As a consequence, the amount of district heating supply increases in spite of the decreasing total heating demand. As the heating supply from small-scale CHP installations and renewable energy systems is also growing for individual heating systems, the share of conventional oil- or natural-gas-fired installations shrinks significantly after 2020 (Fig. 3.6).

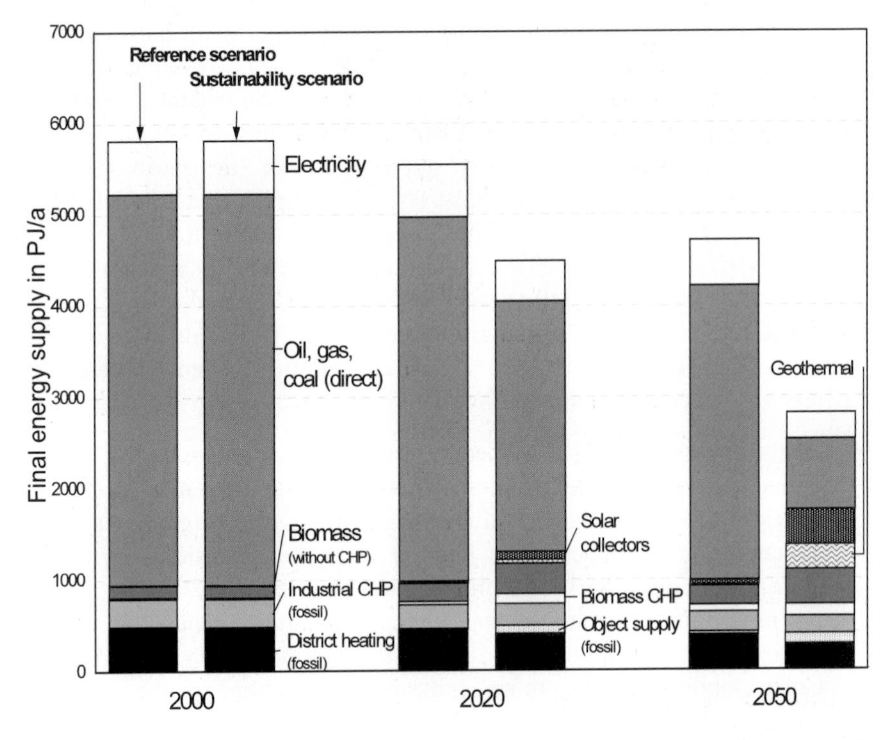

Fig. 3.6. Development of heating supply in two scenarios (Source: Krewitt et al. 2004)

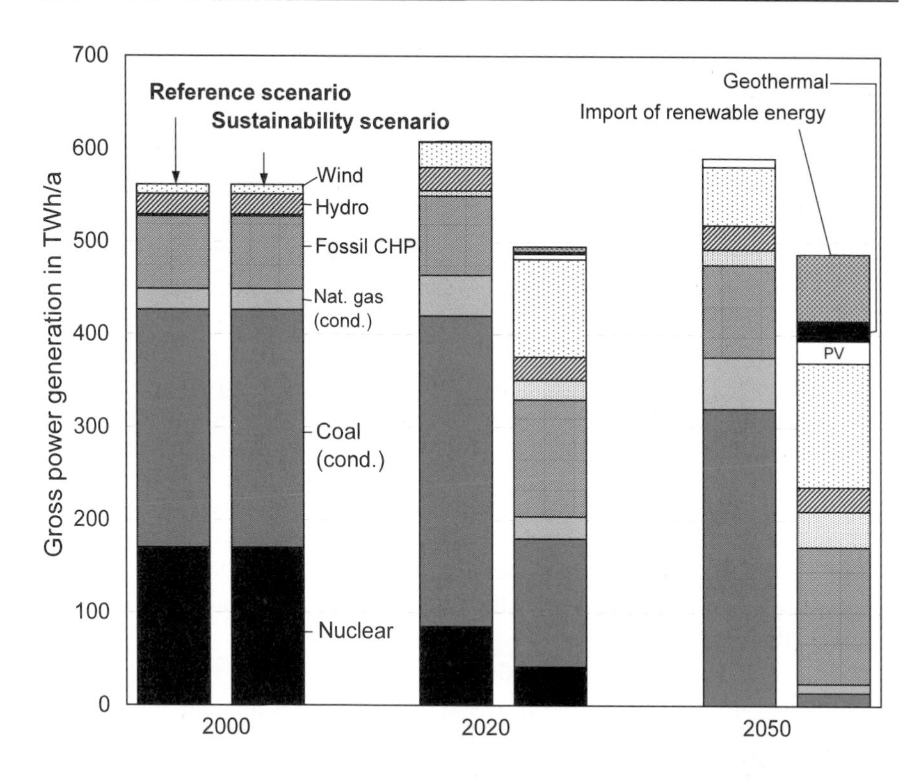

Fig. 3.7. Development of electricity supply in two scenarios (Source: Krewitt et al. 2004). Cond=condensation power station

Even though the heating demand in the reference scenario is much higher than in the sustainability scenario, the cogeneration share in the latter is significantly higher than in the reference case, because political support as well as infrastructural prerequisites (e. g., district heating systems) are assumed to be more favorable for CHP than in the reference scenario. According to the sustainability scenario, with its rather high share of district heating (Fig. 3.7), it is estimated that a micro cogeneration capacity of about 3.3 GW_{el} could be installed in small domestic applications (Table 3.2). It has to be noted that this potential is based on the assumption that district heating will be spread widely. If such a high rate of district heating would turn out to be unfeasible due to barriers against the diffusion of district heating systems, such as low public acceptance or long implementation times, or if market actors such as gas distribution or appliance companies would strongly push domestic CHP systems, parts of the market segment for district heating could also be assumed by micro cogeneration.

Table 3.2. Installed power of decentralized cogeneration systems (< 10 MW_{el}) in Germany, in 2000 and in two scenarios in 2050 (Source: Krewitt et al. 2004)

GW_{el}	2000	2050	
		Reference scenario	Sustainability scenario
Micro cogeneration	≈ 0	0.6	3.3
District heating	1.2	2.8	7.9
Decentralised commercial and industrial CHP	2.7	8.1	10.3
Total	3.9	11.5	21.5

The following can be concluded from the foregoing scenario analysis:

- In the reference scenario, the technical market potential for (micro) CHP is considerably smaller 0.6 GW_{el} than in the sustainability scenario (3.3 GW_{el}), despite the lower heating demand in the latter. Apparently, the climate and CHP policy of the country plays a fundamental role for the diffusion of (micro) CHP.
- The technical market potential for micro cogeneration is relatively small compared to other CHP options, such as district heating and larger commercial or industrial CHP applications. Apparently, district heating is a more favorable option in many cases (see also Chap. 9). However, the split between micro cogeneration and district heating, as shown in Table 3.2, depends on several optimistic assumptions regarding the diffusion of district heat systems. In the sustainability scenario, very high penetration rates of district heating were assumed. If we assume that district heating does not develop so successfully, to a certain degree an additional part of this potential could be transferred to micro cogeneration.
- Only the sustainability scenario meets the expectations of some micro cogeneration industries. Whereas, in the reference scenario, the long-term (2050) micro cogeneration potential of 0.6 GW_{el} is significantly below the future sales envisaged by the micro cogeneration industry[3], in the sustainability scenario, some 3.3 GW_{el} of micro cogeneration systems would approximately meet the industries' expectations.

[3] For instance, Vaillant estimated in 2001 that, already by the year 2010, 100,000 units could be sold per year in Germany and 250,000/a in Europe, corresponding to approximately 0.5 GW_{el} and 1.2 GW_{el}, respectively. For European micro cogeneration, Frost & Sullivan forecast an installed total capacity of 3.5 GW in 2010.

3.3 GW_{el} corresponds to one million systems, taking an average system size of 3.3 kW_{el}. If an increase to a yearly production rate of 50,000 systems within ten years were assumed for the German market, this market size would be reached in 2030. Thus, by 2030 the share of micro cogeneration would be about 50 times larger than today.

- The contribution of micro cogeneration to overall electricity generation can be expected to be rather limited, even in the sustainability scenario. Assuming average full-load hours of 4000 h/a, 3.3 GW_{el} would correspond to an electricity production rate of 13.2 TWh/a. This represents less than 3 % of the 2050 electricity demand in the sustainability scenario.
- The share of renewable energy in the total heating and electricity supply portfolio is strongly determined by the climate policy of the respective country. If there are strong climate-protection goals, high shares of renewable energy carriers will be required to meet them, unless nuclear power and carbon capture and sequestration are options with a significant share. As long as micro cogeneration systems are not readily operationable with renewable fuels (see Sect. 9.4), this would inevitably imply competition between micro cogeneration and renewable energy carriers.
- Nevertheless, even with high shares of renewable heating supply and reduced energy demand, there is ample potential for micro cogeneration in the sustainability scenario . The sustainability scenario clearly indicates that a reasonable and well balanced set of complementary measures – to simultaneously increase energy efficiency, the contribution of renewable energy sources, and the share of CHP in heating and electricity supply –may provide sufficient potential and suitable frame conditions for the application of innovative energy technologies.

4 Economics of Micro Cogeneration

Lambert Schneider[1]

The prospects of a broad diffusion of innovative micro cogeneration technologies depend significantly on their economic performance. In this chapter, the economic viability of micro cogeneration will be assessed from three different perspectives: that of two potential micro cogeneration operators, property owners and vertically integrated utilities, as well as that of society in general.

The operators' perspective should demonstrate in which cases micro cogeneration plants could become an economically attractive alternative to other supply options and which micro cogeneration technologies appear to be particularly favourable under given market conditions. In the operators' perspective, we will include all relevant taxes and subsidies and consider actual tariffs and regulations when calculating costs. The perspective of vertically integrated utilities appears especially appealing since a utility, which supplies natural gas and electricity, while at the same time being a distribution network operator (DNO), may have both synergies and conflicts of interest as regards the operation of micro cogeneration plants.

With regard to the societal perspective, we want to establish whether and in which cases micro cogeneration can be an economically beneficial innovation for society. When calculating costs from the societal perspective, we will exclude all subsidies, taxes and levies. We do not consider external environmental costs in the calculations but we will analyze the range of greenhouse gas abatement costs to economically quantify the benefits for the environment. In all perspectives, the economic performance of micro cogeneration is determined for Germany.

Evaluation of the economic performance of micro cogeneration is undertaken quantitatively for a number of micro cogeneration technologies in different representative buildings under economic conditions in Germany. The economic performance is then compared to other heat and electricity supply scenarios.

[1] The author would like to acknowledge valuable support from Sabine Poetzsch (Öko-Institut), Jens Gröger (Deutsche Energieagentur) and Lars Winkelmann (Berliner Energieagentur GmbH)

4.1 Micro Cogeneration Technologies

The economic performance of micro cogeneration plants is assessed with regard to a number of representative micro cogeneration technologies which are also used for the ecological assessment in Chap. 5 (see Sect. 5.1), allowing a direct comparison of the economic and ecological performance of these technologies.

The selected technologies represent micro cogeneration plants which are either already available on the market or are at an advanced stage of development and are expected to be brought onto the market round about by 2005. Due to the high capital costs of fuel cells and the considerable uncertainty surrounding their achievable target costs (Krewitt et al. 2004), we do not include fuel cells in the comparison of economic performance.

Reciprocating engines with an electrical capacity of about 5 kW and upwards have been available for many years. The Dachs by Senertec is the market leader in micro cogeneration plants in Germany with 10,000 engines having been produced by October 2004 (see Sect. 3.1). Technologies 3, 4, and 5 in Table 4.1 correspond to different versions of the Dachs. The Ecopower by Power Plus Technologies is also a reciprocating engine and corresponds to technology 2 in Table 4.1. Some companies are currently developing smaller reciprocating engines. Since 2003, Osaka Gas in Japan has been marketing a reciprocating engine from Honda with an electrical capacity of 1 kW. The engine is sold together with a boiler and a heat storage tank for about € 5,500 in Japan but is not yet available in Europe. The package costs in Japan are used as the basis for the 1 kW_{el} reciprocating engine, technology 1, in Table 4.1.

Stirling engines are being developed by a number of companies (see Sect. 1.2.2). Several companies (e.g. Enatec, Powerbloc, WhisperTech, MicroGen) are focused on developing engines with a capacity of about 1 kW_{el}, since this size is well-suited to single-family houses. The Whispergen, developed by WhisperTech, is likely to be the first plant of this size to be brought onto the market in 2005. After the completion of field tests in the UK, the British E.ON company Powergen ordered 80,000 WhisperGen units which will be sold for about £ 3,000 in the UK.[2] The WhisperGen is used for specifying reference technology 6. Reference technology 7 corresponds to a somewhat larger Stirling engine of 3 kW_{el}, which is currently being developed by Mayer & Cie. Purchase costs for 2004 are quoted at € 13,000 (ASUE 2004). Reference technology 8 refers to a Stirling engine of 9 kW_{el} that has been developed by the company Solo. About 40 plants

[2] Press release by Powergen from 16 August 2004

Table 4.1. Technical parameters and estimated costs of selected micro cogeneration technologies (for emission factors see also Table 5.1)

		Reciprocating engines					Stirling engines		
		1	2	3	4	5	6	7	8
Capacity									
Electric	[kW]	1.0	4.7	5.5	5.5	5.0	0.8	3.0	9.5
Thermal	[kW]	3.3	12.5	13.9	14.9	12.6	8.0	15.0	26.0
Efficiency									
Electric	-	20%	25%	25%	25%	25%	10%	15%	24%
Total	-	85%	88%	88%	93%	88%	85%	90%	96%
Investment costs									
CHP module	[€]	5,700	13,200	13,300	14,500	14,400	4,300	13,300	25,900
Other	[€]	1,000	3,700	3,700	3,700	3,700	1,200	3,400	5,300
Maintenance costs	€/MWh_{el}	50	36	26	26	26	20	15	10
Economic lifetime	[h]	20,000	80,000	80,000	80,000	80,000	80,000	80,000	80,000

have been sold so far. Purchase costs for 2003 are quoted at € 24,900 (ASUE 2003).[3]

Since the size and performance of these technologies vary, it should be noted that they are not directly comparable. In addition, some cost estimates or parameters are still associated with uncertainty. The results of the analysis can therefore only be indicative; but they do provide an initial assessment of the economic performance of different micro cogeneration technologies. Table 4.1 summarizes the economic parameters for the reference technologies. The price basis of cost estimates is 2005.

4.2 Reference Buildings

Micro cogeneration technologies can be installed in different residential or commercial properties. The economic performance of the plants not only depends on investment, operation and fuel costs but also on the heat and electricity demand characteristics of the buildings. To reflect differences in heating and electricity demand, the economic performance of the eight reference technologies is assessed in five different buildings which represent typical potential applications in Germany.

[3] Currently, the Solo engine is slightly subsidized by the company but short-term target costs are around € 24,000.

Table 4.2. Characteristics of reference buildings for the economic assessment of micro cogeneration plants

		Single-family houses		Apartment houses		Hotel
Heat demand		Low	Average	Low	Average	
Heating surface	[m²]	131	112	457	913	1,263
Heating load	[kW]	7	11	23	67	75
Heating demand	[MWh/a]	9	16	29	109	84
Hot water demand	[MWh/a]	2	3	12	19	38
Electricity demand	[MWh/a]	4	3	13	27	49

We consider two single-family houses, two apartment buildings, and a hotel. The heat demand characteristics of these buildings are taken from a CHP simulation tool (ZSW 2000). The electricity demand is based on a representative survey of households in Germany (FHG-ISI et al. 2004). Key parameters of the five buildings are illustrated in Table 4.2.

In Germany, the conditions for electricity supply from micro cogeneration plants have changed considerably with the liberalization of the electricity market. All electricity consumers, including buildings which are supplied with micro cogeneration plants, may choose their electricity supplier. In the case of single-family houses, micro cogeneration plants usually supply electricity and heat to the property owners who have chosen to install them. However, in apartment buildings an operator of a micro cogeneration plant cannot force the tenants to purchase electricity from the micro cogeneration plant. If a tenant prefers to be supplied with electricity from another supply company, the micro cogeneration plant operator is forced either to produce less electricity or to feed additional electricity into the grid. Feeding electricity into the grid or selling electricity to consumers elsewhere are both less attractive in economic terms. For micro cogeneration operators in properties with several parties, therefore, it is a key prerequisite that a large amount of the generated electricity be sold to consumers directly at the production site. Some operators of micro cogeneration plants offer a discount of 5 to 10 % in relation to the electricity price of the local utility to their consumers in apartment buildings. In practice, this incentive has proven to be sufficient for operators under Third Party Financing arrangements to be able to supply about 80 to 90 % of electricity consumers in apartment buildings. Where several parties are supplied by one operator, we calculate the economic performance on the assumption that only 80 % of the parties are supplied with electricity from a micro cogeneration plant.

In the absence of CHP, heat is usually provided by gas- or oil-fired boilers. In the case of apartment buildings, heat may be generated by a central boiler, which distributes heat to all apartments, or by small boilers for each apartment. In Germany, most apartment buildings are equipped with a central boiler. In this case, the boiler could be replaced or supplemented by a micro cogeneration plant operated by the owner of the building or a third party.

Supplying single apartments in apartment buildings with micro cogeneration is more difficult. In principle, Stirling engines of about 1 kW_{el} could be used in individual apartments instead of boilers since they run rather noiselessly and are not expected to require frequent maintenance. However, in practice, the deployment of micro cogeneration in individual apartments would be difficult for two main reasons: Firstly, connecting a micro cogeneration plant would require modification of the electricity connection. Secondly, most apartments in Germany are rented and property owners do not have economic incentives to invest in micro cogeneration since they do not profit from saved energy costs and can only recoup investment costs to a limited extent via increased rents. In the context of energy-saving measures for buildings, this problem is broadly known and also referred to as the user-investor dilemma. As a result, we only consider centralized heat supply by a single boiler in the case of apartment buildings.

4.3 Heat and Electricity Supply Scenarios

In order to compare the economic performance of micro cogeneration with other supply options, we consider three different scenarios for supplying heat and electricity to the five buildings described above:

- In the *reference case*, electricity is purchased from a utility and heat is generated with a condensing boiler. The boiler is assumed to operate with an average seasonal efficiency of 97% based on the net calorific value.
- In the *micro cogeneration scenario*, the buildings are supplied with heat and electricity by different micro cogeneration plants. Where necessary, an additional boiler covers peak load. Additional electricity is purchased from a utility and, since electricity generation is at times larger than electricity demand, a portion of the electricity generated by the micro cogeneration plant is fed into the grid.
- In an additional scenario, several buildings are jointly supplied with heat and electricity by a *small heat and electricity network with CHP*. In this

case, a CHP plant (with an electrical capacity of about 10 to 50 kW) and an additional boiler supply heat to a small district heating network and provide electricity to consumers in the area through its own small electricity grid. As in the case of micro cogeneration plants in apartment buildings, it is assumed that only 80 % of the consumers within the supplied area would prefer to purchase electricity from the operator of the micro cogeneration plant. In practice, the size of such grids varies and depends mainly on the willingness of property owners to participate in such a system. For this analysis it is assumed that the grid encompasses one block. Based on the settlement structure of large German cities, we estimate 15 single-family houses or 21 apartment buildings to be typical block sizes.

4.4 Economic Parameters

The costs of the different supply options are calculated at 2005 prices without VAT. Inflation is assumed to be 2% per year. Levelized supply costs are calculated with a model. The time horizon is 10 years: from 2005 to 2014. If the technical lifetime of components (boiler, district heat grids) is longer than 10 years, a salvage value is taken into account which is based on linear depreciation over the technical lifetime. The nominal rate of return for capital is 8 % in calculations from an operator's perspective and 5 % in calculations from a societal perspective.

4.4.1 Energy Prices and Subsidies

Future prices and costs for electricity and natural gas are estimated in own scenarios. Electricity prices are based on Eurostat price statistics for different consumption quantities (European Commission 2005a), forward prices at the German power exchange (EEX) and projections for other components of electricity prices, including grid tariffs, the concession levy, and the renewable energy and CHP levies (European Commission 2003; Schlesinger et al. 2000; Deutscher Bundestag 2002; Krzikalla and Schrader 2002). The German concession levy is imposed by the local governments for the use of public grounds. The renewable energy and CHP levies are imposed on consumers for the costs of the renewable energy law and the CHP law (see Chap. 8).

Table 4.3. Average nominal prices and costs for electricity (low voltage) and natural gas, without VAT in €ct/kWh

	2005	2010	2015	2020
Electricity consumer prices (incl. electricity tax and levies)				
Households (3 MWh/a)	17.2	18.2	19.8	21.8
Commercial users (70 MWh/a)	12.9	13.4	14.5	15.9
Feed-in (CHP bonus until 2014)	8.7	9.2	4.9	5,6
Electricity supply costs (w/o taxes and levies)				
Households (3 MWh/a)	11.8	12.5	14.1	15.8
Commercial users (70 MWh/a)	10.1	10.8	12.1	13.7
Natural gas consumer prices (incl. gas tax and concession levy)				
Consumers with 50 MWh/a	3.7	4.1	4.6	5.1
Consumers with 300 MWh/a	3.4	3.7	4.1	4.6
Consumers with 1,200 MWh/a	3.2	3.6	4.0	4.4
Natural gas supply costs (w/o gas tax and concession levy)				
Consumers with 50 MWh/a	2.9	3.3	3.7	4.3
Consumers with 300 MWh/a	2.5	2.9	3.3	3.8
Consumers with 1,200 MWh/a	2.5	2.8	3.2	3.6

Natural gas price projections are based on Eurostat statistics (European Commission 2005b), past border prices (BMWA 2005) and projections (Schlesinger et al. 2000; Deutscher Bundestag 2002). Table 4.3 illustrates the assumptions for prices and costs for electricity and natural gas.

The operation of micro cogeneration plants is promoted by legislation in Germany (see Sect. 8.1.3) with the most important effects being:

- exemption from electricity tax for power plants with an electrical capacity below 2 MW,
- exemption from natural gas tax for CHP plants with an average energy efficiency above 70%, and
- payment of a bonus of 5.11 €ct/kWh for electricity fed into the grid from small CHP plants commissioned before 2009.

This legislation is considered in the economic analysis of independent operators and vertically integrated utilities. In the societal analysis, the costs of the different supply options are assessed without consideration of natural gas and electricity taxation, the concession levy, the renewable energy and CHP levy, or the CHP bonus.

4.4.2 Economic Losses or Benefits for Distribution Network Operators

A broader diffusion of micro cogeneration can result in economic benefits for the electricity network, including savings on transmission and distribution losses, the deferral of upgrades of the electricity network, or the removal of local bottlenecks in the distribution system. The economic benefits or costs for the electricity network depend on the penetration of distributed generation. With low and moderate use of distributed generation, costs can be saved while with very high penetration the network may even need to be reinforced if significant quantities of electricity are fed to the higher voltage level (see Chap. 9 for more details).

Since power from micro cogeneration is mostly consumed on-site and rarely fed into the medium voltage level, it can be expected that micro cogeneration will have economic benefits for the electricity network. Certainly, these benefits are as yet difficult to quantify. As a conservative (i.e., lower-end) assumption, we estimate that overall transmission and distribution costs in a network would be reduced by about 10 % if 30% of low-voltage power were to be generated by micro cogeneration or other distributed generation. We consider this economic benefit in calculations made from the perspective of integrated utilities and from a societal perspective. From the perspective of independent plant operators, we make calculations using existing tariffs in Germany and include the CHP bonus (see Table 4.3.).

4.4.3 Economic Perspectives of Different Operators

Next to these grid effects, there are several other costs or economic benefits that depend on the type of plant operator. We consider three different potential operators of micro cogeneration plants to be particularly important: Property owners, energy service companies, and vertically integrated utilities.

- The property owner owns a plant, purchases natural gas and additional electricity, and usually contracts a company for regular maintenance of the plant. As the plant operator, the owner both bears most economic risks and profits from economic benefits.

- Energy service companies can provide several types of Third Party Financing services. In cases involving small CHP plants, they usually install, own, and operate the plant and sell heat and electricity to consumers. Heat and electricity prices are agreed upon in a service contract with the property owner. In cases with several electricity consumers, the energy service company also establishes electricity supply contracts with each consumer. Thus, the energy service company both bears most economic risks and profits from economic benefits.
- Energy supply utilities can provide services similar to those of energy service companies. For example, a service contract may include the following constellation: The property owner purchases a plant from the utility. The utility installs the plant, conducts regular maintenance at a fixed cost, sells natural gas and additional electricity to the property owner, and offers a special rate for electricity fed into the grid and the purchase of electricity. A vertically integrated utility may also connect micro cogeneration plants to a virtual power plant, increasing the value of electricity generation. In this scenario, the property owner and the utility share economic risks and benefits. This constellation may be particularly attractive for vertically integrated utilities that are simultaneously electricity and gas suppliers and network operators.

Table 4.4 summarizes the economic benefits and barriers for different operators. Energy service companies and utilities can often purchase natural gas and additional electricity at lower costs compared to property owners and consumers. In addition, they can use synergies and gain cost reductions by purchasing, installing, operating, and maintaining many plants. However, energy service companies may run into difficulties when implementing Third Party Financing models since the property rights and access to the micro cogeneration plant have to be registered as easement in land registers. For small micro cogeneration plants, this might be a prohibitive barrier since transaction costs are high and many property owners dislike the idea of such easements being associated with their properties. Thus, it appears more likely that small micro cogeneration plants will be owned by the property owners while they may be installed and/or operated by energy service companies or utilities.

Vertically integrated utilities could benefit from closer consumer relations. Consumers with a micro cogeneration plant may prefer long-term contracts for electricity and natural gas supply and tariffs for electricity fed into the grid. Vertically integrated utilities may also gain additional economic benefits if many micro cogeneration plants are integrated into a *virtual power plant*, enabling utilities to shape power generation and reduce grid congestion (see Chap. 9). In addition, they could increase their natural gas sales. On the other hand, on-site electricity generation by micro cogeneration plants reduces the quantity of electricity distributed and thus the revenues from grid charges. A key question is whether distribution network operators will be allowed to pass on the loss of revenues from network charges for distribution to all consumers. If not, they have strong incentives to generally discourage distributed generation.

Table 4.4. Benefits and barriers for different operators of micro cogeneration plants

	Property owner	Energy service company	Vertically integrated utility
Natural gas purchase conditions	O	+	+
Provision of additional electricity	O	+	+
Plant purchase and installation	O	+	+
Plant maintenance	O	+	+
Contracts and legal issues	+	-	O
Network charges	+	+	- / O
Ancillary network services (e.g. load shaping, minutes reserve, etc.)	O	O	+

"+" = advantageous conditions, "O" = neutral, "-" = disadvantageous conditions

The potential economic benefits for a vertically integrated utility are as yet difficult to quantify. In our economic analysis, we assume that a vertically integrated utility has economic benefits from additional sales of natural gas and better consumer relations which equal a 5% reduction on natural gas prices. In addition, we estimate a net benefit of 0.5 €ct/kWh electricity generation if all micro cogeneration plants are connected to a virtual power plant.[4]

[4] Note that a *virtual power plant* involves considerable costs (e.g. for information technology) and that 0.5 €ct/kWh is understood as a net benefit.

4.5 Micro Cogeneration Electricity Generation Costs

In this section, electricity generation costs of micro cogeneration plants are determined independently of the heat and electricity supply characteristics of the buildings referred to above. Electricity generation costs are calculated for two cases: Operation with a supplementary boiler and stand-alone operation.

A supplementary boiler is regularly installed since it is economically more advantageous for supplying peak heat demand. This allows cost-saving through the installation of a smaller CHP unit which can then operate more continuously. In addition, a supplementary boiler can provide heat during maintenance or failure of the CHP plant. For small micro cogeneration plants, stand-alone operation may also be feasible, given that the costs of small supplementary boilers are relatively high. However, in such cases, the operation time of micro cogeneration plants is limited by the heat demand characteristics of the building. Weather conditions and typical building standards in Germany imply a maximum full load operation time of about 1,500 to 2,000 hours per year for stand-alone operation.

Electricity generation costs are calculated by adding up levelized investment, maintenance, and fuel costs, and subtracting a credit for avoided heat generation costs. Where the CHP plant is operated with an additional boiler, only marginal fuel costs of heat generation are avoided. With stand-alone operation, the micro cogeneration plant replaces the boiler, and thus avoids full heat generation costs, including investment, maintenance, and fuel costs of the boiler.

Fig. 4.1 shows levelized electricity generation costs against full load operation hours for a number of micro cogeneration plants for the two cases: stand-alone operation and operation with a supplementary boiler. The figure reveals several interesting results:

- **Small micro cogeneration plants for single-family houses (about 1 kW_{el}) are only economically feasible in stand-alone operation.** If operated with an additional boiler, electricity generation costs are significantly higher than electricity purchase costs for households. In contrast, in stand-alone operation, electricity generation costs become lower than electricity purchase costs from about 1,000 full load operation hours onwards. Thus, although the heat demand limits stand-alone operation to about 1,500 to 2,000 hours per year, the stand-alone operation of micro cogeneration is economically feasible.

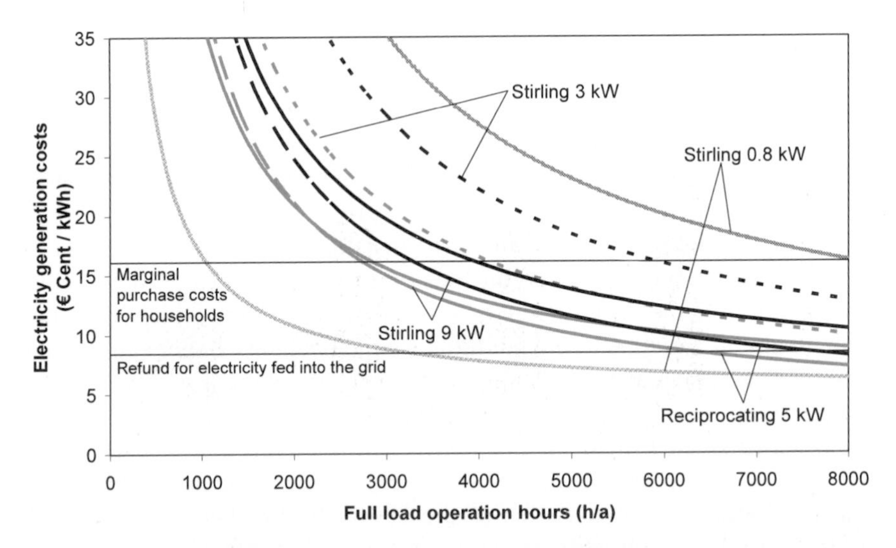

Fig. 4.1. Levelized electricity generation costs (2005 prices without VAT) of different micro cogeneration plants for stand-alone operation (grey lines) and operation with a supplementary boiler (black lines)

- **Micro cogeneration plants for apartment buildings are only economically viable if operated with a supplementary boiler and for at least 5,000 hours or so.** Different from micro cogeneration plants for single-family houses, all larger engines have electricity generation costs far above electricity prices if operated as stand-alone plants for only 1,500 to 2,000 hours per year. With a supplementary boiler, most plants are economically feasible if they provide electricity to on-site consumers for at least 3,500 to 4,000 full load operating hours. Since up to 30% of the generated electricity is typically fed into the grid, a full load operation time of about 5,000 hours ensures an economically attractive operation in most cases.
- **Sufficient on-site electricity demand is a prerequisite for economic operation.** It is not economically feasible to install micro cogeneration plants with the purpose of feeding electricity into the grid, since the total return – including a bonus payment according to CHP law in Germany – is lower than the electricity generation costs in most cases. Consequently, the availability of a sufficient quantity of electricity consumers at micro cogeneration plant sites is a prerequisite for their diffusion. This certainly limits the potential for micro cogeneration.

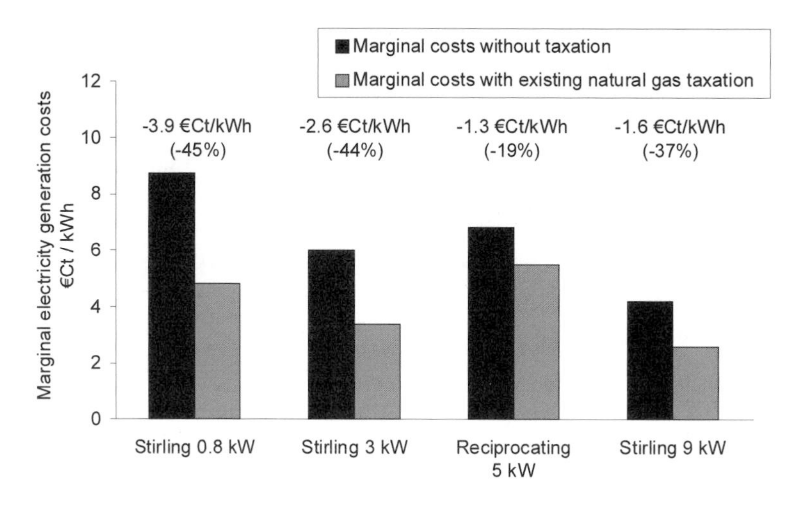

Fig. 4.2. Marginal electricity generation costs of different micro cogeneration plants with and without natural gas taxation (nominal prices in 2005)

The1 kW_{el} reciprocating engine is not included in Fig. 4.1 since electricity generation costs are rather high (about 40 €ct/kWh). This is mainly due to considerable investment and operation costs. It should also be noted that the stand-alone operation of Stirling engines may be more reliable than with reciprocating engines. If the motor of a stand-alone operated reciprocating engine fails, heat supply is disrupted. Repairing the engine in case of failure or regular maintenance activities may take some time, involving a supply risk, particularly in winter. With Stirling engines, heat is generated external to the engine. Some Stirling engines are designed in a manner that allows continued heat supply in case of failure in the Stirling process.

Natural gas taxation presents another important factor influencing electricity generation costs. In Germany, efficient CHP plants are exempt from natural gas taxation, while fuel combustion in boilers is currently subject to a tax of 0.55 €ct/kWh natural gas (see Sect. 8.1.2). Fig. 4.2 shows that the effect of the natural gas taxation on marginal electricity generation costs is noteworthy. The tax on natural gas fired in boilers increases the credit for avoided heat generation costs. As a result, marginal electricity generation costs of the selected technologies decrease by 1.3 - 3.9 €ct/kWh electricity or by 19 to 45 % respectively. Fig. 4.2 also illustrates that the taxation effect differs considerably according to CHP technology. The effect is greatest in the case of the 0.8 kW_{el} Stirling engine (technology 6) due to the high ratio of heat production (and thus avoided tax) relative to electricity generation. Thus, the low electrical efficiency of about 10% is an advantage in terms of natural gas taxation. However, this taxation effect is not backed by the environmental performance of such CHP plants, since plants with lower electric efficiency have poorer environmental performance (see Chap. 5). In this light, it may be necessary to reconsider the criteria for natural gas tax exemption in Germany.

4.6 Economic Performance of Micro Cogeneration in Different Buildings

The economic performance of the selected micro cogeneration technologies (Table 4.1.) in the reference buildings (Table 4.2) is determined by simulating operation of the micro cogeneration plants, taking into account the actual heat and electricity demand characteristics of the buildings. The simulation is conducted with the tool "BHKW Plan", developed by ZSW (2000). It is assumed that micro cogeneration plants mainly follow heat demand and that surplus electricity is fed into the grid. This way of operation is economically viable in Germany because marginal electricity generation costs are usually lower than the refund for electricity fed into the grid. However, this applies only for plants that are eligible for the German CHP bonus of 5.11 €ct/kWh. According to current legislation in Germany, the CHP bonus will only be paid for plants that are commissioned before 2009.

The simulation delivers electricity and heat generation quantities for the micro cogeneration plants as well as the share of generated electricity that is used in each building and the share that is fed into the grid. The economic performance of different supply cases is assessed by comparing total costs for heat and electricity supply for each building. Total supply costs are calculated by levelizing all costs (investment, operation, fuel and electricity purchase) and all revenues (electricity fed into the grid) over a period of ten years.

4.6.1 Independent Operators' Perspective

Fig. 4.3 illustrates the economic performance from the perspective of independent operators for the three supply scenarios: the reference scenario, the micro cogeneration scenario, and the small district heat scenario (see Sect. 4.3). The economic performance is illustrated as total heat and electricity supply costs relative to the reference scenario. For the micro cogeneration scenario, Fig. 4.3 shows the economically most favourable micro cogeneration plant (referred to as "best plant").

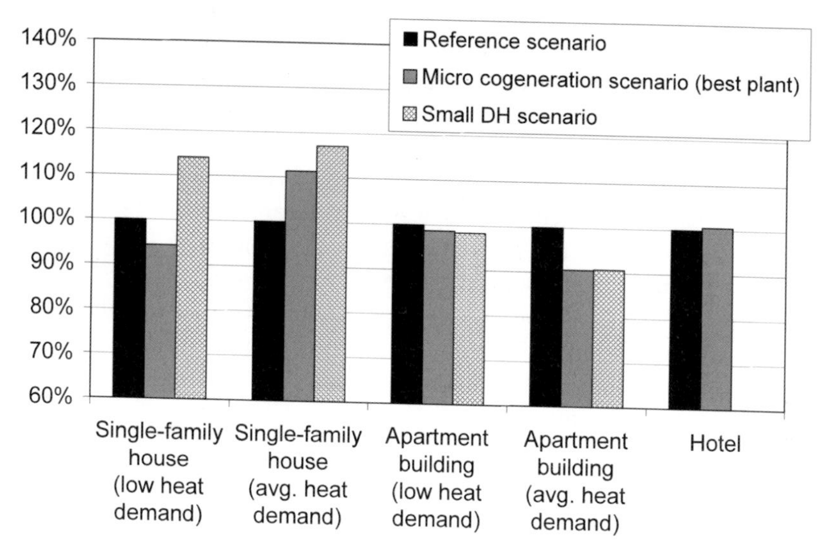

Fig. 4.3. Heat and electricity supply costs for the reference buildings from the independent operators' perspective

For the single-family house with low heat demand, the 0.8 kW_{el} Stirling engine (reference technology 6) is the economically most attractive option for several reasons. Firstly, a supplementary boiler next to the micro cogeneration plant is not required in such a building, since the heat output of the micro cogeneration plant is sufficient to cover the maximum heat demand. Secondly, the investment costs of about € 4,300 for the CHP plant are relatively low. Finally, the relatively low electrical efficiency of 10% is not disadvantageous in economic terms since it better reflects the ratio of heat and electricity demand in single-family houses. The low electrical efficiency leads also to significant tax advantages (see above).

In the case of the single-family house with an average heat demand, the micro cogeneration plants considered are not economically attractive. The 1 kW_{el} reciprocating engine (reference technology 1) has relatively high investment costs and the 0.8 kW_{el} Stirling engine (reference technology 6) would make it necessary to install a small supplementary boiler to cover the peak heat demand, involving significant additional costs. Other micro cogeneration plants with sufficient heat capacity and reasonable investment costs may be economically viable for this building type.

For apartment buildings, both micro cogeneration and small district heat with CHP are more attractive than the reference case, particularly in cases of apartment buildings with average heat demand. For small district heat networks, the heat supply infrastructure is the most important cost factor.

Therefore, such small heat networks are only economically attractive in areas with a sufficiently high density of heat demand. This applies to blocks of apartment buildings in cities but not to areas with dispersed single-family houses. The main advantages of small grids are the significant cost reductions brought about by electricity generation with reciprocating engines. For example, specific investment and maintenance costs of small reciprocating engines with a capacity below 10 kW_{el} are about twice as high as those of engines with a capacity of 30 to 50 kW_{el}. In addition, operators of small district heat networks benefit from purchase costs for additional electricity being considerably lower than regular electricity prices for households. Furthermore, both micro cogeneration and small district heat networks profit from their exemption from natural gas and electricity taxation as well as from the CHP bonus.

In the case of the hotel, the economic performance of micro cogeneration is similar to the reference scenario, despite rather favorable heat and electricity demand characteristics for cogeneration. The main reason for micro cogeneration not being as attractive in economic terms is that electricity prices for the hotel are already considerably lower than for households.

4.6.2 Perspective of Vertically Integrated Utilities

Total heat and electricity supply costs from the perspective of vertically integrated utilities are illustrated in Fig. 4.4. The figure shows that both micro cogeneration and small district heat systems are less attractive for vertically integrated utilities than for independent operators. The loss of revenues from network charges is greater than the estimated benefits in terms of long-term cost savings in the electricity grid, increased natural gas sales (estimated at a 5 % lower price), and economic benefits from connecting micro cogeneration plants in a virtual power plant (estimated at 0.5 €ct/kWh). This result even holds if economic performance is calculated with a natural gas price that is 10% lower and if benefits from the interconnection of micro cogeneration plants are estimated at 1 €ct/kWh.

Consequently, a key question is whether and to what extent distribution network operators may pass on the loss of revenues from network charges to all consumers. Regulation plays an important role here. Only if distribution network operators are adequately compensated for these losses, will they develop initiatives to make use of the potential benefits from distributing micro cogeneration plants. Adequate compensation could also help to overcome the current policy of some distribution network operators of impeding the connection of CHP plants (see Chap. 8).

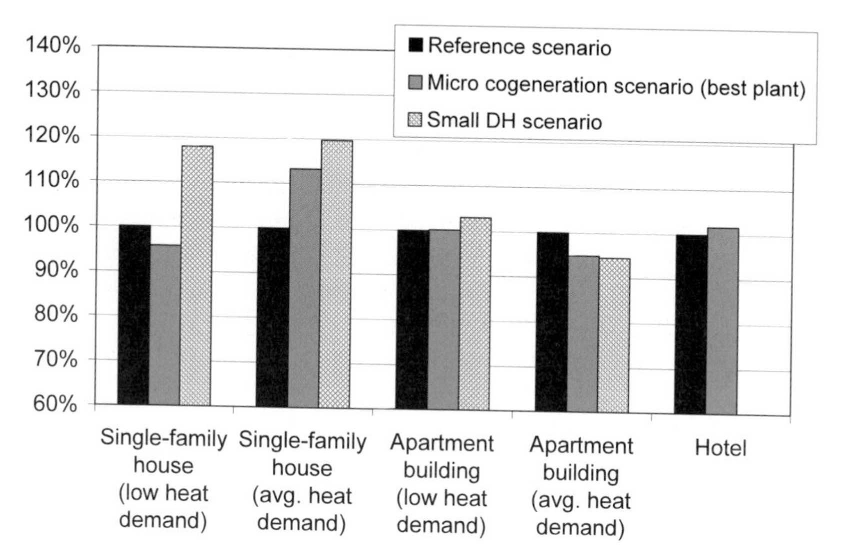

Fig. 4.4. Heat and electricity supply costs for the reference buildings from the perspective of vertically integrated utilities

4.6.3 Societal Perspective

Fig. 4.5 compares heat and electricity supply costs from a societal perspective. Taxes (natural gas, electricity), levies (concession levy, renewable energy levy, CHP levy) and subsidies (CHP bonus) are not included. Please note that external environmental costs are not considered either.

From a societal perspective, micro cogeneration does not yet have cost advantages compared to the reference scenario. For the single-family house with a low heat demand and the apartment building with an average heat demand, the costs of the Stirling engine of 0.8 kW_{el} (reference technology 6) are comparable to the costs in the reference scenario. This plant is also the most attractive micro cogeneration technology in economical terms for other buildings. However, it shows a less favourable environmental performance: GHG emissions are about 11% higher than in the reference scenario, where electricity would be generated by a natural gas combined cycle power plant (see Chap. 5). Consequently, the relatively good economical performance of this micro cogeneration technology is clouded by a relatively poor environmental performance compared to other micro cogeneration technologies.

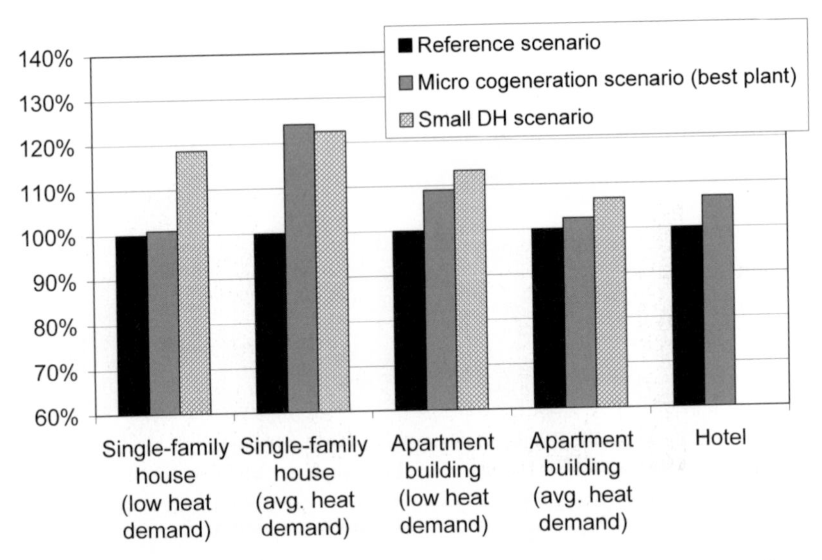

Fig. 4.5. Heat and electricity supply costs for the reference buildings from a societal perspective (without taxes, levies and subsidies)

Other micro cogeneration technologies are significantly less attractive from the societal perspective. This is illustrated in Fig. 4.6 where the economic performance of different micro cogeneration technologies in the apartment building with an average heat demand is compared. Reference technologies 2 to 8 are all favourable for an independent operator but face additional costs from a societal perspective, leading to GHG abatement costs in the range of 100 to 250 €/t CO_2 equivalent. Similar results apply to other building types.

Thus, it is important to take the environmental performance into account when assessing the economic attractiveness from a societal perspective. Currently, micro cogeneration technologies still lack either economic attractiveness or good environmental performance. Low investment costs are traded-off with a less favourable environmental performance. In the further development of micro cogeneration technologies, a key challenge will be to improve electric and total efficiency without increasing unit costs.

Very small district heat systems with CHP are also not yet attractive from a societal perspective. However, larger district heating systems can be expected to have an economically better performance, mainly as a result of larger CHP plants having lower investment costs and electric efficiency being greater while heat losses in the distribution system do not substantially increase.

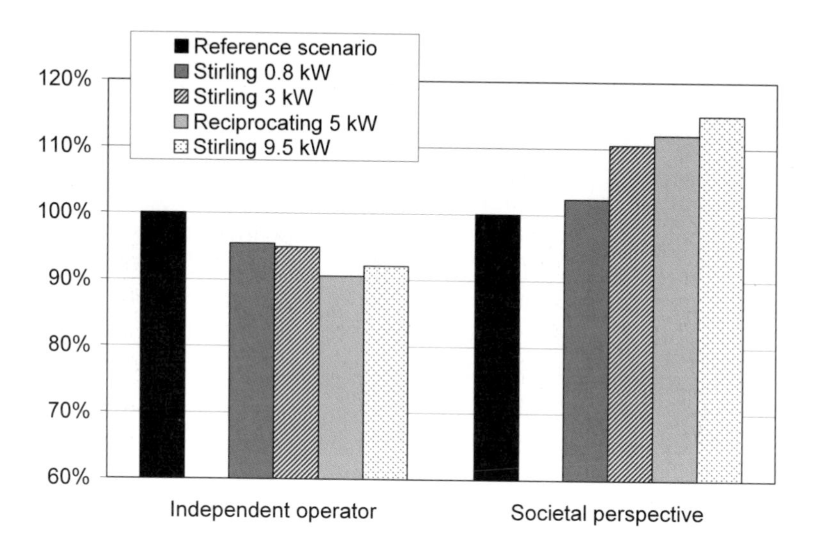

Fig. 4.6. Economic comparison of different CHP technologies in the apartment building with an average heat demand

4.7 Conclusions

Micro cogeneration is becoming an economically attractive option for consumers. Reciprocating engines of about 5 kW_{el} are already on the market and are economically feasible in larger apartment buildings with a sufficient heat and electricity demand. For single-family houses, systems of about 1 kW_{el} that fully substitute the boiler appear particularly advantageous in economic terms. Thus, for the market of single-family houses it is an important prerequisite that engines are designed to cover typical heat loads in the range of 10 kW at reasonable costs.

The economic attractiveness of micro cogeneration for consumers largely rests on a number of regulatory advantages. These include the exemption from electricity and natural gas taxation, the payment of a CHP bonus for electricity fed into the grid and the avoidance of concession levies and grid charges for electricity generated on-site. However, the picture is quite different if these effects are excluded from economic calculations: From a societal perspective, as long as external costs are not included in the calculation, none of the analyzed micro cogeneration technologies is yet economically viable when taxes, levies and subsidies are not considered. In contrast, heat and electricity supply costs are substantially higher for most technologies compared to heat supply by a boiler and electricity purchase from a utility. Investment costs need to decrease further to make micro cogeneration a viable option from this perspective. Sensitivity analyses show that these results hold with varying energy prices, e.g. with natural gas prices being 20 % higher or lower and electricity prices being 10 % higher or lower.

With current micro cogeneration technologies, there is also a trade-off between environmental and economic performance. For example, the Stirling engine with a capacity of 0.8 kW_{el} has a rather good economic performance but a relatively low efficiency, resulting in higher GHG emissions compared to a condensing boiler and electricity generation in a natural gas combined cycle power plant.

The economic feasibility of micro cogeneration also depends on the type of operator. Vertically integrated utilities may benefit from closer costumer relations, increased natural gas sales and the provision of network services by integrating many plants into a virtual power plant. However, they may also face a loss of revenues of network charges which may not be compensated by long-term cost savings in the network due to micro cogeneration. A key question is whether distribution network operators may pass on these losses of revenues to other consumers, thereby increasing the average grid charges. If so, micro cogeneration plants may be an interesting option for vertically integrated utilities. Otherwise, the loss of revenues from grid charges is likely greater than the economic profit from other benefits and, consequently, distribution network operators would have economic incentives to impede the diffusion of micro cogeneration.

Small district heat networks are an economically viable alternative to micro cogeneration in areas with higher heat densities (i.e. apartment buildings). In such areas, micro cogeneration or small district heat networks can realize significant overall cost reductions (>10 %) compared to the separate supply of heat and electricity. However, from a societal perspective, without including external costs in the calculation, this supply option is also not economically attractive. Larger district heat systems may be more favorable in areas with higher heat demand, since relative investment costs for CHP plants are lower, while electric efficiency is significantly higher and distribution losses do not substantially increase.

In summary, micro cogeneration is not yet but could become an economically and environmentally attractive option in areas with low heat density. For its diffusion, reasonable investment costs are a key prerequisite. From a societal perspective, high efficiencies are important to achieve emission reductions at a reasonable cost.

5 Environmental Impacts of Micro Cogeneration

Martin Pehnt, Corinna Fischer[1]

During the past decades, many efforts have been made to improve the environmental performance of energy conversion systems. One of the most prominent examples is the reduction of sulfur dioxide emissions through flue gas desulfurization. Other environmental impacts have also been reduced in the course of implementing modern power plant technologies, power plant modernization or restructuring, and switching fuels.

Nevertheless, the increasing demand for services, mobility, light, communication, and heating is still responsible for significant environmentally damaging effects, which are caused throughout the whole energy conversion life-cycle: from primary fuel extraction, through processing, distribution and conversion into electricity or heat, to energy delivery and use. The related impacts, such as oil tanker spills, the accumulation of radioactive waste, the emission of hazardous substances, and the use of non-renewable energy resources, continue to damage our environment.

Among the most challenging issues today is the anthropogenic emission of greenhouse gases. The scientific community acknowledges that far-reaching measures have to be taken to reduce the increase in CO_2 concentration in the atmosphere, in particular with regard to energy use. However, even the unambitious goals of the Kyoto protocol for the reduction of greenhouse gas emissions will not be met by many countries worldwide.

It has often been argued that, due to its decentralized nature and high total efficiencies, cogeneration, and ultimately also micro cogeneration, contributes toward environmental relief. The present chapter evaluates this assumption and investigates the potential environmental effects of a diffusion of micro cogeneration on the following levels:

1. Foremost, the direct impacts of power plant *operation*. This concerns, for example, the emission of pollutants or climate-altering gases –

[1] CF: Section 5.3.3

which depends significantly on the efficiency of fuel conversion –, the use of lubricants or other auxiliary materials and the like.

2. The upstream environmental impacts of *fuel supply*, e.g. the extraction, processing, and transport of gas or other fuels required for the operation of the system. In cases where micro cogeneration systems have lower fuel demand than competing technologies, the upstream impacts may be smaller as well.
3. The environmental impacts of *power plant manufacturing and construction*. Typically, due to the long life span of the systems, power plant production is of lower importance than the other life cycle stages, unless some very environmentally detrimental materials are used (Pehnt 2002).
4. The impacts of electricity and heat *distribution*. The transport of electricity (in electricity grids) and of heat (e.g. in district heating systems) inevitably leads to loss (see Sect. 5.3.1), which can be partially avoided through local production of electricity.
5. *Connecting* micro cogeneration plants to "virtual power plants" by means of communication devices. This could either create further environmental impacts or reduce them.
6. *Indirect environmental impacts,* which could be created due to the immediate influence of decentralized systems on customer behavior.

5.1 Life Cycle Assessment of Micro Cogeneration

To assess the issues raised in items 1 through 4 for the case of micro cogeneration, and to compare it to competing energy supply technologies, we use the methodology of *life cycle assessment* (LCA). The two key elements of LCA are

- assessment of the total life-cycle ("cradle-to-grave approach") of a given energy conversion technology, including the exploration, processing, transport of materials and fuels, the production and operation of the investigated energy conversion units, and their disposal/recycling, and
- assessment of different environmental impacts on resources, human health, and ecosystems.

An LCA basically consists of four steps: firstly, a *goal and scope definition* serves to describe the investigated product, the data sources and system boundaries, while defining the functional unit, i.e. the reference for all related in- and outputs. This functional unit could be, for instance, the provision of one kilowatt-hour electricity to the customer or the supply of a hotel with heat and electricity for one year. Secondly, an *inventory analysis* "involves a data collection and calculation procedure to quantify relevant inputs and outputs" (ISO 1997). Thirdly, the potential impacts of the in- and outputs of the inventory analysis are determined by an *impact assessment,* which categorizes and aggregates the environmental interventions. For that purpose, impact categories, such as global warming, eutrophication, acidification, or summer smog are defined and characterization factors calculated which describe the contribution of different substances to that particular impact category (e g. CO_2, CH_4 or N_2O to global warming). Fourthly, through *interpretation*, the findings from the inventory analysis and the impact assessment are combined in order to draw conclusions or formulate recommendations.

In the case of micro cogeneration, specific issues are associated with data uncertainty and forecasting, because many of these technologies are not yet fully developed (Pehnt 2003). In addition, cogeneration is a technology path with joint or co-products, which have to be taken into account to obtain a full picture of the environmental consequences.

5.1.1 Considering the Co-product

For the results of an LCA of cogeneration systems, the way that the co-product is considered turns out to be important. As cogeneration systems produce heat and electricity simultaneously, and the comparison to other, non-cogeneration systems often takes place based on the functional unit "1 kWh electricity" or "1 kWh heat", both products need to be taken into account together.

This situation is very similar to that which many companies have to deal with when they have more than one product to sell. Take for instance a farmer raising cows. Each cow produces milk; but eventually it will be slaughtered and converted into meat. How can the farmer allocate the cost of feeding a cow to produce both milk (daily) and meat (at the end of the cow's life)?

There are two main ways to deal with this problem. The first is *allocation* of the cost between the products with an appropriate key. In the case of the farmer, this might be the relative nutritional value of milk and meat, or the prices of milk and meat on the market.

For the case of a cogeneration system, one allocation key could be the energy generated. Take the example of an electric efficiency of 40 % and a thermal efficiency of 40 %. For each kWh electricity, one kWh heat is produced. The allocation key would thus be 50:50. That means that 50 %

of the CO_2 emissions are related to the production of electricity and 50 % to heat.

However, this allocation key does not really characterize the value of the products. Instinctively, one kWh electricity seems more valuable than one kWh low-temperature heat. And, in fact, only a fraction of the heat could be converted into electricity. Therefore, often the *exergy* is used as the allocation basis. Exergy describes the amount of useful energy that is contained within a product. The exergy of electricity is equivalent to its energy, whereas the exergy of heat, in contrast, is given by the Carnot efficiency multiplied by the energy value.

The second way to deal with co-products is to estimate the *avoided burden*. If a cow, for instance, had not produced milk, it would have been necessary to buy some substitute, such as soy milk. The total cost for its meat would then be the sum of feed costs over the lifetime of the cow minus the avoided costs for soy milk.

In the case of a micro cogeneration system, this means that – if we want to compare micro cogeneration to other electrical power plants – we have to identify the heating systems that would actually be replaced by such systems. This depends on a number of factors, for instance the country, the type and age of the houses, the preferred fuels, and the like. It also depends on the perspective of the decision-maker: from the perspective of a home owner who installs a new system, it is the individual, old heating system that might be superseded by a micro cogeneration system. Boiler manufacturers might think of micro cogeneration competing with other modern heating systems. A politician who has to decide which heating system to support financially will have to consider, for instance, a micro cogeneration home energy system compared to a modern condensing boiler.

For the cogeneration systems investigated here, we choose the *avoided burden approach*, where the co-product is credited with an alternative generation route. This approach reflects most accurately real decision-making situations. Also, if we had chosen an allocation procedure, it would be necessary to compare micro cogeneration to other electricity *and* heat systems simultaneously.

We furthermore regard electricity as the main product, while generated heat is credited, and analyze the replacement of a gas condensing boiler by a variety of micro cogeneration systems. The question answered through this substitution perspective is: which environmental impacts do micro cogeneration systems have if we install them instead of a modern gas condensing boiler? Choosing this perspective relies on the following assumptions:

- **Fuel.** We want to determine the effects of a new technology, not of switching fuels. For instance, switching from oil to gas offers environmental advantages. However, this fuel switch can also be accomplished with other technologies. It cannot be causally attributed to micro cogeneration. Therefore, we assume the same fuel for the substituted heat supplying technology.
- **Marginal systems**. Usually, micro cogeneration will not be installed as a substitute for an older technology. It will rather be viewed as an alternative to another modern technology when replacement is required in order to meet future demands for heat and electricity. In this sense, two modern, marginal systems are to be considered as substitutes. Hence, for investors, households, or decision makers, the relevant question is: which technology is to be preferred from an environmental perspective?

Combining these two assumptions, a modern gas condensing boiler is considered as the substituted heat technology. It should be kept in mind that this way of crediting is the most conservative way to assess micro cogeneration. The advantages of micro cogeneration, compared to central generation without it, increase significantly as soon as less modern technologies or fuels with higher carbon contents are considered as the substituted heat supply.

Which *electricity* system should the micro cogeneration systems be compared to? Again, this depends on the perspective of the decision maker. To identify advantages and disadvantages in comparison to the status quo, the current electricity mix can be chosen as a benchmark. Generally, however, we need to take a dynamic, forward-looking perspective. In Germany, for instance, some 40 GW_{el} power plant capacity must be replaced by 2020 (see Chap. 2). Micro cogeneration represents one alternative way to cover this demand for new capacity. Micro cogeneration should, therefore, be compared to modern marginal power plants that would typically be installed. In many cases, this will be a natural gas power plant. This procedure is in accordance with the EU directive on cogeneration, which demands that

> each cogeneration unit shall be compared with the best available and economically justifiable technology for separate production of heat and electricity on the market in the year of construction of the cogeneration unit (EU 2004).

On the other hand, the generation characteristics of micro cogeneration systems tend to contribute to peak load as well. The peak load power mix differs from country to country. In Germany, for instance, it includes low efficiency gas power plants, but also virtually CO_2 free hydropower plants. In our environmental comparison, we include both mix and gas combined-cycle plants to represent the alternative paths for centralized electricity generation.

5.1.2 Input Data for Micro Cogeneration Systems, Fuels, and Conventional Power Plants

The micro cogeneration systems investigated have already been defined, as summarized in Table 4.1, and include several fuel cell, Stirling, and reciprocating engine configurations. These are compared to larger cogeneration systems, such as a large reciprocating engine and a gas combined-cycle plant both feeding district heating systems. As benchmarks for separate generation options, a gas combined-cycle plant without cogeneration, a modern lignite power plant, and the future German electricity mix, as forecasted by Enquête (2002), are assessed.

The reference time of the comparison is the year 2010. We choose this base year because several of the micro cogeneration systems, namely the fuel cell, the very small reciprocating engine, and the Stirling engine, are not yet fully technologically mature. For these systems, we do not evaluate the status quo, but rather extrapolate the technical performance to the year 2010, to take into account further expected technological progress.

We restrict our LCA analysis to natural gas and fuel-oil as fuel. Even though micro cogeneration can, in principle, be operated with other fuels (see Sect. 9.4 for a discussion on fuel flexibility), the greater part of the systems will be operated with natural gas. Furthermore, by 2010, renewable micro cogeneration systems, such as pellet Stirlings or biomass gasification/fuel cell units, are not yet likely to be mature. It is obvious, however, that with respect to greenhouse gas emissions, the use of renewable fuels would be preferable (see Sect. 5.2.2) .

For the purpose of our research project, life cycle data were obtained for various micro cogeneration and conventional systems from the system manufacturers and additional data sources. The LCA data were then implemented using the material flow and life cycle program Umberto (www.umberto.de). This program calculates the impacts caused by the full life cycle of the micro cogeneration systems and carries out an impact assessment in accordance with the international standard ISO 14042.

The fuel cell LCAs are based on our analysis in Krewitt et al. (2004). The materials LCAs used for the various technologies, such as catalyst materials (e.g. platinum group metals), building and structural materials (such as concrete and steel), and materials for the electrical equipment (copper etc.) are mainly based on Pehnt (2002). The natural gas LCA, as modeled in Umberto, represents a standard import mix for Germany, extrapolated to the year 2010 (gas sources: Germany 13 %, Netherlands 19 %, Norway 33 %, GUS 35 %). Electricity and conventional heat production data are taken from the detailed bottom-up Umberto/IFEU database. The district heating systems are based on the detailed model described in Chap. 4, including heat losses depending on the heat density, the length and isolation of the distribution grid, etc.

To account for uncertainties regarding the actual course of future development, but also variations with respect to application, manufacturer, technology, and other parameters, error bars are used to show the bandwidth of results. Some of the important input parameters included in the bandwidth are discussed below with respect to representativeness and validity.

- *Thermal efficiencies* depend strongly on: the application context, e.g. on whether a condensing operation can be achieved; which return temperatures are possible; how many start/stop procedures are carried out; how many full load hours are possible, etc. In addition, it is important whether a condensor is applied to exploit the condensing heat of the exhaust gas. This can, in principle, be used in most micro cogeneration applications. Often, it is rather an economic than an environmental question whether this extra device is used or not. Condensing operations are included in the possible system designs investigated here (Table 5.1).
- The emission of *NO_x* particularly of reciprocating engines varies strongly, depending on the emission reduction concept used ($\lambda = 1$ or lean operation, see Sect. 1.2.1), manufacturer, operation mode (partial load operation, static or dynamic operation, etc.), age of the system, and maintenance interval. The latter two aspects are particularly important for reciprocating engines. The NO_x emissions of $\lambda = 1$ systems can vary by one order of magnitude, depending on whether the catalyst has been newly exchanged or not. Likewise, emissions are very sensitive towards the calibration of engine characteristics. Engines in lean operation typically show more stable, but somewhat higher, emission levels.
  For the small Stirling engine, e.g. the WhisperGen, higher NO_x emission factors are reported. However, it is assumed that future systems will show reduced emissions, comparable to conventional gas burners.
- There is still uncertainty with respect to the bandwidth of unburnt *methane (CH_4) emissions* from reciprocating engines. In the literature on engine CHP, emission factors from 20 to 500 mg per MJ fuel input can be found. A Danish study, for instance, states that “compared to

complete combustion, the global warming potential for gas engines was on average raised by 20 % by the emission of unburned fuel" (Kristensen et al. 2004). Due to the high global warming potential of CH_4 (the impact of one kg CH_4 corresponds to 23 kg CO_2), 500 mg methane per MJ fuel corresponds to some 150 g of CO_2 equivalent per kWh electricity produced; methane emissions thus have a significant impact on overall emissions. Here, we assume a bandwidth of 10 to 320 mg/MJ fuel input, with 100 mg/MJ as default value based on field tests of reciprocating engines (Senertec 1992-2003). For fuel cells and Stirling engines, the emission factors are orders of magnitude lower.

- The condensing boiler as an "avoided burden system" is assumed to have an average efficiency of 97 %. This figure, indicating the somewhat low efficiency of the condensing boiler, was based on a large number of systems investigated in a variety of studies. It has been shown that the potentially high efficiency obtainable from the condensing operation cannot always be achieved, particularly in the summer.

Table 5.1. Input data for the LCA of micro cogeneration systems (see Table 4.1)

		Micro CHP											**Larger CHP**	
		Reciprocating engines						**Stirling engines**			**Fuel cells**		**Recipr. engine**	**Comb. Cycle w/CHP**
Reference number		1	2	3	3a	4	5	6	7	8	9	10	11	12
		1kW	3-6 kW (l=1)	3-6 kW lean (HighNOX)	3-6 kW lean (Fuel oil)	3-6 kW lean (HighNOX cond)	3-6 kW lean (LowNOX)	0.8 kW	3 kW	9.5 kW	PEMFC	SOFC	50 kW	
Capacity (Default)														
Electric	kW_{el}	1,0	4,7	5,5	5,3	5,5	5,0	0,8	3,0	9,5	4,7	1,0	50,0	
Thermal	kW	3,25	12,5	13,9	11,4	14,9	12,6	6,0	15,0	28,5	12,5	2,7	98	
Efficiency (Default)														
Electric	-	20%	25%	25%	28%	25%	25%	10%	15%	24%	32%	32%	30%	45%
Total	-	85%	88%	88%	88%	93%	88%	85%	90%	96%	85%	85% *	88% **	88% **
Efficiency (min)														
Electric	-	20%	25%	25%		25%	25%	10%	15%	22%	28%	28%	28%	45%
Total	-	80%	84%	84%		88%	84%	80%	85%	88%	80%	80% *	85% **	80% **
Efficiency (max)														
Electric	-	20%	25%	25%		25%	25%	12%	19%	26%	32%	32%	33%	49%
Total	-	85%	95%	95%		99%	95%	90%	94%	100%	90%	90% *	90% **	88% **
NO_x emission														
Default	mg/Nm^3	n.a.	125	300	2150	300	135	80	80	80	3	3	125	95
Bandwidth	mg/Nm^3	n.a.	50-400	70-400	n.a.	70-400	60-200	-	20-110	20-110	-	-	50-400	n. a.
Economic Lifetime		80,000 hours or 15 years, whichever is lower												

All efficiencies are seasonal. * plus hot stand-by loss ** plus distribution losses

Lean = lean operation, cond = use of condensor

5.1.3 Impact Assessment

For an impact assessment, characterization factors are required so as to be able to calculate environmental interventions (such as greenhouse gas emissions or acidification) from emission factors. For this study, we take characterization factors in accord with standard LCA procedures (CML et al. 1992; UBA 1995; Franke and Vogt 2000; Giegrich and Detzel 2000). Due to space limitations, we can only present results for two important selected impact categories in this book: the emission of greenhouse gases (GHG) and acidification. The first category represents climate change effects, but also reflects the use of fossil resources, as the impact category is dominated by CO_2 emissions from fossil fuel use. The second assesses the contribution of local emissions of, for instance, NO_x, NH_3 and SO_2 to acidification. Structurally similar results could have been obtained for eutrophication, for example. A number of additional impact categories are included in the model but will not be reported here.

The results of these cradle-to-grave analyses will be described in the following sections. First of all, the contribution of micro cogeneration to climate change will be assessed (Sect. 5.2.1). Then emissions of pollutants are discussed in Sect. 5.2.2, followed by an assessment of immissions (5.2.3), and ending with a comparison on the household level of individual operators of micro cogeneration (Sect. 5.2.4).

5.2 Results of the Life Cycle Assessment

5.2.1 Greenhouse Gas Emission Reduction of Natural Gas Micro Cogeneration

Micro cogeneration versus central electricity production. As far as the prevention of GHG emissions is concerned, micro cogeneration systems are superior not only to the average electricity supply mix in Germany, but also to efficient and state-of-the art separate production of electricity in gas operated power plants and heat in condensing boilers (Fig. 5.1). Compared to separate generation with a natural gas combined cycle plant (without cogeneration) and a natural gas condensing boiler, micro cogeneration reduces GHG emissions by between 10 to 30 %, with the exception of the small Stirling engine. With lower electrical efficiency, its GHG level lies above the one of a gas combined cycle plant. Similar reductions are achievable with respect to primary energy demand.

There is still some uncertainty about the achievable total average efficiency of fuel cell systems. Estimates for future domestic systems range between 80 and 90 %, whereas today the systems are far from this value. When taking the lower bound of 80 % total efficiency for fuel cells only, the achievable reduction for domestic fuel cell systems does not seem very high. However, one can also reverse the argument: based on natural gas as a fuel, domestic systems might reach the same CO_2 emissions as a modern gas combined cycle plant, even though the systems are a factor of 100,000 smaller.

The reduction effects become greater when other fuels enter the comparison (see Sect. 5.2.2). For instance, compared to separate production with more coal-dominated electricity mixes, such as the German electricity mix, or even compared to a lignite power plant, GHG reductions above 50 % can be achieved. Also, comparison with oil-fuelled boilers would lead to even higher GHG reductions. In such cases, the GHG reduction is to a great extent due to a fuel switch from higher carbon-containing fuels to natural gas.

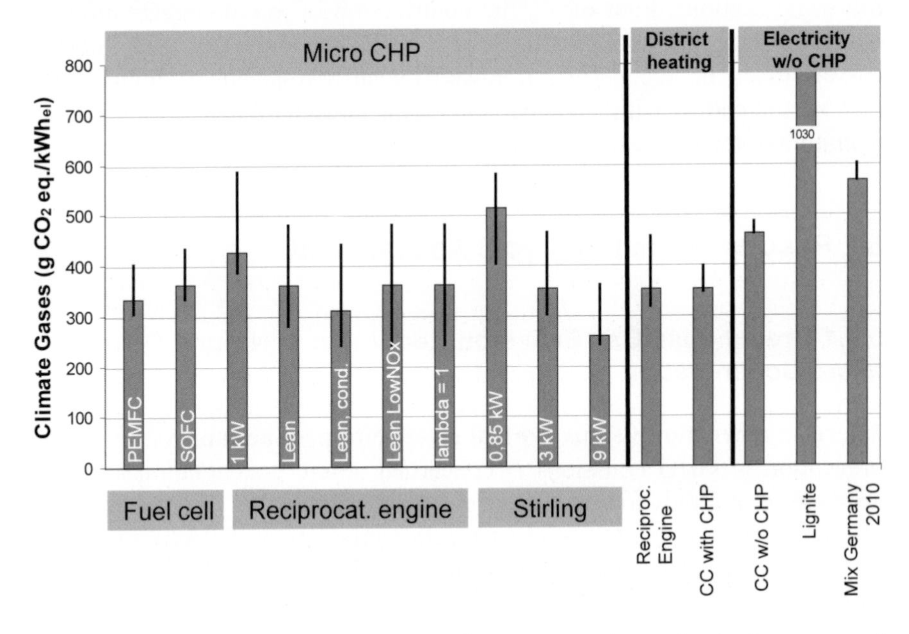

Fig. 5.1. Life cycle GHG emissions of micro cogeneration technologies compared to large CHP and conventional electricity production in the year 2010;
Functional unit 1 kWh electricity at low voltage level, co-produced heat is credited ("avoided burden"; see Sect. 5.1.1)

Which is the best micro cogeneration technology? The determination of the best among the micro cogeneration technologies depends considerably on the efficiencies achieved. Under the assumption that gas condensing boilers are being substituted for, all technologies (fuel cells, reciprocating and Stirling engines) are within a very narrow range, with the exception of the small Stirling engine, which has lower electrical and overall efficiency. The high total efficiency of the large Stirling engine leads to the lowest GHG emissions. Fuel cells reach almost the same GHG emission level as the large Stirling when they achieve a total conversion efficiency of 90 %. However, their thermal management is more complex, and the heat sources inside the fuel cell system are diffused. This makes it more difficult to harvest the heat co-produced in the system than, for instance, in a Stirling or reciprocating engine.

Fig. 5.1 also unveils large variations in GHG emissions for each system. These are particularly caused by varying thermal efficiencies, which depend on a system's context (e.g. return temperatures) and configuration. In addition, for reciprocating engines, the values for methane emissions due to unburned natural gas vary considerably (see Sect. 5.1.2). Therefore, the error bar for this technology is larger. Further research and measurements are needed in this area.

Small versus large CHP. The comparison between the different sizes of CHP reveal that micro cogeneration leads to GHG emissions on a level similar to district heat cogeneration. Whereas the electrical efficiencies of large plants are generally higher than those of smaller units (Fig. 1.3), the total efficiency of large plants is not in general higher than that of the small systems. In addition, heat distribution losses for the district heating systems and electricity distribution losses in the case of large combined-cycle (CC) plants have to be taken into account. Whether large or small CHP is favorable, therefore, depends on the specific situation under investigation.

In any case, district heating systems are not only an efficient way of realizing CHP, but also offer the possibility of a straightforward integration of renewable energy carriers, such as biomass or solar thermal energy, into the heat supply (see Chap. 9).

Table 5.2. Reduction of greenhouse gas (GHG) emissions per kWh_{el} produced, achieved by micro cogeneration technologies with different baselines and heat credits; shown are the values for the default cases in Table 5.1; fuel: natural gas

Technology Status 2010		**GHG reduction (heat credit: cond. boiler)** compared to		
		Comb. Cycle w/o CHP	Mix G 2010	Lignite
Fuel cell	PEMFC	27%	41%	67%
	SOFC	21%	36%	65%
Reciprocating engine	1 kW	8%	25%	59%
	lean	22%	36%	65%
	lean, condensor	33%	45%	70%
	lean, LowNOx	22%	36%	65%
	lamda=1	22%	36%	65%
Stirling engine	0.85 kW	-11%	9%	50%
	3 kW	23%	37%	65%
	9 kW	44%	54%	75%

Example: Production of electricity in a SOFC produces 36 % less GHG emissions than the future German electricity mix when we assume that a gas condensing boiler is replaced.

5.2.2 Pollutant Emissions

It is not only climate change that is an issue of current environmental policy efforts. More generally, the protection of human health and the environment is in focus. In Europe, Directive 2001/81/EC on national emission ceilings for certain atmospheric pollutants, the so-called NEC directive, controls the reduction of sulfur dioxide (SO_2), nitrogen oxides (NO_x), volatile organic compounds, and ammonia (NH_3). The aim is to combat impacts such as acidification, eutrophication, and ground-level ozone. This is to be achieved by limiting annual national emissions so that they do not exceed a ceiling laid down in the annex of the directive. Germany, for instance, is required to further reduce NO_x emissions in the year 2010 by 22 % compared to the year 2000. Even this emission reduction, according to the German Advisory Council on the Environment, will not be sufficient to meet the long-term goals of the NEC directive.

Even though the low level of ambition of the directive is criticized by many institutions, scenario calculations show that many countries will not even be able to comply with its soft ceilings. According to the European Topic Centre on Air and Climate Change, the main problems foreseen by countries relate to the emission of NO_x (ETC-ACC 2004). In the past few decades, emissions of NO_x have generally decreased significantly, particularly due to measures taken regarding power plants and vehicles. In the recent past, however, the *rate* of reduction has decreased significantly, too.

The specific contribution of energy-related technologies to pollutant emission reductions is significant. Table 5.3 summarizes for some important air emissions the contribution of the energy sector to total European emissions. For instance, only 18 % of 1998 European NO_x emissions stemmed from the "energy industry" sector, while the dominant part was emitted by the transport sector. Of course, this does not imply that reducing NO_x in the power sector is of no importance. But it indicates the order of magnitude a possible reduction of impacts implies for the overall European emissions situation. It is remarkable that, for SO_2, the energy industry causes almost two thirds of the European emissions, mainly due to coal power plants. It therefore seems reasonable to investigate options toward reducing these emissions, and to ask whether micro cogeneration may help lead to such a reduction path.

Table 5.3. Contribution of the energy sector to total pollutant emissions in EU15 (Source: EEA 2001, 2004)

%	1980	1990	1998	2001
CO_2		34.6	32.3	33.0
NO_x	25.1	21.3	18.0	
SO_2	57.3	61.6	65.0	
N_2O		4.3	3.8	
CO	1.0	0.9	1.0	
CH_4		0.2	0.4	
NMVOC	0.3	0.3	0.4	
NH_3	01	0.1	0.1	

It is common to all gas-based power plants that the emission of SO_2 is almost completely avoided. Also, in gas-based micro cogeneration and other power plant technologies, the contribution of ammonia to, e.g., acidification, is almost negligible. Therefore, of all exhaust emissions by micro cogeneration systems, NO_x is of particular ecological importance in the context of the NEC directive. Nitrogen oxides are formed during the combustion process in most of the micro cogeneration technologies and contribute to many environmental impact categories, such as acidification, human health impairment, and eutrophication. Furthermore, NO_x assists in the formation of Ozone and thereby contributes to summer smog and global warming; it also is a precursor material for the formation of small particles.

Whereas the total efficiencies of micro cogeneration technologies are similar, the various technologies differ considerably with respect to their emissions, particularly of NO_x. Reciprocating engines, which are based on discontinuous combustion at high temperatures, exhibit the highest emission factors. Stirling engines, due to their continuous combustion and the possibility of applying modern burner technology (such as flameless oxidation burners), lead to very low emission factors. This is also true for fuel cells, due to the electrochemical nature of their combustion, the comparatively low temperatures involved in the reforming reaction, and the requirement to clean up impurities such as sulfur and CO. The direct NO_x emission factors for the operation of micro cogeneration systems, as derived from analysis of existing technologies and field tests, are given in Table 5.1.

In the following, we will concentrate on acidification as an important impact category, mirroring particularly the differences with respect to NO_x emissions of micro cogeneration technologies and, when compared, for example, to coal power plants, SO_2 emissions.

In Germany, micro cogeneration systems have to be "state-of-the-art" in terms of pollutant emissions (see Sect. 8.1.7). This standard is regulated by the administrative instruction "TA-Luft"; the values of the TA Luft will be easily fulfilled by all micro cogeneration systems (Table 5.1). However, as pointed out in Sect. 8.1.7, in some other European regions, emission legislation is much stricter than in Germany with respect to small CHP.

Fig. 5.2 displays the results of the life cycle assessment of the micro cogeneration technologies for acidification. It shows that fuel cells offer great advantages with respect to environmental impacts that are caused by local pollutants, such as acidification (mainly due to NO_x and SO_2). This also holds for other impact categories not shown in the figure. Unlike engine CHP plants, fuel cells couple the advantages of reduced energy consumption with low direct emissions. The remaining acidifying emissions stem mainly from their natural gas operating supply and, to a much lower degree, from system production (for the PEMFC, 13 % of the total life-cycle emissions stem from system production, for the SOFC 9 %).

Similar advantages can be achieved by means of Stirling engines, which have very low direct emission factors as well, at least in the case of the innovative FLOX burner. For the small Stirling engine, we assumed that NOx emissions of 80 mg/Nm^3 will be achieved in future systems from high-volume production.

Acidifying emissions from small reciprocating engines (considering the heat co-product) are somewhat higher than those of centralized gas power plants, due to more efficient emission control in large power plants.

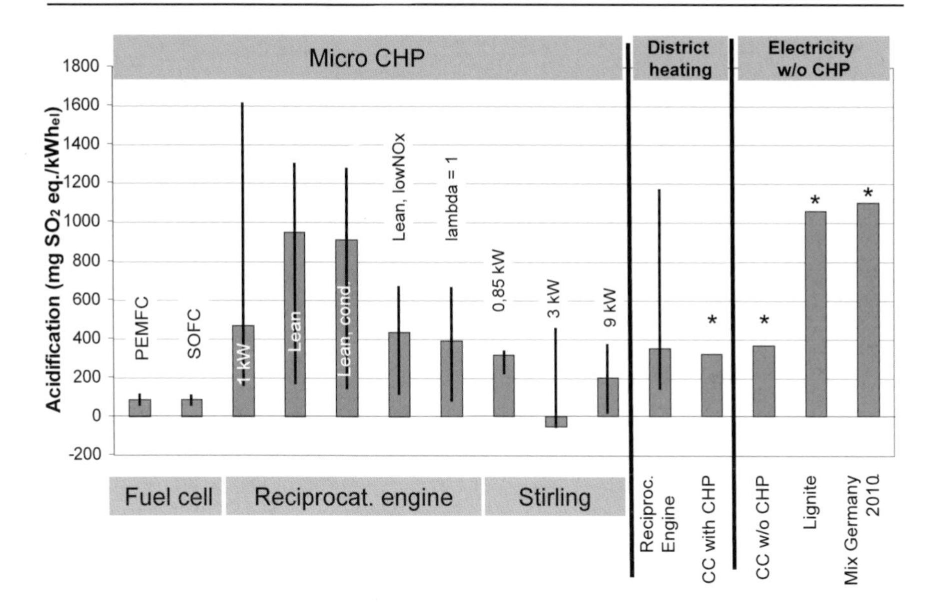

Fig. 5.2. Life cycle assessment of acidifying emissions (NO_X, SO_2, NH_3) of micro cogeneration technologies compared to large CHP and conventional electricity production without CHP

5.2.3 Local Impacts from Air Pollutant Releases

When interpreting the LCA results for pollutant emissions, it is equally important to evaluate the impact of these emissions on the recipients side: how does the *local air quality* change due to the emissions? To address this question, the environmental relevance of the additional NO_X emissions of reciprocating engines will be assessed. For this purpose, a dispersion calculation for a virtual residential area was carried out, based on the software package AUSTAL 2000 (AUSTAL2000 2003). For this calculation, it was assumed that, in a residential area based on multi-family residences, one third of the houses would be equipped with a reciprocating engine. This corresponds to 100 systems in a 1 km^2 area. In addition, avoided air pollutant releases due to the substitution of an alternative gas heating system are accounted for.

With respect to technical parameters, we conservatively assumed 5000 hours of full load operation per year, an NO_X emission factor of 300 mg/Nm^3 (expressed as NO_2, see Table 5.1), and a share of 10 % NO_2 of primary NO_X emissions. This latter assumption is of great importance, because NO_2 is of primary environmental concern. A site with flat topography with relatively critical weather conditions was selected: non-

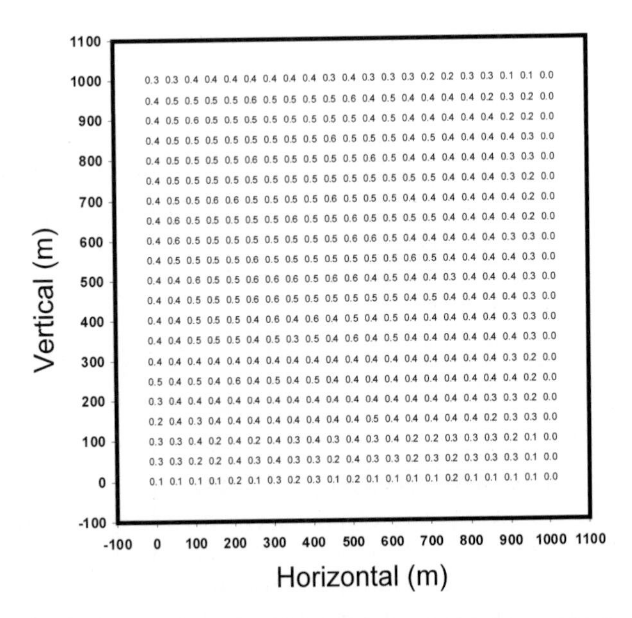

Fig. 5.3. Annual average concentration of NO_2 in ambient air (µg/m^3) in a residential area due to installation of 100 reciprocating engines per km^2

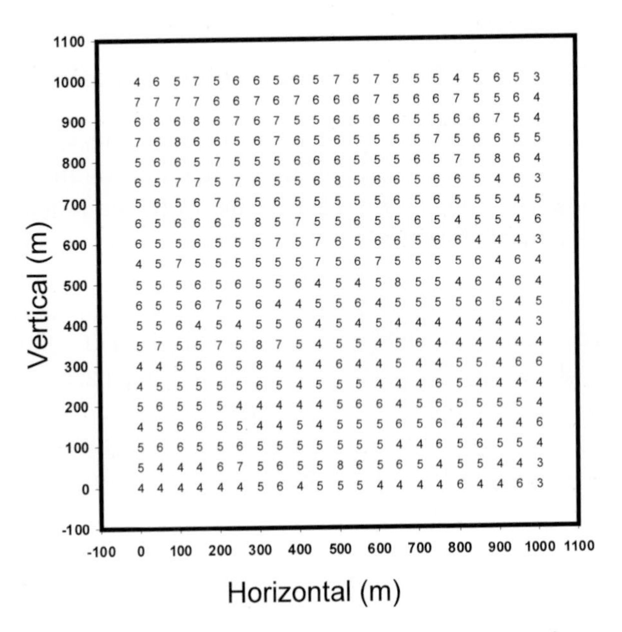

Fig. 5.4. Short-term concentrations of NO_2 in ambient air (µg/m^3; indicating the concentration that is not exceeded in more than 18 hours/a) in a residential area due to installation of 100 reciprocating engines per km^2

uniform wind distribution, large share of stable weather situations, and low wind speed that tend to lower the dispersion of pollutants. The surface texture corresponds to that of an urban area.

The results of this calculation are shown in Fig. 5.3 and Fig. 5.4. Fig. 5.3 depicts the additional annual average concentration of NO_2 in the residential area due to net emissions from reciprocating engines. The annual average of 0.6 $\mu g/m^3$ can be compared to the permissible level for the annual average NO_2 concentration; the German limit being 40 $\mu g/m^3$. The air concentration associated with the installation of the 100 units is very low – below a level of irrelevance, which is by German law defined as 1.2 $\mu g/m^3$.

In addition, the maximum short term concentrations of NO_2 in air (i.e. not to be exceeded in more than 18 hours) are well below 7 $\mu g/m^3$ (Fig. 5.4), compared to the allowable level which by German law is 200 $\mu g/m^3$.

With respect to the impact on ambient air quality, reciprocating engines are not critical under the given circumstances. It has to be reiterated, however, that a flat topography was chosen for this analysis. Complex terrain (e.g. a narrow valley) might cause higher pollutant concentrations in certain places as calculated here.

From a national perspective, of course, every measure to reduce NO_x emissions is welcome. This is why, from our perspective, low-NO_x reciprocating engines should be further developed and marketed. The market leader, Senertec, for instance, offers a high- and a low-NO_x system. The specific capital cost of the latter is approximately 10 % higher. Thus, only 4 to 5 % of the Dachs systems sold belong to the low-NO_x class.

5.2.4 From Specific Impacts to Supply Objects

In the preceding sections, we have analyzed the environmental impacts per unit of electricity or heat produced, i.e. we have looked at the *specific* impacts. In addition, it is instructive to investigate the impacts per *supply object*, e.g. per single-family house, per apartment house, per hotel, etc. The question we want to answer here is: how much environmental impact reduction can be realized in concrete supply objects with concrete micro cogeneration systems? To answer this question, the specific environmental impacts (per kWh energy supplied) are multiplied by different full load hours, giving the annual environmental impacts.

This analysis involves a further step: we need to analyze the amount of electricity and heat produced by the CHP unit, depending on the object and the CHP system (size, partial load characteristics, etc.). This investigation was begun in Chap. 4 to determine the economic viability of micro cogeneration. We can build on the results of that chapter, i.e. the full load hours, the share of own electricity consumption and grid electricity, the amount of heat produced in the peak boiler, etc. As the technologies were defined identically, we can directly use those results for further calculations.

It is important to note that the calculations in Chap. 4 are based on real available or planned systems. For instance, the calculations for a single-family house utilize an 850 Watt Stirling engine (like the WhispherTech), a 1 kW solid oxide fuel cell (like the Sulzer Hexis system) or a 1 kW reciprocating engine (Honda) (Table 5.4). The differing efficiencies and power-to-heat ratios also imply different full load hours.

We have chosen this approach because it provides additional information; we can not only estimate the impact reduction per supply object, but also how well a particular system fits the need of a supply object.

Table 5.4. Investigated technologies for the LCA of supply objects

	Reciprocating engine	Stirling engine	Fuel cell	District heating
Technologies				
Single-family (low)	1 kW	850 Watt	1 kW SOFC	20 kW rec.eng. w/ 17.8 % heat loss
Multi-family (average)	5.5 kW	3 and 9 kW	4.7 kW	50 kW rec. eng. w/ 2.4 % heat loss
Hotel	5.5 kW	3 and 9 kW	4.7 kW	
Full load hours (h/a)				
Single-family (low)	3373	1940	4786	4800
Multi-family (average)	5526-5639	5509-4160 [a]	6432	6456
Hotel	5555-5730	5505-3939 [a]	7015	-
Share of electricity produced in CHP (%)				
Single-family (low)	29	13	41	39
Multi-family (average)	24	13-31 [a]	24	17
Hotel	26	14-31 [a]	24	17
Share of heat produced in CHP (%)				
Single-family (low)	95	100	68	62
Multi-family (average)	61-65	65-93 [a]	39	30
Hotel	65-68	68-92 [a]	45	-

[a] First value: small Stirling, second value: large Stirling

Beside the LCA of the specific impacts (per kWh) of each CHP unit, which were derived in the preceding sections, we need to model the energy flows of our supply objects (Fig. 5.5). Therefore, for each object, we need to include the following information:

- The net electricity demand of the supply object. The supply object may require additional electricity, and the CHP unit may also feed electricity into the grid. We assume that the substituted electricity and the electricity fed into the grid have the same "quality". This is, to a certain extent, a simplification because micro cogeneration systems are very often thermally led; the electricity production of micro cogeneration units is therefore at maximum in periods of high space-heating demand, which are often also times of high electricity demand. Therefore, micro cogeneration might avoid the need for electricity production from environmentally more problematic peak load power plants (see Sect. 5.3.2).
  Assuming the identical quality of grid electricity bought by the supply object and of electricity fed into the grid, we therefore calculate a net electricity demand. It can be either positive, if less electricity is sold to the grid than bought, or negative, if more electricity is fed into the grid than bought.
- The CHP unit does not cover the full thermal load of the object. Therefore, peak load demand is considered to be covered by a modern gas condensing boiler.

This supply combination of micro cogeneration, grid electricity, and condensing boiler can then be compared to the reference, which assumes 100 % coverage of electricity demand via the German electricity mix from the grid and heat demand from a condensing boiler. Again, our reference time is 2010. The functional unit for this type of analysis is the supply of an object with heat and electricity for one year.

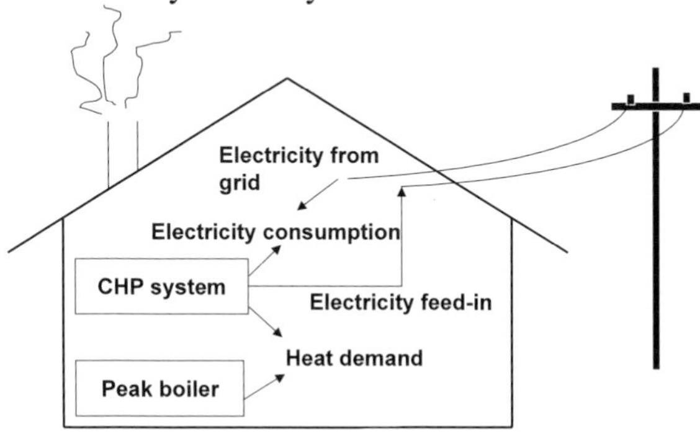

Fig. 5.5. Analysis of supply objects

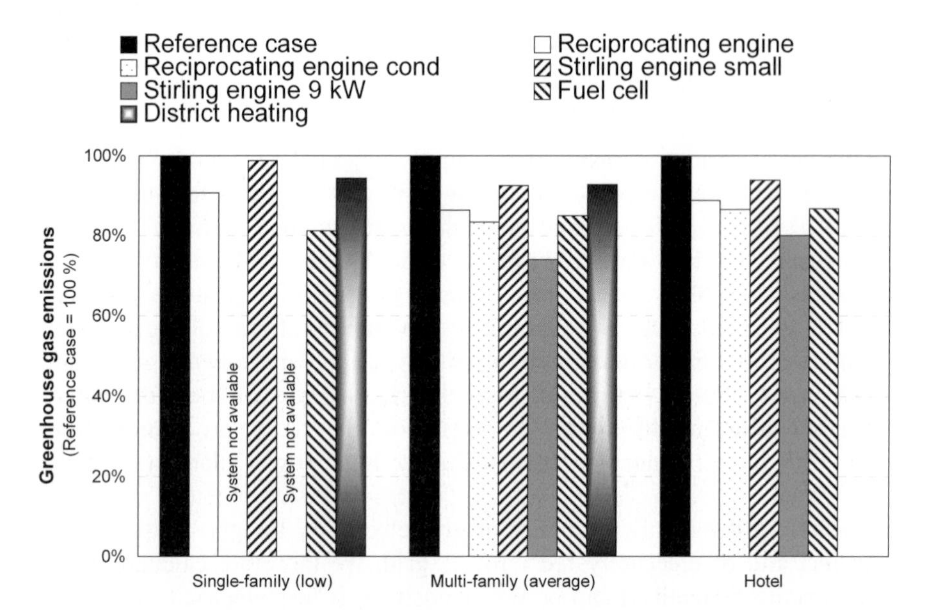

Fig. 5.6. Greenhouse gas emissions of different supply objects

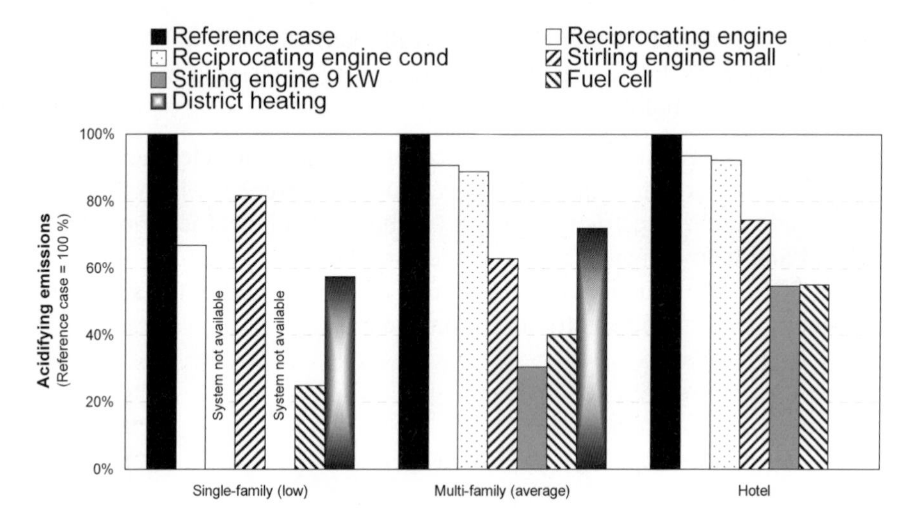

Fig. 5.7. Acidifying emissions of different supply objects

Selected supply objects (see Chap. 4): single-family house with low average heat demand, multi-family house with average heat demand, and small hotel; functional unit: supply of an object with heat and electricity for one year; reference case: grid electricity and gas condensing boiler

The results of this analysis for the impact categories GHGs and acidification are shown in Fig. 5.6. In all cases, the default technology assumptions listed in Table 5.1 are used for the assessment. Several important conclusions can be drawn based on this analysis:

- Unsurprisingly, the advantages of the individual technologies, as assessed in Chap. 5.2, are mirrored in this analysis. This means that CHP technologies offer climate advantages, the fuel cell offers additional acidification advantages, and so on.
- On the other hand, in an economically optimized system configuration, the CHP units cover only between 10 and 40 % of the electricity demand, and between 30 and 100 % of the heat. This implies that the rest has to be supplied by the electricity mix and the conventional heat system. Therefore, the GHG emission reduction of around 20 to 30 % in Fig. 5.1, for instance, is reduced to between 10 and 20 % when regarded per object supplied, due to the constant base demand of peak boiler heat and electricity.
- From this perspective, the fuel cell can increase its competitive advantage particularly in the case of the *single-family house*. The reason is that these systems are limited by the prevailing heat demand. Because the fuel cell has a higher power-to-heat ratio, it produces more electricity for a given heat demand and can thus substitute higher amounts of grid electricity. The district heating system, in contrast, has rather high losses of 17.8 % in the case of single-family houses (and concomitant low heat demand density).
- In the *multi-family house* connected to the district heating grid, in contrast, the heat losses drop to 2.4 %; thus, the difference between district heating and micro cogeneration shrinks as well. In addition, however, district heating has a lower share of the overall heat supply.

In absolute terms, GHG reductions from one ton per year (single-family house) to 11 tons per year (hotel) can be achieved by operating micro cogeneration compared to the electricity mix. Carrying out the same analysis for gas combined cycle plants as suppliers of grid electricity reduces these advantages.

5.2.5 From Gas to Renewables: Other Micro Cogeneration Fuels

In the preceding section, only natural gas as a fuel was investigated, because we believe this will be an important fuel for many European applications (see Chap. 9).

However, particularly for areas outside of larger population agglomerations, other fuels will be of high importance. Fuel oil can be more easily distributed in rural areas than gas. The picture changes further when we move to other regions worldwide. In Japan, for instance, liquefied petroleum or natural gas (LPG and LNG), fuel-oil, or other fuels will be important (see Chap. 13). Also, renewable fuels can, and in the mid-term will, also be used for micro cogeneration applications.

It is not the place here to carry out a number of further LCAs. However, because of the relevance of fuel oil, it was included in a sensitivity analysis (Fig. 5.8). A reciprocating engine operated with fuel oil, on the one hand, combusts a fuel with higher carbon content and, hence, emits more CO_2 per MJ fuel than gas. On the other hand, the efficiency of such a unit is slightly higher (by some 2 to 3 % points; see Table 5.1) than that of a gas unit. In addition, such units will not be replacing gas condensing boilers, but rather modern fuel oil heating systems.

In Fig. 5.8, we compare a 5.5 kW gas- or fuel-oil operated engine (corresponding to the Senertec Dachs) with the large non-CHP power plants from Chap. 5.2.1 ff. We observe that in terms of GHG mitigation, the oil reciprocating engine achieves almost the same level as the gas engine for the reasons given above. However, the acidifying emissions are substantially higher in the case of the oil CHP unit. This is primarily due to the much higher NO_x emissions of the fuel-oil CHP unit.

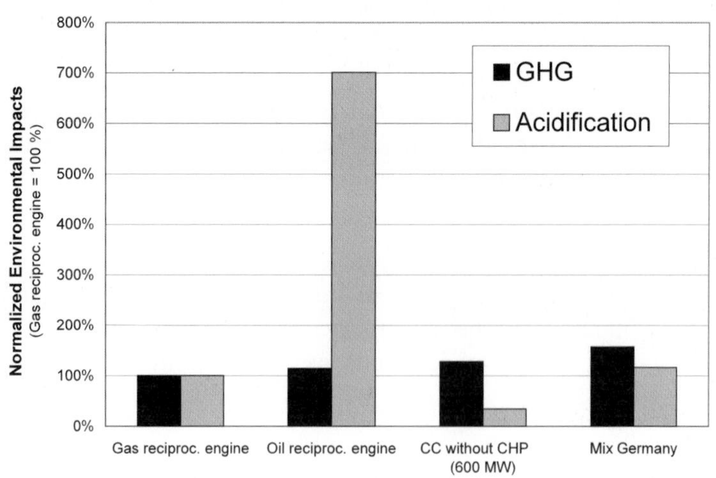

Fig. 5.8. Selected life cycle impacts of micro cogeneration technologies with gas and fuel oil compared to a combined cycle (CC) plant without CHP and conventional electricity production in the year 2010

Functional unit 1 kWh electricity at low voltage level; co-produced heat is credited with a gas condensing boiler (gas reciproc. engine) and an oil condensing boiler (oil reciproc. engine)

When using *renewable fuels*, e.g. biogas or wood pellets, the overall GHG emissions for micro cogeneration systems will be drastically reduced. Even for rapeseed oil, the LCA will show significantly lower overall GHG emissions, despite its production requiring fossil fuel inputs and leading to further greenhouse gases due to the emission of N_2O from the agricultural system (Gärtner and Reinhardt 2003). This positive GHG balance is, however, a general characteristic of most renewable energy systems (Pehnt 2005) and not only of renewable micro cogeneration, implying that also on the side of centralized production of electricity, large CHP, or decentralized heat production without cogeneration, renewable, and thus virtually GHG-free, systems could be identified. In other words: when we discuss renewable fuels, the GHG (and energy) assessment per kWh final energy delivered will be excellent for almost all conversion systems, regardless whether they are small or large. In the context of that discussion, aspects such as technical maturity, logistical feasibility and cost are very important to keep in mind (see Sect. 9.4).

5.3 Beyond the Single Micro Cogeneration Unit

So far, we have discussed environmental impacts arising from installation of single micro cogeneration power plants. But there are also impacts on a more abstract system level, which are either caused by the number of systems, or which are of a more indirect nature. These impacts shall be qualitatively discussed in the subsequent chapters.

5.3.1 Avoided Grid Losses

Micro cogeneration units produce electricity in direct proximity to the customer. Either the produced electricity is directly used within the object, or the CHP electricity is fed into the low-voltage grid, but typically not transformed to the medium-voltage grid.

Grid losses are reduced not only by the reduced haul length, but also the lower net currents required and, thereby, the lower Ohmic losses and lower resistance of conductors and transformers, due to the lower temperatures of these devices. However, it is very difficult to determine actually avoided grid losses, because they "depend not only on the grid load displaced but also on the time, weather, load conditions, loadshapes, and – especially – physical placement in the grid" (Lovins et al. 2002). For example, on the low-voltage level, electricity losses in Europe range from 6.1 % in Germany and 9.6 % in France to 24.2 % in Croatia (Pehnt 2002;

Frischknecht and Emmenegger 2003). For the United States, Lovins et al. 2002 report distribution losses ranging between 3 and 13 %.

The possibility of avoiding grid losses also depends on the capacity of a system to reduce reactive power consumption and on the load following capabilities of the system. If, for instance, an electricity-led CHP device does not react instantaneously to changes in the electricity demand of the household, the excess electricity is fed into the grid. For thermally-led micro cogeneration units, in periods of low general consumption, grid losses might increase, whereas in periods of high consumption, the losses might decrease (Lehmann et al. 2003). Distributed generation might also enhance grid losses in cases where generation is located at the end of a network or connected with long feeding cables.

Arndt et al. (2004) calculate the impact of micro cogeneration (fuel cell) systems for a given Bavarian low-voltage grid and determine a reduction of some 0.26 %-points. Converted to, for instance, GHG emissions, this would translate to only 2 g CO_2 equivalents/kWh_{el} for the German electricity mix. From an environmental point of view, this is only of minor importance. In terms of economic viability, however, the congestion relief of the grid can be of importance, especially in cases where adding grid capacity can be deferred or even avoided (see Chap. 9).

5.3.2 Connecting Micro Cogeneration to Create Virtual Power Plants

Starting from the conventional micro cogeneration application, it is often argued that connecting many micro cogeneration units to create a "Virtual Power Plant" would offer additional economic advantages. But the virtual power plant application could also have additional environmental benefits:

- **Avoiding more "dirty" marginal power plants.** When one central operator co-ordinates the operation of micro cogeneration units, one objective could be to minimize the demand for peak load electricity or even to use the devices for regulating energy for other fluctuating energy sources, such as renewable energy carriers. Regulating energy is not only more expensive but also, in many cases (exception: hydropower from reservoirs), more environmentally intensive with respect to GHG emissions and other impacts. In Germany, peak load power plants are particularly old fuel-oil and gas power plants. Even in cases where hydropower is used for peak load purposes, peak shaving would set free hydropower capacity for other purposes, such as integration of renewables.

- **Enhanced full load hours of CHP systems.** A higher load factor due to more optimized system operation could lead to more CHP electricity and heat generation, thereby substituting for more conventionally supplied energy. Also, heat losses due to stand-by operation of CHP plants could be reduced in this way.
- **Enhanced acceptance by utilities of fluctuating energy systems.** A more "company-psychological" factor is that, by employing virtual power plant communication, municipal utilities will more easily be prepared to accept fluctuating energy feed-ins, such as wind or solar power, because they have the possibility to partially balance these fluctuating energy contributions by controlling the electricity production of their individual systems. This aspect, however, is more focused on renewable-based virtual power plants and is less relevant for micro cogeneration.

5.3.3 Indirect Ecological Impacts through User Behavior

Ecological impacts of technologies are not determined solely by their technological characteristics nor by their being embedded in technological and economical systems. New technological options may stimulate new behaviors, habits, and routines that counteract, or reinforce, the environmental effects of the technology (see also Chap. 6).

It is difficult to predict the indirect effects of micro cogeneration, as no systematic research on its users has been done so far. However, we can make inferences from experience with other energy-related technologies. Among them are energy-efficient household appliances, efficient conversion technologies, and the use of renewable energy sources.

In efficient appliances and conversion technologies, a *rebound effect* is often found (Gottron 2001; Menges 2003). This means that energy savings achieved by more efficient technologies are at least partly compensated, and sometimes overcompensated, by an increase in energy demand. The rebound effect can be traced to different causes.

First, new technologies may stimulate energy demand by offering new comfort features. It is a well-known effect, to give an example, that the switch from single coal or wood stoves to central heating in residential buildings leads to increases in energy consumption of up to one third, even though central heating is more efficient. The reason is that, given the possibility to heat all rooms easily and comfortably, users increase the number of heated rooms as well as the average temperature (Kleemann et al. 2000, p.13; Wilhite and Norgard 2003).

Secondly, energy-efficient technologies reduce expenditure for energy services and, therefore, set free disposable income, which may in turn increase the expenditure on energy-consuming appliances (direct rebound effect) or on consumer goods which need energy for their production (indirect rebound effect).

Thirdly, psychological research on environmental consciousness and action has shown that environmental behavior is selective, and people draw up a "subjective ecobalance" by offsetting certain symbolic types of environmental action against behavior in other areas. Performing a specific "environmentally friendly" behavior salves the conscience and legitimizes abandoning environmental considerations elsewhere. Possessing an energy-efficient technology can therefore provide an excuse for using it excessively or using other energy-consuming equipment (Diekmann and Preisendörfer 1992 and 1998, Bilharz 2003).

Similar considerations apply to the use of *renewables*: the sensation that energy is, economically and ecologically, "for free", invites liberal use.

Finally, if a resource is used less due to more efficient applications, the resource price declines, thereby stimulating demand for the resource again (market effect).

However, new eco-effective technologies may also stimulate environmental consciousness. This is particularly true for energy conversion technologies. The installation and monitoring of a new conversion system implies an involvement with topics of energy conversion and use. In some cases, detailed feedback on performance is provided that can be compared with consumption patterns. Research shows that timely and detailed feedback on both electricity and heat consumption can greatly reduce them (Jensen 2003; McCalley 2003). Information on production, especially when combined with information on consumption, is likely to have a similar effect.

Whether a new technology stimulates environmentally beneficial or unfavorable behavioral changes or not, depends heavily on the respective technology and on the context. Economic incentives for increased consumption are only present when the new technology indeed "pays off" and financial savings are not overcompensated for by rising energy prices. Direct rebound effects depend on whether increased appliance use or number of appliances indeed promises improved comfort – room temperature, for example, will not be raised indefinitely. Psychological offsetting of environmental benefits and damage depends on many aspects: how strongly is ecologically sound behavior perceived as a social or personal norm? Are we dealing with a conscious decision or habitual behavior? How much does the person know about the actual environmental effects of various behaviors? Whether information and feedback can help to change consumption patterns depends on subjective motivation and the actual amount, detail, and form of information provided.

Various technologies and contexts lead to an array of different empirical findings. Van Elburg (2001) finds a strong rebound effect in cooling appliances and compact fluorescent lamps. Cooling appliances are replaced by bigger ones, and lamps are increasingly used to light previously unlit areas – for example, the garden. This is for several reasons. Apparently, both types of appliances provide noticeable financial savings that can be invested in additional appliances. Furthermore, bigger fridges allow comfort gains in terms of better planning of household purchases and supply, and illuminating the garden satisfies aesthetic and status needs. Nadel (2003), in contrast, reports very weak rebound effects in appliances, with the exception of air conditioning and compact fluorescent lamps. Unfortunately, no more details are given. Henderson et al. (2003) report that rebound effects in heat energy consumption occur after the installation of insulation especially in households with low prior energy consumption and in houses which had greater heat loss. Similar effects are found by Genennig and Hoffmann (1996) and by Haas et al. (1999) for home electricity generation via photovoltaics: electricity consumption rises in households with low prior consumption and decreases in households with high prior consumption. Apparently, the "free" energy is used to raise the comfort level of users who were previously deprived of such comfort. In contrast, Haas and Biermayr (2001) find no difference in electricity consumption between households using renewable energies and conventional households. A temporal perspective on *changes* in consumption, however, is lacking here. What is interesting is the fact that most renewable energy users subjectively *perceive* themselves as energy conscious and economical (Karsten 1998; Haas et al. 2001). Another interesting effect is observed by Haas et al. (1999) and Reif (2000): photovoltaics users are willing to adapt their temporal consumption patterns to the production patterns of their panels, in order to maximize self-reliance. In Reif (2000) this is supported by financial incentives.

What can be inferred from these findings for micro cogeneration?

1. Micro cogeneration is targeted at home owners who, in Germany, form a relatively wealthy section of the population. It is safe to assume that the comfort levels of their dwellings are already high. Therefore, a significant increase of energy consumption is improbable.
2. A positive effect of feedback and increased involvement with energy topics is possible, but depends highly on the specific form, timing, and

amount of detail of feedback, and on the presence of other incentives – e.g. price incentives, importance of independence, or ecological motives. It seems advisable to provide information both on production and consumption patterns to allow users to compare and match them.
3. Similar to photovoltaics, there is potential for adapting temporal consumption patterns. This opens up interesting possibilities for load management.
4. One potential rebound effect may result from the German Energy Savings Decree (*Energieeinsparverordnung*, EnEV). The decree sets a maximum level for the overall energy demand in new buildings. Building owners who install a micro cogeneration plant are permitted to apply less stringent heat insulation measures. However, poor insulation of buildings has very long-term implications on energy demand, while a micro cogeneration plant that has reached its technical life span may be replaced by a conventional boiler, increasing the energy demand of the building.
5. Finally, it is difficult to predict to what extent micro cogeneration possession may be used as a justification for otherwise ecologically unsustainable consumption patterns. This depends, among other things, on the relevance of ecological norms to the user, on the consciousness of behaviors, on the degree to which micro cogeneration possession is perceived as ecologically relevant, and on knowledge of its effects. This is a question that needs to be empirically explored.

5.4 Conclusions

Most micro cogeneration systems are superior, as far as the reduction of GHG emissions is concerned, not only to average electricity and heat supply, but also to efficient and state-of-the art separate production of electricity in gas power plants and heat in condensing boilers. This claim is the result of our Life Cycle Assessment on the technology level. It is true, despite the strong dependence of the results on the electrical and thermal efficiency of micro cogeneration technologies and further parameters such as methane emissions from reciprocating engines. Effectively, at an electrical capacity of up to five orders of magnitude smaller than large gas combined-cycle power plants, lower GHG emission levels can be achieved. The GHG advantages of micro cogeneration plants are comparable to district heating with CHP.

The performance of micro cogeneration technologies with respect to climate and resource protection depends mainly on the total conversion efficiency (including the thermal output of the system) that can be achieved. In some cases, an unfavorable heat integration of micro cogeneration systems may lead to operation at the lower end of the assumed bandwidths of total efficiency. In such cases, CHP systems come rather close to electricity production in modern combined cycle plants without CHP. The optimization of micro cogeneration implementation thus involves careful integration into the supply object. Generally, the systems should not be oversized, to ensure that the full amount of heat is actually used in the system.

Under the assumption that gas-condensing boilers are the competing heat-supply technology, all technologies are within a very narrow range, with the exception of the small Stirling engine, with its lower electrical and total efficiency.

Looking at the GHG reduction potential on the level of a supply object (e.g. a single-family or multi-family house), the mitigation potential is somewhat lower, because the micro cogeneration systems do not supply the whole energy demand. Rather, additional electricity from the grid and heat from a peak boiler have to be taken into consideration. Here, fuel cells offer the advantage of a higher power-to-heat ratio, implying that, for a given heat demand, more electricity can be generated that displaces conventionally produced electricity.

Environmental impacts other than those related to climate and resource protection relate more specifically to technology. Whereas for fuel cells and Stirling engines (as long as these use innovative flameless burner technologies) emissions of air pollutants are extremely low, reciprocating engines emit more significant amounts of NO_x, CO, and hydrocarbons. Furthermore, the emission factors of reciprocating engines depend heavily on operation characteristics, age and maintenance of the systems. Thus, larger bandwidths characterize this system. As a consequence, acidifying emissions of small reciprocating engines (considering the heat co-product) are somewhat higher than those of centralized gas power plants, due to more efficient emission control in the latter.

In addition to investigating the emissions side, analysis of the immissions situation of a residential area supplied by reciprocating engines was carried out. The analysis shows that for selected, rather unfavorable, meteorological conditions, with a flat topography and urban housing structure, the additional immission of NO_x due to the engines is below the value of significance according to German legislation, and does not create severe additional environmental impacts.

The emissions reduction and resource protection potential of micro cogeneration could partially be offset by a “rebound effect”, implying that energy savings achieved by a more efficient technology are at least partly compensated, and sometimes overcompensated, for by an increase in energy demand. On the other hand, use of innovative energy systems in the household may rather sensitize its users regarding energy consumption and stimulate more energy conscious behavior.

Which of either of these effects prevails is an empirical question. We suspect that increasing comfort will not occur to a relevant degree, because micro cogeneration is targeted at a part of the population which has already reached satisfactory comfort levels. The effects of additional income and psychological offsetting cannot be exactly determined. The former depends on to what extent micro cogeneration indeed produces financial savings. The latter is influenced by many factors, including environmental consciousness and knowledge about the environmental impact of various behaviors.

A positive effect of feedback and increased involvement with energy topics is possible, both concerning total energy consumption and temporal adaptation of electricity use. The exact effect depends highly on the specific form, timing and the level of detail of feedback, and on the presence of other incentives.

When interpreting the overall environmental achievements of micro cogeneration, it has to be kept in mind that, as long as the fuels for micro cogeneration units are based on fossil resources, greenhouse gases are emitted and finite energy resources used. Therefore it is necessary to embed the use of micro cogeneration units into a strategy of renewable energy carriers, energy efficiency, and energy saving.

6 From Consumers to Operators: the Role of Micro Cogeneration Users

Corinna Fischer[1]

Technology does not develop autonomously, nor is its course solely determined by the quest for technological and economic optimisation. Rather, technology development is shaped by the actions and interactions of various societal actors. Technology users are an important constituent of this tightly woven network of actors. With an emerging technology like micro cogeneration, user influence begins with *technology acceptance,* which determines the chances of micro cogeneration to be introduced and to find widespread diffusion. This is particularly true when it is not being sold to housing developers or landlords, but directly to the end user, as in the case of single-family homes. Furthermore, users do not only accept or reject a new technology, but also *participate* in its development by giving feedback and interacting with producers. Finally, users can help or hinder the *diffusion* of a new technology by communicating about its advantages and disadvantages with other potential buyers.

Apart from their role in technology development, users also have an impact on the sustainability of a technology by their *patterns of use* – be it through differences in handling, timing and intensity of use, or through indirect effects, such as the "rebound" effect, meaning that energy efficiency gains are nullified by increased consumption in other areas (see Sect. 5.3.3). Since micro cogeneration is in an early phase of the innovation cycle, little data is available with respect to patterns of use and sustainability. Therefore, we will focus in this chapter on technology acceptance and on user motives and demands that could inform further technology development.

Since micro cogeneration is still in the introductory phase, a specific group of users is important to look at, whom Villiger et al. (2002) called

[1] The author would like to thank a number of people for their valuable contributions. Raphael Sauter conducted the research and analysis of the literature. Martin Pehnt, Raphael Sauter, Katja Schumacher and Julia Werner gave detailed feedback on earlier drafts of this chapter.

"innovators" and which we for conceptual reasons will name "pioneers".[2] Pioneers take up a new product in a very early stage of development, some even before general market introduction. Their role is to test and help develop the new technology and also to propagate it, paving its way into a broader market. Pioneers have specific characteristics. Rogers (1995) describes them as "venturesome"; interested in new ideas; financially well-established, which allows them to compensate for potential losses on an investment; and well-educated which enables them to understand the innovation. While the very first pioneers may be social outsiders due to their originality, pioneers in later stages have a high social status and substantial communication skills which help them communicate the new technological options.

Micro cogeneration technology blurs the boundaries between energy consumers and producers. Consequently, there is a broad spectrum of individual or corporate users who are motivated by different mixtures of producer and consumer logic. On the "producer" end of the spectrum, there are municipal utilities testing micro cogeneration as a potential option to broaden their portfolio of supply technologies and services. On the "consumer" end, we can find micro cogeneration in private homes, primarily used as a heating system, while the electricity may be consumed in the household or fed into the grid to reap extra benefits. Between those two extremes, small industries and handicraft enterprises, service enterprises like hotels or swimming pools, public bodies, educational and social institutions like schools or hospitals, and nonprofit or business organisations use micro cogeneration, each with a specific mix of motives, including immediate economic benefits, future business opportunities, public relations, and educational or idealistic motives. Furthermore, operators may differ from end users of the products, such as, for example, when a municipal utility tests a micro cogeneration installation for district heating of private homes.

The empirical evidence for current micro cogeneration types suggests that a variety of different user types and user / operator combinations indeed exists. We can find commercial users, private households,

---

[2] The term "innovators" is ambiguous because it is used by Villiger et al. (2002) in a slightly different sense from Rogers (1995) on whose framework they build. In Rogers' concept, "innovators" are entrepreneurs and engineers who develop an innovation, while users step in at a later stage, as "early adopters". Villiger et al. (2002) term the very first users "innovators". To avoid ambiguities, we describe such initial users as "pioneers", which also matches well the terminology used in innovation and diffusion research in the political sciences.

educational institutions, business associations, hospitals, municipalities or municipal energy suppliers as end users of all types of micro cogeneration, be they based on reciprocating engines, fuel cells or Stirling engines. The plants are being used for own supply, for the supply of third parties or for demonstration or educational purposes. Figures for the Senertec Dachs, which has to date sold over 10,000 units, suggest what typical shares might be like. Here, commercial users dominate (about 60 % of installed capacity), private households come second (about 32 %), complemented by a few municipalities (8 %). However, shares for private households seem to be growing, as in 2004 55 % of sales went to private homes (Karl Kiessling, Senertec Managing Director, personal communication).

There is also a variety of operating arrangements. With a mature technology, such as reciprocating engines, most users seem to operate their plant themselves, as the experience with the Senertec Dachs demonstrates. By contrast, with an emerging technology, such as fuel cells, field tests are being conducted in which energy suppliers (e.g. RWE or EnBW) operate the fuel cell plants on behalf of end users. Experience with Stirling engine-based plants shows a variety of different operator / user combinations, for example, a municipal energy supplier running a plant to supply municipal buildings, another municipal supplier running one for the supply of private homes, or a commercial customer using one for his own supply (for an overview of projects see http://www.stirling-engine.de/referenz1003.html)

In this chapter, we aim at describing micro cogeneration pioneers in more detail. After a short overview of the data and literature on which the analysis is based, we describe the target group in socio-demographic terms and discuss some of its relevant attitudes. The bulk of the chapter consists of a discussion of their relationship to micro cogeneration technology: their motives for testing micro cogeneration, what they expect from it, and how they evaluate it. After a discussion of the role of pioneers as potential multipliers, we add some remarks on barriers to micro cogeneration adoption.

6.1 Data and Literature

At the outset, it is important to note that our endeavour is subject to considerable data restrictions, meaning that results are not available for all types of users or micro cogeneration technologies alike. The information stems from several sources. First, we have conducted an extensive *study on pioneer users of fuel cell-based micro cogeneration.* In 2001, the power utility “Energie Baden-Württemberg” (EnBW) initiated a field test of fuel-cell-based Sulzer Hexis micro cogeneration plants in German "pioneer households". Potential participants were found via a newspaper advertisement campaign in spring 2002. About 6000 persons responded, of which almost 1000 were shortlisted for the test because their homes fulfilled the necessary requirements. To date, 15 plants have been installed, with plans to install about 55 by the end of 2006. The plants are operated and maintained by EnBW with users paying average tariffs for the heat and electricity generated by them. Of note is that the pioneer households had to pay a one-time "innovation contribution" of 2000 Euro, so no financial incentive was involved. To interpret the findings, it should be pointed out that the marketing strategy concentrated on the "fuel cell" aspect and did not elaborate on the notion of micro cogeneration in general.

Our study is comprised of various components. Qualitative and quantitative data were collected to complement each other. First, we conducted a focus group discussion in December 2003 with the first seven actual users who had installed the device at that time. On the basis of an interview guide, group members were encouraged to discuss the reasons for their application, their experiences, advantages and disadvantages of fuel cell micro cogeneration as compared to other electricity or heat technologies, and their willingness to help in the diffusion of the technology. This discussion was complemented by three more group discussions and a postal survey, centered on persons who had applied to take part in the field test but who had not been chosen to actually test the device. However, we reasoned that non-chosen applicants' motives and attitudes would come close to those of actual users.

The postal survey was sent out to a random sample of 462 applicants, that is half of the applicants who had been shortlisted. 142 of them responded, making a satisfactory response rate of 30 %. For the focus group discussions, all the applicants within the region around Stuttgart were invited. The aim was to keep travel distance to the discussion location below 50 km. There were 94 applicants in the region who were contacted by mail, 35 of whom volunteered to take part in a discussion. 28 were finally chosen on the basis of availability at the scheduled time, of which 26 actually showed up. Participants were randomly assigned to three groups. Discussion topics were similar to those of the actual users group, though instead of experiences we discussed hopes and fears regarding the fuel cells. Furthermore, a discussion of preferred ownership models was added.

Additionally, we infer some conclusions from a *literature study* on analogous cases. We define "analogous cases" as technologies for home production of electricity and/or heat which are, in comparison with established ones[3], innovative and advanced regarding efficiency and environmental effects. On the heat side, the investigated technologies were small biomass plants and solarthermal collectors. On the electricity side, most studies focussed on photovoltaics. The geographical focus was on Germany and Austria. Twelve studies dating from 1992-2003 were found which dealt with consumer aspects (Table 6.1. Literature basis of the study of analogous cases

Across all data sources, the user types included comprise primarily private individuals, a few representatives of public administration or non-profit associations, and a few business associations. Enterprises and energy suppliers are clearly under-sampled. The fuel cell micro cogeneration study focuses on end users and not on the operator (the EnBW). In the literature covered, end users and operators are usually identical[4]. Therefore, results presented in this chapter are valid for individual end users, and less so for associations or public bodies. To discover motives of business representatives, additional research will be necessary.

Where possible, we compared our sample to representative studies of the German population. For socio-demographic characteristics, we used the "General Population Survey for the Social Sciences" *(Allgemeine Bevölkerungsumfrage der Sozialwissenschaften ALLBUS)*, a representative survey concerning socio-demographic data and a number of current political and social issues that is conducted every two years (see http://www.gesis.org/Dauerbeobachtung/Allbus/). To analyse attitudes, a number of studies on attitudes regarding technology, energy or the environment were consulted.

[3] Conventional German heating systems are gas- or oil-based central heating, gas-based apartment heating, district heating, and, in a few cases in East Germany, individual coal stoves. Electricity is usually taken from the grid.

[4] There is one exception, the study by Reif (2000), in which houses have been built and equipped with photovoltaics by a non-profit cooperative who made them available for hire-purchase to low-income families.

Table 6.1. Literature basis of the study of analogous cases

Study (by year)	System	Topics	Sample
Hackstock et al. 1992	ST	Information, motives, user evaluation of the Austrian ST Do-It-Yourself program	238 ST owners
Greenpeace 1996	PV	Success of campaign (= number of PV systems installed), motivation, user evaluation of state support programmes	1662 persons interested in the campaign
Genennig 1996	PV	Motivation, satisfaction, user evaluation of the support programme, electricity consumption behaviour	1445 PV owners in survey, 48 in interviews.
Katzbeck 1997	PV	Possession of other energy technologies; attitudes towards electricity consumption, consumption data	32 PV owners
Rohracher and Suschek-Berger 1997	BM	Economic, technological and social conditions, motivation, trigger, sources of information, user evaluation of state support programmes	25 BM owners, 28 multipliers, control group of 116
Karsten 1998	PV	Motivation, experience, suggestions	123 PV owners: 106 households, 17 firms, associations and public bodies
Haas et al. 1999	PV	Characteristics of the users: are they "early adopters" as characterized by Rogers (1995)?	60 PV owners, 17 merchants, 660 experts
Reif 2000	PV	Output and consumption data, ratio of output/consumption, degree of self-reliance	23 PV users: 22 households, 1 kindergarten
Haas et al. 2001	ST,PV BM, HP	Investment and consumption behaviour, influence of structural aspects, attitudes on energy consumption	101 owners, control group of 177
Hübner and Felser 2001	ST, PV	Motives, barriers, desired characteristics of the technology	Secondary study
Polzer 2003	ST, PV	Number and size of systems in a town, motivation, satisfaction, suggestions	147 ST owners, 22 PV owners
DENA 2003	ST	Motivation, attitudes, image of the technology, barriers, user evaluation of support programmes	ST users, sample size not provided

PV=Photovoltaics, ST=Solarthermal collector, BM=Biomass boiler, HP=heat pump

6.2 Socio-demographic Characteristics

Users covered in the literature as well as in the fuel cell micro cogeneration study have a distinctive socio-demographic profile. They are usually families living in their own home, predominantly in rural areas and small towns. In addition, participants in the fuel cell micro cogeneration study were almost exclusively men[5] in their mid-fifties (54.9 years on average), and thus well beyond the average age of 46.4 years of the German population over 18, as given by a representative sample (ALLBUS 2002).

The findings on professional and educational status are particularly interesting. In the literature, user groups seem to differ slightly, depending on the technology studied. On the one hand, the users of solar thermal collectors (Hackstock et al. 1992; DENA 2003) and biomass (Rohracher and Suschek-Berger 1997) are predominantly non-academic, with a large share being farmers and skilled manual workers. On the other hand, users of photovoltaics (Genennig 1996; Karsten 1998, Haas et al. 1999; Haas et al. 2001)[6] tend to be high-income academics with a special interest in socio-political issues in general and energy issues in particular.[7] Participants in the fuel cell pioneer campaign resemble, with some modifications, the latter group. Due to peculiarities of the recruiting process, they are almost exclusively home owners (131 of 142) of a more rural and, in terms of lifestyle, traditional sample than the PV pioneers: almost all are married, about 20 % live in villages with populations of less than 5,000, with almost another 50 % living in small towns with 5,000 to 20,000 inhabitants. But, like the PV users, they are of an educated high-income strata. Almost 20 % have a net monthly household income of more than 4,500 EUR. In the representative ALLBUS sample (average Germans over 18), this group makes up only 11 %. However, the high household income does not necessarily translate into a high per capita income. The fuel cell pioneers are not only especially well off, they also have especially big families. In their average household, there are 3.15 people, as compared to 2.59 in the general population. That means that the additional income must also feed additional people. It is, then, not so much an indicator of a high level of disposable funds, but rather of social status.

---

[5] Of the 142 respondents of the postal survey, only 7 were female.

[6] There is no information about education in Greenpeace (1996).

[7] The only exception is the study by Reif (2000) mentioned in footnote 4.

Figure 6.1 illustrates fuel cell pioneers' education levels. As they are almost exclusively male, and because in their generation, education is gender-related, their school degrees are compared to degrees of the male part of the ALLBUS sample. The dark-shaded section shows the highest degree available at German schools, *Abitur* or *Fachabitur* (high school graduate; 12-13 years of schooling). The middle section shows an intermediate degree (*Mittlere Reife*, 10 years of schooling), and the light section the lowest degree *(Hauptschule*, 9 years of schooling).

Figure 6.2 shows the fuel cell pioneers' vocational training[8] and current job position. The numbers indicate the frequency of mention (multiple answers were possible). It is apparent that many of the pioneers have technical training and/or work in a technical profession – and also, that many of them are already pensioners, corresponding to their relatively high ages.

How can the findings be interpreted? The striking fact that the sample is almost exclusively male will be discussed below, when motives are presented. The educational and professional difference between owners of different technologies found in the literature evokes some considerations that might also be relevant for different types of micro cogeneration.

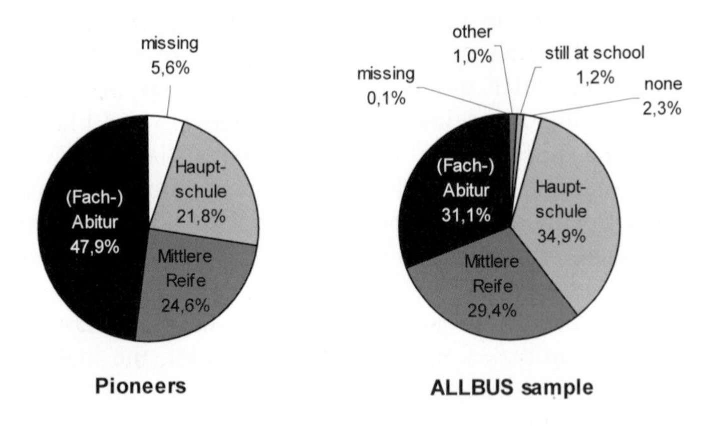

Fig. 6.1. Education of fuel cell pioneers compared to a representative (male) sample

[8] In German social science surveys, "vocational training" includes all higher education beyond school that qualifies one for a job, including university education.

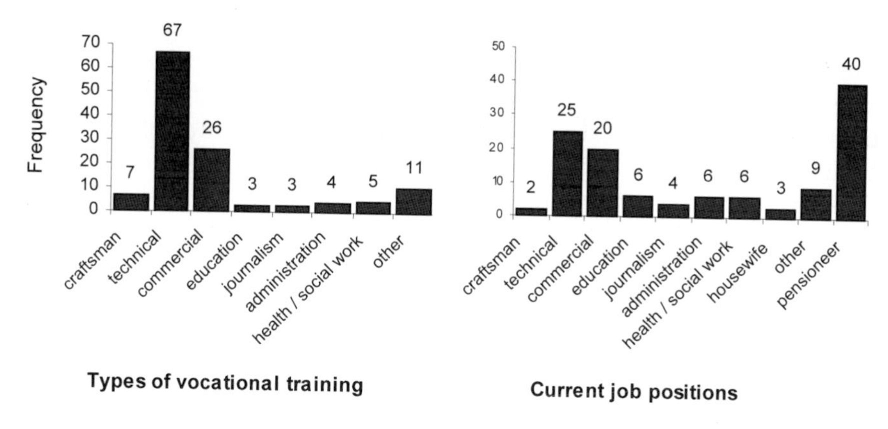

Fig. 6.2. Vocational training and current job positions of fuel cell pioneers

We suspect that solar thermal collectors (as well as biomass), in comparison to photovoltaics, are perceived as rather low-tech applications. They therefore address "Do-It-Yourself" individuals and attract users with technical and manual skills who may perceive the mounting and maintenance of the system as a hobby. Photovoltaics has a high-tech image, is expensive and, without state support, not economical. It is therefore easily perceived not to be a technology for broad application, but a hobbyhorse of technology nerds or environmentalists. It may therefore appeal rather to people with "green" political orientation, post-materialistic values, specialised technical interests, or to individualists willing to take risks. These groups can traditionally be found in the academic, high-income strata. The same argument may hold for fuel cell micro cogeneration – and more so, because it is even less well known and has not been tested extensively in practice. In the following section, we will reveal some further evidence for this hypothesis.

In contrast to fuel cells, reciprocating engines are an established technology that seems feasible and economically sound. Micro cogeneration based on such engines may therefore rather appeal to similar groups as biomass and solar thermal collectors. Stirling micro cogeneration, being a rather old but not too well-known technology, is likely to be found somewhere in between.

6.3 Attitudes

In the fuel cell micro cogeneration study, we explored pioneers' attitudes on technology and the environment. Furthermore, we investigated their interest in social and political affairs, because we suspected that their interest in a technology perceived to be innovative and environmentally benign might be related to a broader interest in social progress. We will discuss the various issues in turn, focussing on the results of the postal survey. Focus group results are generally in line with them and back the findings.

6.3.1 Political Attitudes

One initial result is that the pioneers indeed show a keen interest in the community, both at a local level and with respect to "big politics". 21.3 % do volunteer community work between once a week and once a month (as compared to 11.3 % in the male part of the ALLBUS sample), 31.9 % do it between once a day and once a week (as compared to 20.2 %) and 5.7 % do it even on a daily basis (as compared to 4 %). The pioneers also report high levels of political activism.

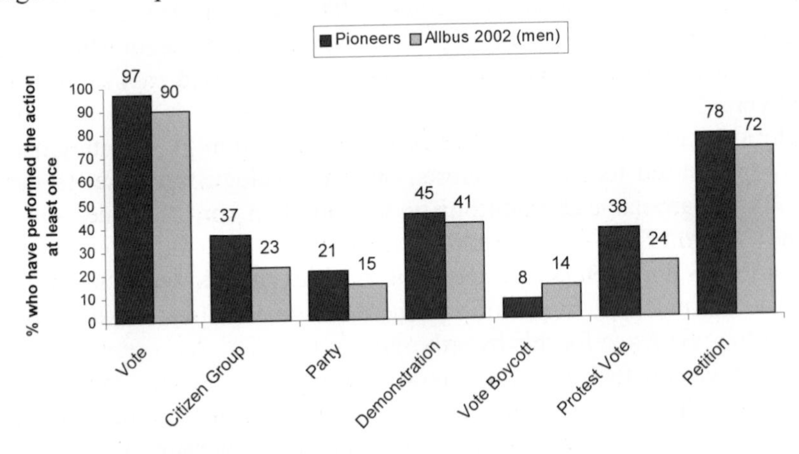

Fig. 6.3. Political activism of fuel cell pioneers. Activities are: voting in an election, being an active member of a citizens' advocacy group, being an active member of a political party, taking part in a legal demonstration, refusing to vote (as a method of expressing protest), voting for a party one would normally not consider (as a method of expressing protest), and signing a petition

Figure 6.3 shows the percentages of that report having taken part in various types of political activism at least once, again compared to the male part of the ALLBUS 2002 sample. It turns out that the pioneers have performed every action more often than the male members of the ALLBUS sample, with the single exception of vote boycotts. Apparently, they feel an obligation to express their political views via voting. All differences are significant at a 0.05 level.

To summarize, the pioneers evoke a picture of persons with a keen interest in public affairs, rooted in the community, and willing to do their share for the common good, but also to stand up for their aims and interests. Focus group members confirm this view: among them were mayors, school directors, and leaders of local industry or craftsman associations.

6.3.2 Technology Attitudes

As expected on the basis of their professions, the pioneers also have a positive picture of technology, and feel competent in dealing with technical issues. Table 6.2 shows their responses to several technology-related items, to be rated on a scale of 1 to 7 (1 equals "fully disagree", 7 "fully agree"). The pioneers' mean responses are compared to the means of a representative sample of the population, surveyed by the Institute for Technology Assessment and Systems Analysis (ITAS) (Hocke-Bergler and Stolle 2003).[9]

Significant differences have been highlighted in grey. An interesting pattern shows up. The pioneers do not differ much from the general population in assessing the blessings of technology. With the exception of the item "Technology is the basis of our living standard", they are very close to the ITAS sample on all positive items. However, they differ significantly when it comes to judging risks and disadvantages of technology. The pioneers are much less cautious than the general population, detecting few risks and problems in "modern technology".

The pioneers also score high in a test measuring subjective technical competence, that is, the confidence in one's own ability to solve technical problems. The short version of the KUT test (*Kompetenz im Umgang mit Technik*, meaning "technology-use competence"; Beier 1999) comprises eight items, such as "Usually, I cope with technical problems successfully", to be rated on a scale from 0 ("fully disagree") to 5 ("fully

[9] Unfortunately, the data does not allow us to differentiate for gender in this case.

Table 6.2. Attitudes of fuel cell pioneers towards technology

Item	Pioneers	ITAS 2003	Signif. (T-test)
Technology is the basis of our living standard	6.04	5.56	.006
Technology inevitably leads to pollution	3.09	4.08	.000
Technology simplifies everyday life	5.81	5.78	.839
Technology needs stricter control	4.70	5.09	.049
Without technology, everyday work could rarely be mastered	5.72	5.67	.766
Technology is opaque and menacing	2.37	3.50	.000
Without technology, life would be more humane	3.07	3.61	.017
Technology enslaves people	3.13	3.79	.004
Technology makes life more comfortable	6.03	5.84	.149
Today, technology is used without investigating its consequences properly	4.26	4.23	.879
Technology helps to prevent catastrophes (plagues, famines)	5.18	4.92	.189
Technology endangers humankind and the environment	2.90	3.43	.006
Technology is needed for the survival of a growing world population	5.35	5.34	.944

1 = "fully disagree", 7 = "fully agree".

agree") with the maximum score being 40. The pioneers show an unusual pattern. Of the 142 respondents, 130 score more than 20, 72 (that is half of them) more than 30 and 37 (about a quarter) even more than 35. The mean score is 30.54. Unfortunately, information about a representative sample is not available for comparison, but the sample of test persons used in constructing the test shows a mean score of only 26.8 (Beier 1999, p. 690).[10]

6.3.3 Environmental and Energy Attitudes

A third dimension of interest was attitudes towards the environment and energy. Table 6.3 shows a comparison of environmental attitudes between the pioneers and a representative German sample studied in 2002 (Kuckartz and Grunenberg 2002). The items were rated on a scale from 0 ("fully disagree") to 4 ("fully agree"). Again, significant differences are highlighted in grey.

[10] The original score is 34.8 because a scale of 1 to 6 was used while in our survey, for consistency reasons, a scale of 0 to 5 was employed. Transposed to a 0 to 5 scale, a mean of 26.8 emerges.

Table 6.3. Environmental attitudes of fuel cell pioneers

Item	Pioneers	Kuckartz and Grunenberg 2002	Signif. (T-test)
Even today, most people still do not act very environmentally concious.	2.72	2.84	.327
People like me cannot do much for the environment	1.31	1.99	.000
There are limits to growth that our industrialised society has already reached or will soon reach.	2.66	2.56	.528
The environment should be protected even if it means a loss of jobs.	2.11	1.83	.069[a]
Science and technology will solve many environmental problems without us having to change our lifestyle.	2.08	2.05	.841
Most products of science and technology damage the environment.	1.29	1.79	.000
If we go on like this, we are approaching an environmental catastrophe	2.38	2.51	.443
It worries me to think about the environmental conditions our children and grandchildren will have to face	2.34	2.69	.025
Media reports about environmental problems often fill me with outrage	2.32	2.53	.150
Even today, politicians do by far too little for the environment	2.61	2.76	.293
In my opinion, many environmentalists strongly exaggerate the significance of environmental problems	1.56	2.30	.000

0 = “fully disagree”, 4 = “fully agree”.
[a]The difference becomes significant (.048) if a Mann-Whitney-U-test is employed instead of a t-test.

It becomes clear that the pioneers take environmental problems seriously. They strongly reject the notion that environmental problems might be exaggerated, and even tend to consider the environment to be more important than jobs. However, they do not react to the problem emotionally. In the item expressing worry, they score lower than the general population. A possible reason is that the pioneers have high confidence in their own ability to contribute to a solution. The item “People like me cannot do much for the environment” is rejected categorically. And of course, the pioneers would never assign science and technology a role in creating the problem. Survey participants do not appear to be more enthusiastic about the potential of science and technology to solve environmental problems than the general population. In the focus group discussions, however, this idea was expressed clearly. To the pioneers, what was technically advanced was also environmentally benign. Both notions often melted in the notion of *Zukunfts(weisende) Technologie*, maybe best translated as "forward looking technology "or "technology of the future".

Our picture so far is that the pioneers care about the environment, have a positive view of technology, and think they can contribute to solving environmental problems as well as dealing competently with new technology. So we might conclude that they themselves take measures to protect the environment and that technology plays an important role in those measures.

We have explored this idea by examining examples of self-reported environmental behaviour and of investment in "green" technologies, focusing on energy. Figure 6.4 shows self-reported environmental behaviour in the energy domain, as compared to the Kuckartz and Grunenberg sample.[11]

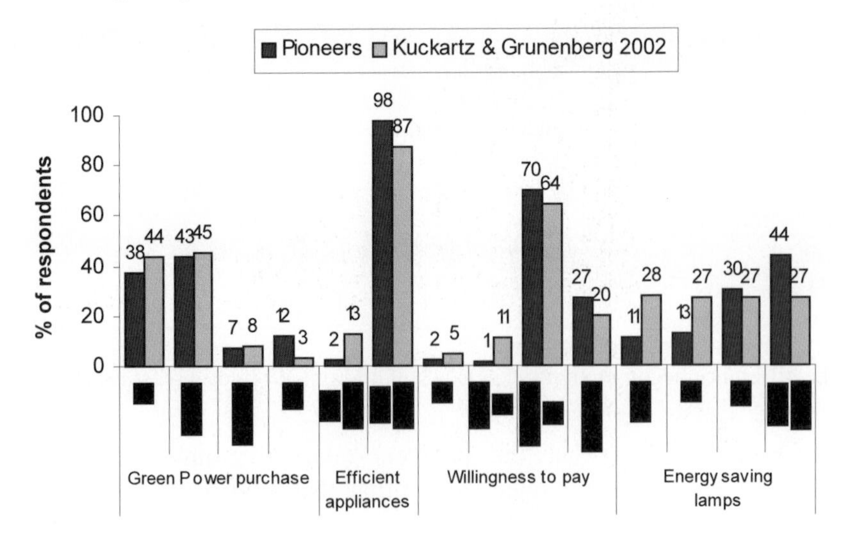

Fig. 6.4. Energy-related behavior of fuel cell pioneers

[11] Self-reported behaviour does not exactly mirror real behaviour, but assuming that the rate of misreporting is about the same in both samples, the comparison is instructive.

The actions examined were: purchasing Green Power, buying efficient household appliances, willingness to pay more for such appliances, and number of energy-saving lamps in the household. We can see that the pioneers rate better than the general population on any of these items, though the differences do not become statistically significant in the Green Power case. In the other items, significance is at the 0.01 level.

Furthermore, it is evident that the pioneers have already invested in various environmentally sound technologies (Fig. 6.5). This does not only demonstrate their willingness to invest in "green" technologies. What is more, these "trigger technologies" provide them with some experience and knowledge, lowering the threshold for considering an innovative technology like fuel cells. The role of knowledge is confirmed by the fact that many of the pioneers have actively sought out information on different energy- and environment-related issues (Fig. 6.6). Most have done so before learning about the field test, but learning about it has spurred additional interest, especially in fuel cells.

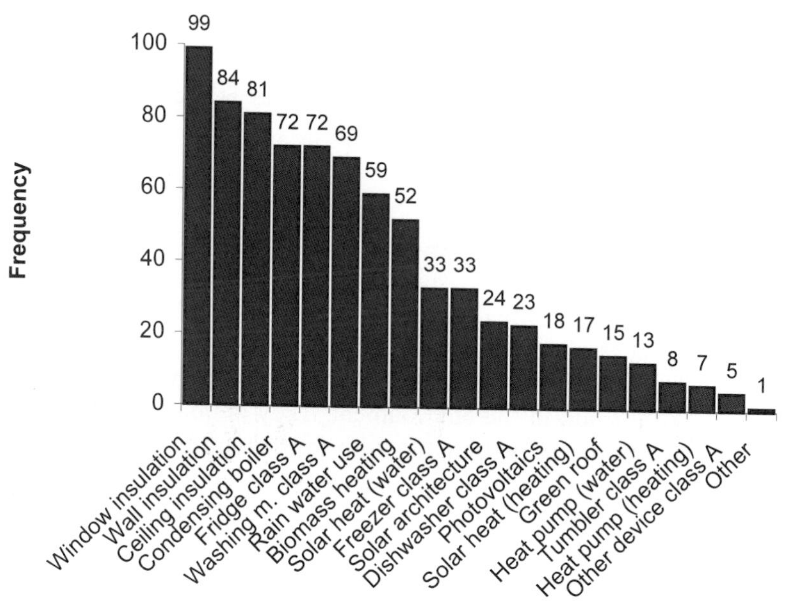

Fig. 6.5. Possession of environmentally sound technologies

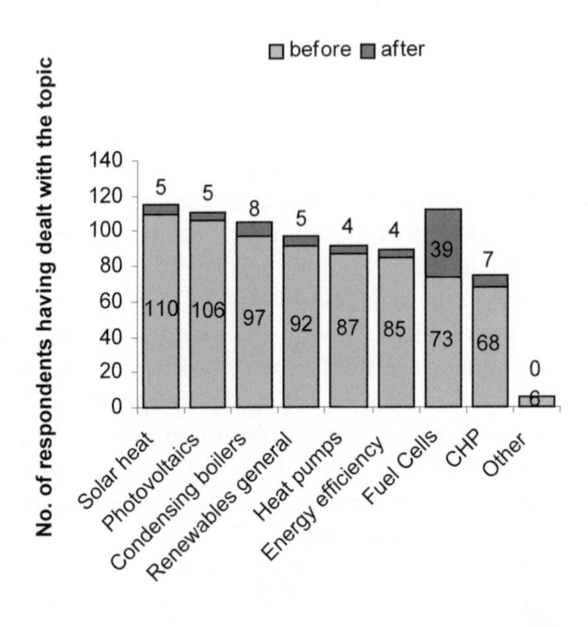

Fig. 6.6. Active information on energy technologies before and after learning about the fuel cell micro cogeneration field test

6.3.4 Conclusions on Attitudes

Now, what have we learned about the pioneers so far? Results about pioneers' attitudes fit well with the hypotheses drawn from their sociodemographic profile. Obviously, they are people with an interest in community affairs and in the environment. They are positive that existing problems may be solved, particularly with the help of new technology, and they are willing to contribute their share to the solution. Their enthusiasm about technology means that this kind of contribution is not perceived as a sacrifice, but rather contributes to personal fulfilment.

In the next section, we will deal with the pioneers' relationship to their respective energy technology and see whether this fits the picture sketched here. We will look at the events that prompt them to become interested in an innovative energy system, at the reasons they give for showing an interest, at requirements they think such a system must fulfil and at their opinion on whether their system meets those requirements.

6.4 Motives, Goals, and Interests

6.4.1 Reasons for Considering Fuel Cell Micro Cogeneration

Our literature study shows that innovative technology users report mixtures of different motives. Among them are: independence, interest in the new technology and a desire to promote it, the wish to help the environment, and economic benefits.

But as much as the general set of motives is constant, just as much does their relative importance differ. Sometimes environmental motives are most prominent; sometimes it is independence or technical interest. Economical motives, however, are never primary. The differences are probably to a great extent due to methodological reasons. The studies analyzed word questions differently, present different sets of answers to choose from, pose them in different order, and elaborate topics in different detail.

The fuel cell micro cogeneration study helps to clarify some of these aspects with respect to micro cogeneration. Survey participants were asked in an open question about their reasons for considering the fuel cells; the answers were categorized later. Figure 6.7 shows the results.

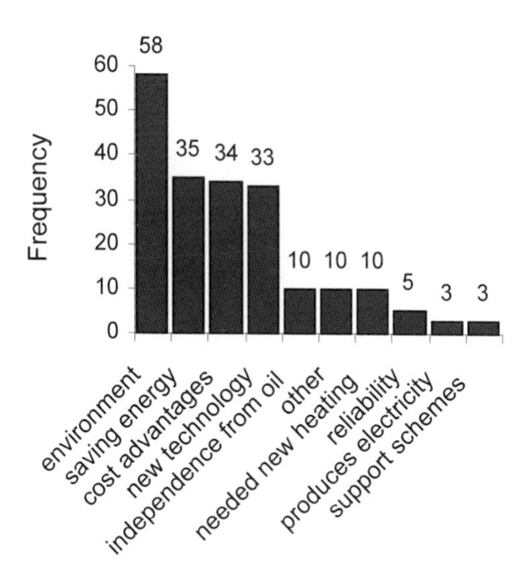

Fig. 6.7. Reasons for considering a fuel cell micro cogeneration as a home heating system

We can see that environmental protection (comprising answers like "environmental protection", "lowering emissions" or "protecting resources") is by far the most frequent motive, followed by energy saving (including energy efficiency), cost considerations, and interest in the technology. The desire to save energy, by the way, is a hybrid motive; it can stem both from environmental and from economical reasons.

A similar picture emerges from the focus group discussions. Technical interest is inseparably interwoven with an interest in everything "new" and "forward-looking". Interviewees state that they are "open for everything that's new", they want to "be in on the ground floor with this technology". For this purpose, they are willing to invest some money and incur certain risks. Fuel cell systems attract them because they are perceived as especially modern in comparison to other generation techniques. Visions articulated in this context dealt with a hydrogen economy, import of solar hydrogen from desert areas, and self-sufficient home systems based on a combination of fuel cells, photovoltaics, and hydrogen storage. Interviewees have an optimistic outlook, assuming that technology will solve environmental problems. At the same time, new technologies are expected to be more "efficient" and "economical", thus linking environmental and economic benefits. As one participant put it:

> "It is my aim to rearrange the building in terms of optimum [energy] use. Better heat supply and heat insulation – a combination of solar thermal water heating and a fuel cell would be ideal. Later, this could be combined with photovoltaics, and I'd achieve some form of self-sufficiency."

Other than we supposed for the PV users in the literature, but similar to the suspected motives of the biomass and solar heat users, technical interest of focus group members mixes with a desire to "do it yourself", to "tinker" with one's home energy system. Such systems are perceived as a space for self-actualisation:

> "Home technology provides room for creativity. It is something you can shape, you can unfold your talents on, find solutions. It's not only 'It's working, and that's all I'm interested in'. No, I want to be ahead, find better solutions than others."

6.4.2 Requirements for Home Energy Supply

So far, the results fit into the picture of the environmentally concerned and technologically enthusiastic citizen. However, there are also contrasting findings. We asked survey participants which aspects generally they find important when thinking about their home energy supply. They were given various choices, each to be rated on a scale from 0 to 4 (0 = not important at all, 4 = very important). When respondents are confronted with such a general question which not explicitly mentions fuel cells, the picture changes dramatically (Fig. 6.8 shows the mean rating of the various items).

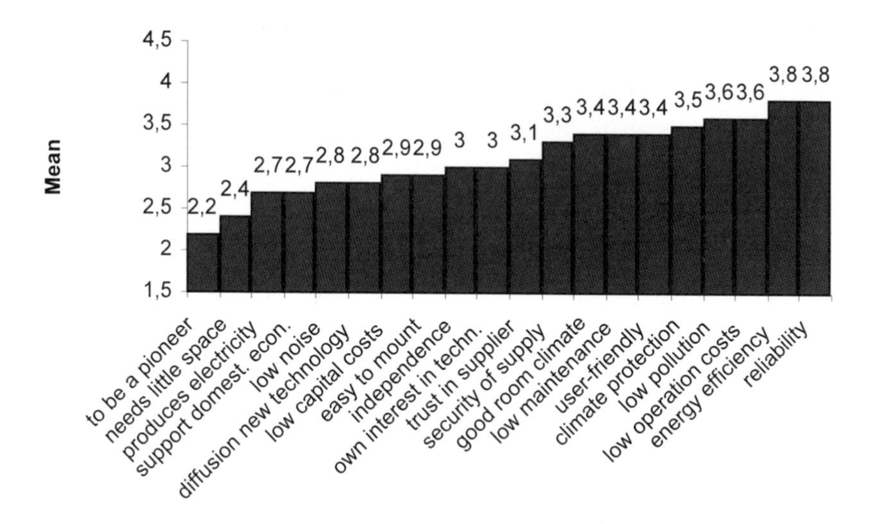

Fig. 6.8. Pioneers' requirements for home heating systems. 0 = "not important at all"; 4 = "very important"

Suddenly, reliability, energy efficiency, and low operation costs become the main concerns. The environment ranks thereafter, and to "own an interesting technology" or "support the diffusion of new technologies" follow much later. How can we explain this apparent incoherence? To answer this question, we will first have a look at some additional information: opinions on fuel cell performance and further results from the focus group discussions. After voicing their requirements for home energy supply, respondents were given the same set of items and asked how well the fuel cell micro cogeneration fulfilled them in their opinion. Figure 6.9 gives the results, again presenting the mean rating of each item, which had been rated on a scale from 0 ("does not fulfill the requirement at all") to 4 ("fully fulfills the requirement").

According to Fig. 6.9, pioneers see the fuel cell micro cogeneration plant's strength in the domains of environmental benefits and innovative technology. In contrast, performance with regard to "practical" concerns like user-friendliness, reliability and low maintenance as well as cost aspects is judged rather sceptically.

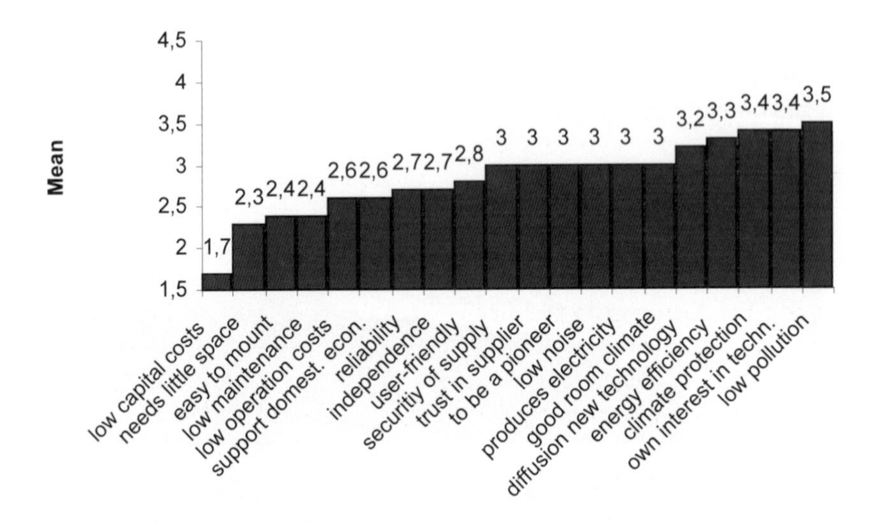

Fig. 6.9. Expected performance of fuel cell micro cogeneration. 0 = "does not fulfill the requirement at all"; 4 = "fully fulfills the requirement"

This is a realistic picture. In fact, fuel cell CHP systems promise a good environmental record. However, they still have a long way to go to become fully operational. At the moment, the prototypes do not yet function reliably, there are problems with the installation due to the size of the devices, and production and maintenance costs are prohibitive.

These problems were also reflected by the focus group participants. In spite of their sweeping technological visions, they also feared failures due to the immature stage of the technology and, in exchange for their willingness to contribute, demanded that they be protected from economic and technological risk. What is more, there was general agreement that new solutions needed to become economically feasible in the long run. Interviewees did not demand high financial returns, but required that the investment had to "pay off" after some time. They were also aware that aversion to risk as well as economical considerations are even stronger in the group of mainstream users. They voiced clearly that, in order to win a mass market, fuel cells must become more reliable and, first of all, much cheaper:

> "It is a small, idealistic minority who think like [us]. Most people think differently. The system needs to pay off after 10, 12 years – otherwise it won't succeed."

> "I'd like to comment on the topic 'convincing other sections of the population'. In my case, one part of that population was my wife! Women have a more

practical perspective. [...] She has doubts about new developments which pose a certain risk – because you don't want to install a new heating system every few years!"

The last statement points to the importance of the gender issue. Being responsible for the household's functioning, the family's well-being, and usually lacking the technological enthusiasm of their male counterparts, women with a traditional socialisation are definitely more difficult to reach if a new technology suggests possible inconvenience, hassle or even failures, threatening to leave the family in the cold. In this light, we can interpret the seemingly contradictory results (Fig. 6.7 vs. Fig. 6.8). Like everybody, the pioneers have some very practical, down-to-earth requirements for their heating system. It needs to work reliably, provide comfort, and be economically feasible. On the other hand, they are well aware that fuel cell CHP is a novel, experimental technology. Because of their keen technical interest and their environmental concern, they would like to give it a chance. They are willing to contribute to the experiment as long as they are protected from the biggest risks. However, this does not mean they are willing to accept high costs or unreliable quality for the sake of the environment once and for all. They expect that fuel cell development will advance, and that the deficits will be remedied.

6.4.3 Micro Cogeneration Technologies in Comparison

How about other micro cogeneration technologies? Which arguments can be brought forward in favour of them? Do they differ from the arguments for fuel cells – and in which respect? Are more established micro cogeneration technologies like reciprocating engines better suited to accommodate users' needs? To find out about these questions, focus group participants in the three applicant groups were asked to rank different technologies for the provision of heat and electricity services with respect to which of them they would most like to invest in. They were also requested to give reasons for their ranking. Heat and electricity production technologies were ranked separately, with micro cogeneration present both on the heat and on the electricity side. Three micro cogeneration technologies were included (fuel cell, Stirling engine and reciprocating engine) which had been introduced to the participants in a brief presentation prior to the ranking task.

The resulting picture is quite mixed. The rankings constructed by the three discussion groups differ, the only commonalities being that solar technologies (photovoltaics and solar heat) generally rank relatively high, and purchasing "green power" from the grid always ranks lowest. The three micro cogeneration technologies are always close together, somewhere in the middle, differing in their exact rank.

However, as the exact ranking result is very much dependent on group dynamics and on the amount of information present in the specific group, what is more important are the various arguments and evaluation criteria brought forward and discussed by group participants. Some of these arguments were similar in all three groups. Differences emerged due to both the relative weight group members gave the various criteria and different perceptions as to what extent the technologies would meet those criteria. Rather than contradicting each other, the various arguments each add additional aspects, so that a consistent picture can be constructed.

All micro cogeneration technologies generally were viewed as being energy efficient, which positively influenced their ranking. With regard to other aspects of environmental performance, feelings were mixed. They were criticized for still running on fossil fuels, which made them less attractive than solar technologies. Therefore, it was seen as a positive vision for the future to run a Stirling engine on renewable energy sources or a fuel cell on hydrogen. The reciprocating engine performed worse in this respect. Furthermore, it was seen as "dirty" in operation, probably also loud, and old-fashioned. On the upside of that "old-fashionedness", however, participants regarded it as an approved and reliable technology.

One of the core criteria, and more so than in the survey, was independence. Independence primarily meant control over one's own energy supply. One aspect was independence from the grid. Technologies that allow the storage of fuels at home or solar technologies were therefore favoured (which also meant that oil was preferred over gas). Micro cogeneration's performance in this respect depended on the fuel used. Another aspect was independence from fossil fuels, which were seen as subject to shortages and price fluctuations. This aspect privileged wood-pellet heating and solar technologies over most micro cogeneration. However, fuel flexibility was seen as a clear asset of the Stirling engine. Finally, independence also meant control over how one's energy is produced. Participants desired to verify potential ecological benefits and efficiency gains themselves. Therefore, all micro cogeneration technologies were preferred to green power from the grid, which encountered much distrust.

While solar technologies ranked better in environmental performance and independence, their supply security and their ability to completely meet energy demand was doubted. They were therefore seen as "additional" technologies, which could gainfully be combined with micro cogeneration.

Finally, practical considerations and cost arguments were discussed. Among the practical considerations were lack of space for fuel storage (which might, in contrast to the arguments discussed above, benefit grid-bound technologies) and the size of the device, which at the moment is a problem for the fuel cell. As to prices, no clear advantage for one or the other system was found, which was also due to lack of information.

6.5 The Pioneers as Multipliers

The role of a pioneer is not only to be the first in testing a new technology, but also to support its further diffusion. Can we expect this kind of pioneer to be available for micro cogeneration?

Positive hints come from the literature on renewable energy pioneers. Virtually all participants in all studies are highly satisfied with their systems. Satisfaction is even higher when participants monitor their own system, receiving detailed feedback on its performance. Given the fact that the innovative energy systems were actively sought out by the users, we assume that satisfaction stems from having realized personal goals and aspirations, and from observing the positive environmental and, at times, economic effects of this realization. Therefore, we assume that micro cogeneration users will also be satisfied, provided the systems work reliably. This satisfaction could make them successful multipliers.

Satisfaction was explored in the fuel cell micro cogeneration study via the focus groups, both with applicants and with actual users. During the group discussion, applicants got the opportunity to see a micro cogeneration system in action and get up-to-date information on its state of development, so they could make an informed judgement. Both applicants and users were indeed generally satisfied. However, they showed some disillusionment about the system's technological maturity. Some applicants had hoped to be able to buy a fuel cell-based micro cogeneration system in one or two years via mainstream sources, not knowing that there are still a number of technical difficulties to overcome and that the product will certainly not go into mass production within this decade.

Nonetheless, pioneers were interested in helping to diffuse the technology. In the focus groups, a desire for showing and explaining the technology to others was voiced over and over again. Teachers wanted to show their students; craftsmen articulated they would like to set up informational events for their colleagues. The pioneers welcomed the oportunity to discuss in the focus groups, and demanded from the energy supplier to "go public" much more strongly; some focus group members teamed up and started drafting a PR campaign. As one member put it:

> "I look at my solar thermal panel every day. It would be the same with that [fuel cell] system. I'd tell everybody: 'Hey, come on, look, I show you something.' "

In the postal survey, 26 % claimed to be "very interested" and 44 % to be "interested" in passing on information and showing their fuel cell to others, if they had one. Only 2 % were "not very interested" and another 2 % "not interested at all".

However, pioneers do not want to simply promote the technology any which way. They consider it important to provide fair and realistic information, pointing out advantages and disadvantages. They deem it necessary to accurately inform the public about the state of development in order to avoid raising false expectations.

6.6 Uncertainty as a Barrier

So far, we have identified motives of pioneer users and evaluated their potential of aiding diffusion. However, it is also important to identify potential barriers that may inhibit adoption and diffusion. A substantive barrier identified in the literature is *uncertainty* about the reliability and maturity, and also about the cost and profitability, of technologies. These problems are reflected in the fuel cell micro cogeneration study. Many respondents had difficulties assessing how well the fuel cell might perform in relevant areas. About one third marked "don't know" for each of the following items: reliability, user friendliness, ease of mounting, and need of maintenance. About 25 % "don't know" about operation costs. Focus group members feared immaturity and possible "teething troubles" of the systems. They expected frequent failures and need for repair, and were sometimes concerned about safety issues. Though they were willing to incur a certain degree of risks, they did not want to function as "guinea pigs". The uncertainty of costs also seems deterring. People cannot assess either investment costs or operation and maintenance costs and, therefore, find it difficult to soundly calculate whether the installation may be economically feasible. For reciprocating and Stirling engine CHP systems, we expect these concerns to be less pressing.

What strategies can help to overcome such uncertainties? First and foremost, reliable information is crucial. A second helpful tool to protect users from financial and technical risks is a contract arrangement with a contractor bearing the financial risk and guaranteeing heat and power supply for an agreed-upon price. Focus group members welcomed such a contracting system for the development and testing phase. It was particularly appreciated that, in their case, the contractor was a well-known, sizeable, and reputable energy supplier. However, with few exceptions they hesitated to accept contracting in the long run. Once the system is mature, most interviewees preferred to own it themselves. Ownership is an important dimension of the independence that is sought.

Thirdly, financial risks could be eased by support schemes. A substantial number of the users described in the literature claimed they would have installed their system even without any support programmes. But even for them, public support schemes played a significant symbolic role, honouring their initiative and willingness to take risks and emphasizing that they are using a promising technology in which investment is worthwhile. Users favoured a combination of investment subsidies and feed-in bonuses.

The topic was not studied in the fuel cell micro cogeneration study because the development stage of that technology is too young to draft support schemes. Nonetheless, the concerns found in analogous cases should apply to micro cogeneration as well.

6.7 Summary and Conclusions: Pioneers for Promoting Micro Cogeneration

We can now sum up some preliminary findings about pioneers for promoting micro cogeneration, their motives and expectations:

Pioneers of micro cogeneration come from a well-educated, established middle-class population with good income. Yet the fuel cell micro cogeneration study shows that they are not the urban academic ecologists we find in the photovoltaic case, but rather a more traditional group: families living in their own houses in rural areas or small towns. Their education is very often of a technical nature, spurring interest in new technologies. What is striking is their relatively high age and the almost complete absence of women.

They show environmental consciousness and a keen technical interest with a hobby component. Combining these two features, they hope to be able to solve environmental problems by means of innovative technology. Both concerns converge in the notion of the "forward-looking technology". The pioneers trust their own ability to solve problems – be they of a technical or environmental nature. Therefore, they want to make their own contribution. Many of them possess "green" energy technologies like solar heat or heat pumps, or efficient household appliances. These technologies might serve as a "door opener" for micro cogeneration.

Environmental protection and the desire to be the first in testing a novelty thus spur their interest in micro cogeneration. However, like everybody else, the pioneers require their home energy system to be cost-effective, reliable, and user-friendly, at least in the long run. In these domains, they spot deficits in fuel-cell-based micro cogeneration. They are willing to give the immature technology a chance, but strongly point to the necessity to remedy the deficits in order to reach a mass market. Other micro cogeneration technologies might perform better in this regard; but, on the downside, especially reciprocating engines are perceived as being "dirty".

The pioneers have a desire for independence and perceive micro cogeneration as a suitable tool for "self production" of electricity, though reliance on grid-based gas or fossil fuels in general might pose a problem. The Stirling engine, with its fuel flexibility, could be a promising technology in this respect.

Provided that their technology works reliably, pioneers tend to be very positive about it. The fuel cell pioneers investigated in our study are even fascinated. Some of their fascination is specific to the fuel cell (especially the visions of a hydrogen economy), but their more general motives also make this group a target group for other micro cogeneration types. Feedback on system performance and possibilities to self-monitor performance may increase satisfaction.

The pioneers are willing to promote the new technology and function as communicators and multipliers. And, other than predicted by Rogers, they are certainly able to do so. By no means are they outsiders. Owning a home, having a family and a good job or well-earned retirement, they have what they need in order to be well-respected in their communities which they appear to be firmly rooted in, doing volunteer work and engaging in political affairs.

Restrictions to the adoption and promotion of micro cogeneration are uncertainties about cost and performance. Helpful tools to overcome these uncertainties are reliable information, especially personal experience, contract arrangements, and support schemes.

Pioneer users can play an important role in promoting micro cogeneration – firstly, by their willingness to experiment with a novel, immature technology, and secondly, by their enthusiasm about sharing their experience and spreading the word, thus promoting the technology's diffusion. At the same time, pioneers are anything but uncritical mouthpieces. They critically follow the development process, demand to be kept informed, make recommendations, and communicate successes as well as failures to others.

Our analysis also shows that pioneers are a special group in some, but not in all, respects. Technology pioneers are not necessarily lifestyle pioneers, they need not be – and in this case, definitely are not – concerned with sufficiency issues or political ecology. They share the concerns of mainstream users with respect to cost effectiveness and comfort, but are willing to put them aside for the sake of an experiment. In sharing these concerns, they build an important bridge to the mass market. In their willingness to overcome them, pioneers are the relevant multipliers and valuable partners technology developers need in order to make their products pass the reality test.

7 Micro Cogeneration – Setting of an Emerging Market

Barbara Praetorius

Over the past few years, the German energy market has witnessed the slow but steady emergence of a novelty in energy service delivery: with the liberalization of the electricity market in Germany, a handful of technology firms and energy service companies started to focus their business in the area of small, and very small, cogeneration units, to be implemented on the level of individual households or larger single buildings like apartment houses, but also hotels, hospitals and the like.

In the German electricity system, large replacement investments are currently moving onto the agenda. Until 2020, up to 40,000 MW of capacity need to be substituted (see also Sect. 2.1). This window of opportunity is a chance for promising novelties – afterwards, it may close again for decades. Economically attractive technologies could be expected to make their way: even more so in a liberalized energy market, as textbooks would predict. In earlier chapters of this book, we have demonstrated that, up to a certain level of energy supply and in certain applications, micro cogeneration has indeed the potential to contribute to a sustainable transformation of the electricity system and, at the same time, would be an increasingly economically interesting investment.

In the German reality, this potential has not been explored "automatically" by the market – yet. And there is presently not much optimism that this situation will change quickly enough to make use of the upcoming reinvestment cycle. This chapter explores the reasons for the unreceptive environment to date. It looks at markets and actors and asks whether the setting is favorable for introducing micro cogeneration on a larger scale from their perspective.

Real markets are a complex issue. Institutions, market structures, and economic and other interests affect the introduction of an innovation. Also, micro cogeneration has the potential to be a "disruptive" innovation: introduced on a larger scale, it would substantially change a number of technological and structural features of the long-established system and markets for electricity and heat supply. It would also affect the commercial interests of a number of players on the market to a more than marginal extent. Their individual attitudes towards micro cogeneration will depend on whether they perceive themselves as winners – or as losers – from a larger-scale introduction of micro cogeneration. Vice versa, the ultimate impact on the individual organization or firm largely depends on its strategies. For example, it makes a difference for its market share, income, and customer relations if an energy supply company invests in micro cogeneration itself, or if it ignores the opportunity and leaves it to other players, such as energy service companies. It is therefore worthwhile to assess the motivations and strategies of relevant actors towards micro cogeneration.

To answer the underlying questions – not only theoretically, but also in a real world context – a set of interviews with relevant actors was carried out and evaluated, with particular attention being paid to their innovation strategies, their attitudes towards decentralized energy technologies, their degree of cooperation with other actors of the innovation cluster "micro cogeneration", and external factors impacting on their attitudes and strategies.[1]

The structure of the chapter is as follows: after a brief overview on potential drivers for innovation and change in the electricity sector, the setting of the relevant markets for implementing micro cogeneration is sketched out. On this basis, the actual actors involved in its implementation are assessed with respect to their interests, motivations, and strategies to foster or to hold back this innovative technology.

7.1 A Changing Environment for Micro Cogeneration

The European energy market has faced substantial changes in its institutional framework during the last decade. The traditional structure had a technical architecture which was based on large central power stations and an institutional structure based on regulated monopoly. This structure is currently undergoing a fundamental transformation. In industrialized countries, the sector is being substantially restructured by the interdependent processes of liberalization, globalization of markets,

[1] The interviews took place from June to December 2004, in the form of face-to-face or telephone interviews on the basis of a standardised interview guideline (qualitative interviews). The interviewees were senior employees from the micro cogeneration industry, large electricity companies, local and regional energy suppliers, grid operators, the gas industry, contracting companies, and independent experts in Germany.

and the development of new technologies for electricity generation, network control and energy use.

Market liberalization. The EU directive on energy market liberalization led to significant changes in the structure of the European electricity market; effects in the gas market are less momentous so far. In Germany, monopolistic regulation of the electricity supply industry was replaced by competition on all levels in April 1998. Since then, customers have been allowed to choose their electricity supply company, and thus have taken on a completely new role. New power producers entered the market; and new trading forms, such as electricity exchanges, emerged. At the same time, a major trend towards mergers and the creation of large-scale international companies has been observed.

With the introduction of competitive markets, risk and uncertainty have become a decisive criterion in investment decisions. Traditional investments in large, capital-intensive power plants, with 30 or more years calculated for capital recovery, are now high-risk adventures with no guarantee that they will eventually pay back.

Environmental concerns. The second stimulus – as recognized internationally – is to lower the negative environmental impacts of energy generation and use to a sustainable level.

Technology developments. The third important stimulus for change stems from recent technological developments. New technologies on all levels – power generation, system control and electricity end-use – have emerged. To date, distributed systems of electricity and heat could be integrated into the grid by means of modern joint control and operator interfaces, thereby possibly replacing large conventional power plants. Renewable technologies are still being improved, and learning curves still show upward slopes with regard to efficiency improvements and cost reduction.

Indeed, market liberalization led to substantial changes in the structure of the German energy industry. The development of key indicators for achieving a functioning state of competition, however, is rather disillusioning: less than 5 % of household electricity customers have used the opportunity to change their supplier, together with 7 % among commercial customers (VDEW 2004). Prices are back to their original levels, and concentration – both, vertical and horizontal – has increased substantially, with the consequence that competition is stagnating.

Shortly after liberalization, as many as 40 % of customers announced their willingness to change their supplier. Newcomers bet on these potential customers and on a decrease of transmission fees. Neither took place, as the small supplier change rates and still high transmission fees (compared to other European countries) show. Since liberalization, the Federal Competition Office (*Bundeskartellamt*) has been continually suing a number of energy companies for exaggerated transmission fees and inadequate price increases. The ex post self-regulation of grid access, as implemented in 1998, was incapable of inhibiting fraud and made it difficult for new market entrants to succeed. The rules established for this so-called negotiated Third Party Access (nTPA) model were not legally binding, and there was no regulatory authority to enforce them. Often, new entrants and distributed generators had to rely on court decisions to enforce the rules (Leprich and Bauknecht 2003). From 2005 on, electricity and gas grids will eventually be regulated by law and by means of a newly established regulation unit for electricity (*Regulierungsbehörde für Telekommunikation und Post*, RegTP). It remains to be seen if the RegTP will be able to achieve more competition on the electricity and gas markets.

7.1.1 Energy Prices

After 1998, electricity prices decreased significantly, which was interpreted as a result of market liberalization. This trend, however, did not persist – prices reached their original levels again within less than 5 years (Fig. 7.1). One might argue that price increases have not been limited to electricity only. Energy prices in general have risen strongly. This is due to a number of factors: turbulence on the international oil market and its effects on other energy markets, price increases for environmental reasons, and delays in liberalizing and regulating electricity and gas supplies. International oil prices reached a record level in October 2004. Coal prices followed, as did gas prices after a delay of a few months. By European comparison, however, the level of electricity prices in Germany is high, for both private households and industrial customers. This difference can partly be attributed to higher taxation. However, a significant share is also likely to be a result of the imperfect liberalization of the German electricity market, as will be discussed below.

The viability of micro cogeneration is partly determined by the ratio of prices for electricity and gas. A high electricity price is advantageous, whereas high gas prices discourage investments into gas-fuelled micro cogeneration technologies. The current situation of increasing crude oil and thus gas prices represents a certain economic risk, which is lowered by the fact that electricity prices are also rising again. The future ratio, however, remains a subject of speculation by experts.

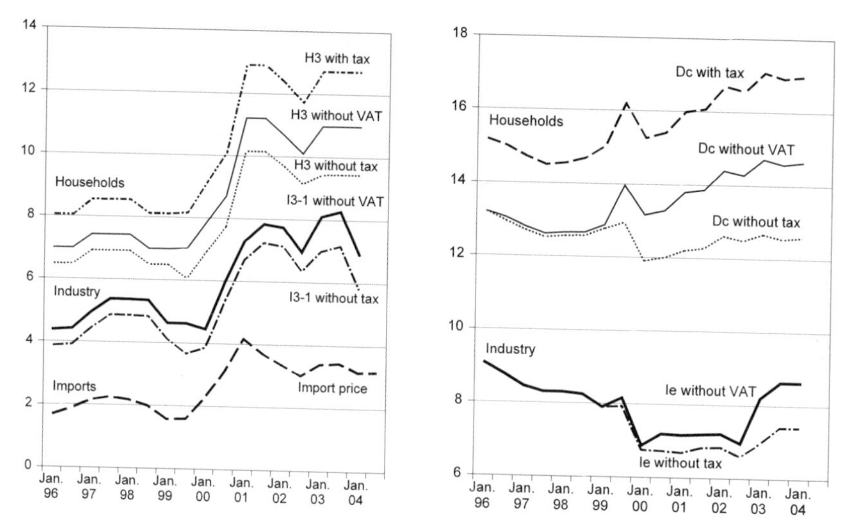

Fig. 7.1. Gas prices (left, in €/GJ) and electricity prices (right, €cents/kWh) in Germany, 1996-2004 (Sources: Gas: Import prices according to BAFA, Household (type H3) and industry (type I3-1) prices in Dortmund, according to Eurostat; Electricity: Prices for households (type Dc) and industry (type Ie), according to Eurostat)

7.1.2 The Electricity Market

As a result of major mergers, the German electricity market is now dominated by four large integrated utilities: E.ON, RWE, Vattenfall, and Energie Baden-Württemberg (EnBW). These "big four" supply 80 % of German electricity; they also own parts of numerous regional and local energy supply companies (Table 7.1). Moreover, the Federal Cartel Office has stated that vertical concentration increased after liberalization. In 2002, RWE and E.ON disposed of more than 210 major and minor (> 10 %) shareholdings in regional and local utilities (Bundeskartellamt 2003). The Herfindahl-Hirschman Index (HHI), a conventional measure for market power, increased to about 2500, as compared to a HHI of 1807 in the year 1994 (Brunekreeft and Twelemann 2004).[2]

[2] The Herfindahl-Hirschman Index is defined as HHI = $\sum_i S_i^2$, where S_i is the market share of firm i. HHI is bounded between 0 (perfect competition) and 10,000 (monopoly).

In a move towards diversification and new markets, the large companies also purchased many small and large shares in other energy companies, as well as in companies in the non-electricity sector and in other countries. RWE acquired Thyssengas in 2004; E.ON and Ruhrgas merged in 2002; and E.ON took over Powergen, now E.ON UK, to mention only a few. For the prospects of micro cogeneration, this tendency towards mergers between gas and electricity utilities is an important trend: electricity and gas markets are now more closely interwoven than they used to be.

On the local level, the initial number of local utilities – around 570 today – has decreased only slightly since 1998; they have rarely merged, but rather have cooperated (BMWA 2003). Many local utilities bundled their activities in purchasing electricity and are thus now in a position to negotiate better conditions and, hence, to be more competitive.

After April 1998, new players such as independent power producers, energy traders, Third Party Financing institutions and the like started entering the markets. However, the number of newcomers decreased again after 2000/2001. There is no statistic that captures the precise numbers. However, the Federal Association of New Energy Suppliers (*Bundesverband neue Energieanbieter*, BNE) states that, after 1998, roughly 100 new companies started offering electricity to private households on a national level. According to BNE, only six survived, together with a handful of "only local" household suppliers (BNE 2004). The number of independent local suppliers to commercial and industrial customers has been estimated to be higher (around 100), but is also decreasing (BMWA 2003).

In addition to liberalization, energy companies have also been requested to separate their generation, transmission and distribution/sales activities by 1 July 2007. This unbundling into independent business units is likely to change the conditions for disseminating innovative technologies, as the interests of the separated levels of the energy/electricity system may differ, for example with regard to distributed generation and energy efficiency. Prior to unbundling, the vertically integrated distribution network operators (DNOs) may view distributed generation as competitors to their own generation (Connor and Mitchell 2002; Leprich and Bauknecht 2003). After unbundling, this disincentive for micro cogeneration should evaporate. Yet, as long as ownership structures still reflect the previous vertical integration, it may be doubted whether independence in strategic decision making exists.

Recapitulating developments since 1998, analysts state that competition on the German electricity market has stagnated (Schmidt 2004). The underlying reasons are manifold: little market transparency, negotiated grid access with high transmission fees, a belated standardization of the supplier-change procedure, and an ex-post competition control have made it difficult for new market entrants to compete with established companies and to gain a large enough number of customers.

Table 7.1. Key figures in the German electricity market (Sources: Monopolkommission 2004; BMWA 2003; Leprich and Bauknecht 2003; annual reports of companies)

	Companies	Electricity generation (share)	Electricity distribution (share)	Shareholdings in regional / local utilities >10 % (number)	Ownership
National level / transission grid	RWE	32 %		~ 20	Mainly private stakeholders; partly German *Bundesländer* and municipalities
	E.ON	30 %		~ 210	
	Vattenfall	12 %		~ 10	
	EnBW	7 %		~ 40	
	Σ	81 %	~ 33 %		
Regional	~ 40	about 7 %	~ 36 %		Partly municipal, partly national level companies
Large local utilities	~ 25	about 10 %	~ 31 %		Partly municipal, partly national level companies
Medium and small local utilities	~ 545	mostly no generation or small CHP			Mostly municipal
Independent suppliers	~ 100	less than 2 %	less than 4 %		Private
Total number	~ 740	100%			

7.1.3 The Gas Market

The German gas market is equally subject to liberalization, albeit with less dynamic. So far, reactions to the relevant EU directive have been more of a strategic nature. Competition has not yet been fully established; a few independent companies that entered the market disappeared again. Regulation of grid access for gas transmission purposes is under way, so this situation may change in the future.

Presently, four fifths of Germany's natural gas supply is imported, mainly from Russia, Norway, and the Netherlands. Gas prices are not yet an indicator for competition in the gas industry; the price of imported natural gas follows developments in crude oil prices, with a time lag of about six months. Long-term supply contracts for natural gas couple the price of gas to the price of oil in accordance with the 'viability' principle traditional in the gas industry. It is unlikely that this will change in the near future.

The German gas supply branch consists of approximately 750 companies. The level of imports and long-distance gas transport is dominated by six large companies, namely Ruhrgas, Thyssengas, Wingas, EWE, BEB, and VNG. The two biggest firms are owned by large electricity utilities: Ruhrgas and Thyssengas belong fully to E.ON and RWE, respectively, while BEB – which also accounts for 50 % of national gas production – belongs to the oil companies Shell and Exxon-Mobil. Wingas is owned jointly by the Russian Gazprom and Wintershall, a subsidiary of the chemicals company BASF. EWE is a regional multi-service company, owned by the administrative districts it supplies (Table 7.2).

Micro cogeneration is a technology employed on the level of final consumers and, thus, a potential tool to increase gas sales. Around 700 gas companies deliver gas to these consumers. Roughly 30 companies act as suppliers from the regional level; they either belong to the big six or are owned by local gas companies. Half of the remaining local companies belong to a multi-utility company, supplying electricity, gas, and water. Only about 17 % are pure gas suppliers.

Table 7.2. Key figures in the German Gas Market

Level	Largest companies		Ownership
Long distance / gas grid	1.	E.ON Ruhrgas AG	E.ON (100 %)
	2.	RWE Gas AG / Thyssengas	RWE (100 %)
	3.	VNG Verbundnetz Gas AG	Wintershall (16 %), EWE (48 %), VNG Beteiligungsgesellschaft (26 %), misc. (10 %)
	4.	Wingas	Wintershall/BASF (65 %), Gazprom (35 %)
	5.	BEB GmbH	Shell (50 %)/Exxon (50 %)
	6.	EWE AG	WEE (82 %), EEW (18 %) [a]
	Total	~ 13	
Regional supply	~ 30		Many ownership linkages between all levels
Local supply	~ 700		
Total	~ 750		

[a] WEE (Weser-Ems-Energiebeteiligungen) and EEW (Energieverband Elbe-Weser Beteiligungsholding) are both local, county-level associations from the EWE supply area (Source: Annual reports of the companies)

7.1.4 Tackling High Transaction Costs and Small Margins

We have shown that, despite the existence of a large number of companies, the energy market remains determined by the formerly established industry, and by mergers between the electricity and gas industry. Retail competition is not as vibrant as originally envisaged, based on the UK example, and prices – in particular, retail prices – are rising again.

Nevertheless, there are also some reasons for the established industry to foster distributed generation. Despite the current stagnation of competition, the market is still on the move. The new German energy regulation office, RegTP, will introduce new rules for grid access; and the upcoming reinvestment cycle should prove to be a more advantageous occasion for new market entrants than shortly after liberalization, when they had to compete with the overcapacities of the existing industry. The uncertainties and risks implied by competition thus remain an important criterion for investment decisions. Low capital costs and the short lead-time of investments in distributed generation technologies are other characteristics that are attractive to the electricity industry.

In parallel, the retail market – albeit less competitive than expected – nevertheless changed its character. Consumers are now being supplied with a *quality* product – not merely "power". Utilities offer complimentary services like information centers, call centers, online customer support, and diverse integrated services, including Third Party Financing solutions.

They provide "environmentally friendly" cogeneration or "green" electricity. They promote themselves as innovative, environmentally responsible utilities. Offering micro cogeneration "technology and service" packages would fit in well with this new identity.

Compared to the separate delivery of heat and electricity, however, micro cogeneration is not yet off the ground. It is only rarely employed in private households. On the level of professional or commercial users, such as apartment buildings, hotels, and small firms, there is relatively more dynamism in the market (see Chap. 3 for details). Despite the subsidy granted to electricity fed into the grid from small micro cogeneration units, and even with the diverse tax exemptions, the size of micro cogeneration implemented in practice mostly exceeds the 15 kW definition applied in this study. In the majority of cases, they are implemented by firms specializing in Third Party Financing solutions, such as local energy agencies, contracting sub-units of the gas industry or of regional energy suppliers.

This situation is essentially a consequence of high transaction costs which are hardly compensated by the small margins that can be gained with micro cogeneration to date. These transaction costs add to the conventional scale economies as reported in Chap. 4.

Transaction costs for micro cogeneration are composed of costs entailed by information and search processes, as well as negotiation costs that run until the actual supply and/or service contracts are signed, as follows:

- **Search and evaluation costs.** The first step includes an assessment of the respective heating and electricity needs and choosing the appropriate micro cogeneration technology.
- **Implementation and maintenance costs.** Micro cogeneration technologies require specific skills and, hence, supplementary training for technicians. At present, the share of appropriately skilled technicians is likely to be small, even more so since micro cogeneration technologies differ in their technical and system characteristics.
- **Negotiations with the grid owner.** Connecting a CHP plant to the grid costs a fee which is to be paid to the grid owner. This also implies negotiations about feed-in remuneration as well as disconnection options for critical load situations and grid bottlenecks. The owner of the micro cogeneration unit negotiates with the owner of the electricity grid about the details of grid connection, including remuneration of electricity fed into the grid, reserves, and compensation for avoided network grid use. Fortunately, these procedures have already been

more standardized for small CHP units; however, the complexity is still considerable.

- **Authorization from the Federal Agency for Economy and Export Control** (*Bundesanstalt für Außenhandel*, BAFA). This is needed in order to qualify for the bonus payments for electricity fed into the grid, as granted by the CHP law (administration fee: € 75 per unit).
- **Reporting demands.** To receive the ex post tax exemption from energy taxes, the owner of the micro cogeneration unit has to collect detailed information about the unit, including yearly and monthly input and output data. If not generated automatically, this is a time-consuming task.
- **Contracts with electricity customers.** In the case of apartment buildings, for example, supplemental contractual issues become apparent, as the electricity and heat generated by the micro cogeneration unit is supposed to be sold to the tenants. Long-established tenants, however, cannot be obliged to change their supplier in order to buy the "home-made" power. Besides the related contractual and legal issues involved, this also poses significant risks to the viability of the micro cogeneration unit.
- **Legal aspects.** In the case of Third Party Financing companies, property rights of, and access to, the micro cogeneration plant have to be listed as easement in land registers. For small micro cogeneration plants this might be a prohibitive barrier, as transaction costs are considerable and as many property owners dislike the idea of easements associated with their properties. Hence, it appears more likely that small micro cogeneration plants are owned by the property owners, though they may be installed and/or operated by energy service companies or utilities. Also, the negotiation of a heat and electricity supply contract with the property owner may cause significant transaction costs for Third Party Financing companies.

Altogether, compared to a conventional heating and electricity supply agreement, micro cogeneration involves a complex and costly procedure with very little reward. Also, the feed-in bonus and the other compensations do not cover the electricity generation costs, with the effect that micro cogeneration operator's can install plants only in buildings with sufficient electricity demand. As a result, many heating potentials for micro cogeneration without sufficient electricity demand cannot be used. This applies, for example, also to apartment buildings where Third Party Financing companies or property owners cannot bear the risk that tenants in the building prefer to be supplied by other energy supply companies. High transaction costs and the related unfavorable institutional setting are therefore an area of major concern to proponents of micro cogeneration. "All inclusive" service packages, as offered by Third Party Financing companies, may lessen the underlying problem. However, even then, some decisive incentive problems remain active, as shall be confirmed below.

7.2 The Actor's Perspective: Strategy, Motivation and Institutions

Despite the rather uninviting setting of the market, as described above, micro cogeneration is currently receiving growing attention. Key players on the markets related to micro cogeneration are carefully observing its development and analyzing its deployment. The mid- to long-term potential for introducing micro cogeneration depends on the strategic interests and attitudes of these and other actors towards decentralized generation. Economic viability is one important factor that determines their interests. However, to explain real world decisions about investment in technology development and diffusion, this is not sufficient. Motivations – and also disincentives – stem from several sources.

A first, core, aspect is the *institutional context*, as described above. Institutions matter; this proposition, raised by New Institutional Economics theorists, is still valid, yet often ignored, when trying to explain the function and outcomes of market interactions (Williamson 1985). Indeed, innovative behavior cannot be explained in a satisfying manner without taking into consideration the institutional framework which is shaping it. Institutions motivate or discourage action. Vice versa, the actors involved with the respective innovation may also motivate changes of the institutional framework.

A second aspect is *cooperation*. The implementation of micro cogeneration entails cooperation between the actors. Some of them are crucial for its implementation, others may be marginal or replaceable. In any case, the functioning, style, and forms of cooperation may motivate – or discourage – further engagement with micro cogeneration.

A third aspect is *intrinsic motivation and strategic attitudes*. Decision making on future investment and innovation is not purely "rational", but also a function of expectations and risk assessment of individuals within their respective company and industry environment.

These aspects are all interlinked; they form the focus of the following assessment. To derive a prospective outlook for diffusing micro cogeneration, a closer look at the settings, motivations, and resulting strategies regarding micro cogeneration is taken. For this purpose, key actors were asked about micro cogeneration activities which they plan or have realized, the future share of micro cogeneration they expect to see, an appreciation and critique of the existing institutional framework, and improvements that would positively stimulate their attitude towards micro cogeneration. The key actors considered here include

- technology developers,
- the gas supply industry,
- large electricity companies,
- local energy companies,
- customers in both private and commercial areas,
- energy contracting and service companies (ESCOs), and
- distribution network operators (DNOs).

7.2.1 Technology Developers

Measured by the number of units sold since liberalization of the energy markets, micro cogeneration is actually booming, at least in its niche. The handful of German technology developers and engine building firms engaging in this market are experiencing an increasing interest in their products (see Sect. 3.1). Senertec has promoted its *Dachs* since 1996 and sold its 10,000th unit in 2004. Power Plus started marketing the *Ecopower* reciprocating machine in 1999 and has implemented some 1000 motors since. The *Stirling* engine built by Solo has been tested in about 30 locations in Germany so far. Foreign Stirling technology developers, like WhisperTech, Microgen and other firms, are also inquiring the German market in search for new sales potentials. A steady increase in sales volumes from the mid-1990s until now can be observed, with about 50 per cent more plants being installed in 2004 compared to 2002 (see Chap. 3).

Last but not least, the diverse fuel cell developers are still in the process of testing and improving their respective fuel cell designs. A number of units have been installed as part of field trials (see Sect. 1.2.3).

The strategies of these developers are substantially different. Senertec is the most successful and well-known vendor of micro cogeneration units so far – so successful that its production capacities seem undersized. In 2004, the delivery time lag after ordering a Dachs amounted to about 6 months. Senertec aims at the production and delivery of 3000 units per year. For this purpose, a sophisticated distribution system was set up, with 30 so-called regional centers all over Germany, and around 280 sales partners with cooperation agreements, as well as framework contracts concluded with energy and Third Party Financing companies. The Dachs is marketed to household customers through lifestyle brochures, a Dachs fan club, and so-called Dachs parties, i.e. information evenings in the house of a Dachs operator for "fans" and potential buyers.

Power Plus ventured later into the market of reciprocating engines and has not yet been able to copy the Dachs success with its Ecopower. In the future, it hopes to benefit from the established distribution system of Vaillant, one of the large boiler companies in Germany, which purchased Power Plus in early 2004. In parallel, Vaillant also develops its own small-scale fuel cell.

Solo operates in a different (non-household) market segment, and is still in a testing phase of its technology during which local utilities and other prospective customers are being carefully approached. It also remains to be seen whether the foreign Stirling firms will enter the German market. The first WhisperGen machines are currently being tested by local energy companies in Germany. Their developers are keenly observing the German market, in search of the best starting points for their marketing initiatives.

In terms of their ownership, it is interesting to note that originally independent technology developers like Senertec and Power Plus, along with a number of fuel cell developers equally, have been purchased by other boiler or CHP technology firms. In the case of Senertec, the British Baxi Group – also a fuel cell developer – is interested in the elaborated German distribution network of Senertec. Just like Senertec itself, the Baxi Group may argue that the more micro cogeneration is being disseminated – regardless the respective technology or firm – the greater its recognition in society, thus increasing the chances to sell further units of its own original technology.

The cooperation between traditional boiler companies and micro cogeneration manufacturers appears promising, as micro cogeneration plants could be marketed as a "better" boiler which also produces electricity. Such a marketing strategy may simplify the deployment of micro cogeneration for various reasons. Firstly, consumers are mostly unaware of CHP and often do not properly understand it. However, they could be informed on micro cogeneration plants easily when they need to replace a boiler, and boiler manufacturers also offer micro cogeneration plants. In addition, micro cogeneration is economically particularly promising, and much easier to install, when the micro cogeneration plant fully replaces the boiler.

7.2.2 Gas Supply Industry

Micro cogeneration is primarily fuelled by natural gas. A likely ally and driver for introducing micro cogeneration to the market could hence be expected among gas suppliers. Before liberalization, the gas industry had solely focused on the heating market, with modern gas-fuelled technologies like the high-efficiency condensing boiler allowing them to gain an increasing share. The image of gas as a clean heating energy, combined with the advantages of grid-connected energy delivery (no need to order and bunker energy), offer good marketing arguments for acquiring new customers. Today, 46.6 % of German private households use natural gas for heating; in new residential buildings, the market share is as high as 75 % (BGW 2004; see also Sect. 3.2). In order to further increase their gas kilowatts sold, gas companies have also invested in the development of cogeneration since the 1980s. At the same time, this became a means to enter the market for electricity generation.

Indeed, many gas companies are closely observing innovation activities with respect to small cogeneration technologies. They check promising technologies – like the small WhisperGen Stirling engine or fuel cells – on location and negotiate framework contracts with engine developers, with the objective of obtaining better conditions in terms of unit prices and maintenance service packages. Both sides benefit from this means of cooperation. Some gas utilities make use of their own contracting subsidiaries to implement small- and medium-size reciprocating engines for industrial customers and larger domestic buildings. In other cases, external contractors support implementation.

However, not every gas company encourages small cogeneration. In fact, the ownership structure seems decisive. With its acquisition of the natural gas importer and long-distance transport company Ruhrgas, for example, the electricity company E.ON is pursuing a strategy to gain access to the gas grid and to the local heating market. The electricity market, however, is not part of this focus, as it is supplied by other E.ON subsidiaries. Other gas-supply units are likely to face similar internal strategic decisions. On the municipal level, the picture is more mixed, again depending on the respective business concept (see below on local energy companies). Integrated companies will compare the potential advantages of increased gas sales with the losses through decreasing electricity sales.

All in all, the gas supply industry is indeed a “natural ally” and invests in implementing micro cogeneration – as long as it is not constrained by its stockholders or, in case of multi-utility companies, by the other business areas like electricity generation. Besides, similar to the local electricity industry, the gas industry is currently tied up with other issues: liberalization and regulation of the gas transmission grid are the real issues at stake; micro cogeneration plays a rather subordinate role.

7.2.3 Large Electricity Companies

Traditionally, the focus of the established large-scale electricity industry has been on larger generation units and a centralized supply system. A certain path dependency in strategic decision management in favor of large power stations may thus be presumed. In an increasingly competitive environment, however, electricity companies would need to carry out substantial changes in their operations and make use of technological innovations. The response of the German electricity industry to liberalization was hence to reduce operating costs, including manpower, capital expenditure, and routine maintenance. At the same time, electricity prices dropped to almost historical lows – albeit only for a limited time.

In terms of the upcoming capacity replacement investments, micro cogeneration could offer a number of advantages to large electricity companies, yet it would need better overall economic performance to be attractive to energy utilities and customers equally. In the mid to long term, learning effects and economies of scale could be expected to improve performance and release advantages. From an investment and operational perspective, micro cogeneration involves lower incremental and strategic risks because the initial investment is small. Regarding the customer side, price reduction, an on-site full-service package, added value services, branding, and diverse tools for consumer retention could be competitive benefits. In this regard, micro cogeneration could be a strategic tool and also offer attractive business opportunities to an energy company when it shifts its profile towards being an energy service company.

This argumentative route is currently being followed by E.ON UK – the former PowerGen and UK's second largest energy retailer – which ordered 80,000 Stirling machines with the ultimate objective of equipping (and thus binding) up to 30 % of the UK households with micro cogeneration in the year 2020 (WhisperTech 2004). However, the features of the UK retail market differ substantially from the German market, in particular with respect to heating needs – German houses are much better insulated – and to the quality of maintenance services – in Germany, the service network functions well, so a full-service package from an energy company is not as attractive to private households as in the UK. This is a major reason for E.ON Germany not to follow the UK example. Also, E.ON UK considers the Stirling initiative as a means to customer retention. This is reasonable for the UK, where about 30 % of household customers have changed their supplier since liberalization. In Germany, with its low change rates, this rationale has little relevance yet.

Also, the advantages of short lead-time and low investment risk entailed by micro cogeneration are currently more than offset by the cost of convincing customers and of managing a system of multiple, decentralized generation points with fluctuating feed-in to the electricity grid – a system that would approach the visions of a “virtual power plant”.

At the same time, the development of distributed generation technologies is being closely watched by German electricity companies. All major and many medium-sized energy companies are involved in developing or at least observing and testing possible micro cogeneration technologies.

Fig. 7.2. “Imagine”: RWE image campaign for fuel cell micro cogeneration

These activities overwhelmingly focus on fuel cells as the relevant micro cogeneration technology. RWE has its own subsidiary, “RWE fuel cells”, E.ON tests small units on location in cooperation with its regional and local subsidiaries, and EnBW tests fuel cells with pioneer operators. Image campaigns support the vision of self-sustaining energy cycles and decentralized power supply (Fig. 7.2). However, they all advertise decentralized generation as a scenario which will only become relevant in the remote future, while ignoring those micro cogeneration systems which are nearly or already commercially available.

From the perspective of large utilities, this strategy seems consistent. First and foremost, they are not likely to be interested in implementing decentralized structures; they rather focus on their traditional path of central generation plants. In parallel, observing the market allows them to participate in the development of a potential niche and to receive signals of an emerging trend towards decentralized generation as early as possible. For the time being, the utilities avoid the transaction costs which would occur in the case of a too-early commitment to decentralized generation structures. Equally, this strategy protects them from business risks such as losses of revenue to new (decentralized) market entrants and stranded assets in the form of under-utilization of their existing assets. In any case, distributed generation is not in the focus of the major companies with regard to the upcoming reinvestment cycle.

7.2.4 Local Energy Companies

Other potential allies for micro cogeneration could be expected to be among the large number of local energy companies in Germany. Competition with the large, cross-regional energy companies is an explicit challenge for them. Customer retention activities are hence an effective objective on the local level of the German electricity supply system, in particular with respect to medium-sized customers like hotels, public baths, or small businesses.

Local energy companies with both a power and a natural gas grid, but little in the way of self-generation capacities, should have good reasons to promote micro cogeneration in order to increase their sales in natural gas, as gas offers a higher margin. However, the margin is still comparatively small, so that the economic incentives are low for engaging in such a "new" technology. Some invest for the reasons mentioned above, in particular for customer retention reasons. Altogether, however, only a small share of local energy companies is presently active in the area of small-scale cogeneration.

In a survey of investment activities in the field of cogeneration, only 20 % of the local utilities answered that they intend to invest in cogeneration smaller than 50 kW_{el}, while some 11 % also consider fuel cells as a future option (VKU 2003). An investigation by the consultant Ernst & Young among German municipalities, done in cooperation with the electricity industry association VDEW, suggests that 65 % consider fuel cells to be a particularly important generation technology of the future, along with cogeneration (58%) (Ernst & Young 2003). Conversely, the future relevance of distributed generation in general is estimated to be comparatively small (Table 7.3). This probably reflects a down-to-earth assessment of the market potential of fuel cells and other small-scale electricity generation technologies. Local energy companies regard the operation of fuel cells on a local level as a complement to their bulk energy suppliers, so as to be slightly more flexible. They do not yet see it as a building block for their own future distributed generation.

Table 7.3. Future relevance of distributed generation for local energy companies (Source: Ernst & Young 2003)

in %		Size of community (residents)		
	Total sample	20,000-50,000	50,000-100,000	>100,000
N=	*43*	*21*	*14*	*8*
very high	4.7	9.5	0	0
rather high	27.9	23.8	35.7	25.0
rather small	44.2	47.6	35.7	50.0
very small	0	0	0	0
don't know	7.0	9.5	7.1	0
n.s.	16.3	9.5	21.4	25.0
total	100.0	100.0	100.0	100.0

Moreover, for local energy companies, their attitude towards micro cogeneration strongly depends on the prevailing local situation. For instance, in some cases, staff was underemployed but could not be dismissed. Here, the maintenance of micro cogeneration systems offered additional tasks, thereby increasing the work load of staff members without increasing the cost for the municipality. In other cases, local energy companies, which are typically also DNOs (see below), could defer grid extension. Similarly, micro cogeneration and other forms of distributed generation may offer an alternative to grid modernization investments in the case of cities with shrinking populations.

Last but not least, local energy companies are currently focussing on issues other than micro cogeneration – upcoming demands related to the new regulation entity RegTP and the unbundling directive keep them more than busy for the time being.

7.2.5 Customers

To justify their disinterest in promoting distributed generation systems, energy companies argue that small and private household customers, albeit fascinated by the fancy or "high tech" fuel cell, are not yet interested in small cogeneration.

Indeed, compared to the vivid retail market in the UK, the German electricity and heating retail markets follow different routines. Supplier change rates are much smaller; and full-service packages, as offered by E.ON UK, are not as appealing to German households. Moreover, they are usually not particularly interested in installing a new technology as long as the old technology is still running well, especially when economic advantages of the replacement are not significant. Only a comparatively small number of consumers who are environmentally conscious and critical about the market power of large energy companies may value micro cogeneration as a contribution to a decentralized and environmentally sound energy supply. In general, however, the size and the respective economic attributes of micro cogeneration technologies are not yet suitable to individual private households.

Today, large apartment buildings, hotels, public baths, and other small businesses appears to form a more promising market for micro cogeneration. Here, energy and heating demand reach levels at which the potential advantages energy service packages are attractive for both sides – the energy supplier and the customer – Third Party Financing subsidiaries and independent contractors and energy agencies also find here an attractive – albeit niche – market for implementing micro cogeneration. Contractual and legal issues, however, are even more a barrier on this level, in particular in apartment buildings, where electricity and heat would need to be sold to the tenants for the micro cogeneration unit to be viable. Here, the complexity of related procedures will often remain a barrier to implementation.

7.2.6 Energy Contracting and Service Companies

In terms of diffusion of micro cogeneration, energy Third Party Financing or energy contracting and service companies (ESCOs) play a crucial role – their business models may even represent the most advantageous way to introduce it to the market. While most of the actors assessed above gain little economic advantages from introducing micro cogeneration, a number of additional commercial opportunities opens up for ESCOs as the operating companies of micro cogeneration units. ESCOs – in the denotation used here – are companies that offer energy services to final customers by means of implementing a technology on site and acting as a link to the original energy supplier (mostly gas or electricity). They are able to overcome typical problems, such as information and skill shortages, delivery or operation and maintenance risks, and the like. They benefit from bundling knowledge with contacts to relevant administrative and financing institutions, they apply standard contracts and are able to negotiate quantity rebates with the technology and the energy industry. All in all, they are thus able to realize higher margins than individuals can in implementing micro cogeneration. It is therefore likely that ESCOs will not only be initial market entrants, but also the principle long term players. This may also be an underlying reason for medium- and large-sized energy companies to have subsidiary ESCOs.

On the level of technical realization of micro cogeneration, ESCOs cooperate with technical firms. Here, the traditional education system poses a problem. The existing system of education of technicians is usually focused either on heating systems or on electricity supply. For the implementation of micro cogeneration, they need to handle both sides. This problem is about to be overcome in education systems, yet it will take some more time for the integrated service installer to be a standard job description.

7.2.7 Distribution Network Operators

Distribution Network Operators are the lynchpin for a broader diffusion of micro cogeneration, as they occupy the points connecting individual units to the electricity network. In Germany, the high-voltage transmission grid is owned by the "big four" electricity companies. The unbundling directive requires network operation to become legally independent. In fact, the large companies have already unbundled their businesses.

However, as long as the ownership structures remain unchanged, the incentives for DNOs to act purely independently are small. In fact, DNOs, and vertically integrated electricity supply companies in general, have often used technical connection requirements to discourage distributed generation (IEA 2002). Jörß (2003) additionally reports procedural prolongations, unjustified financial requests, and other negative experiences of distributed generation operators with their respective DNOs. These "policies", however, vary considerably from company to company and over time (Jörß et al. 2003).

With regard to network operation as such, DNOs are concerned about an increasing share of distributed generation, as this may imply a substantial change in the operation of networks. On the other hand, they may also experience a reduction in investment needs: micro cogeneration does not make use of the high-voltage net, and may reduce the transformer capacity needed, so that reinforcement can be avoided or delayed (see Chap. 9).

Technologically, a single micro cogeneration unit is too small to cause any significant problems for, or changes to, the management of the electricity grid. This may change when micro cogeneration is implemented in large numbers and area-wide. The recent advances in power conversion and information technologies allow for the coordination of a large number of small feed-in points to the electricity grid. Appropriate changes in the technological systems applied by network operators are currently under way. A widely decentralized electricity supply may physically disburden the electricity grid; the diversification of generation feed-in points would theoretically help to improve the quality of supply and to reduce the risk of brown- or blackouts. Network operators may thus see the potential for avoiding investments in grid expansion and upgrading in the mid to long term. Supplemental costs, however, may arise from the need to maintain voltage levels and other technical problems related to large numbers of micro cogeneration in an area.

Economically, the DNOs are affected by CHP proliferation because decentralized power generation decreases the volume of electricity that is distributed to consumers, thereby reducing the income flow and, thus, increasing the fixed cost per kWh distributed. As the regulation of network tariffs will in principle be based on cost recovery, these increased costs per kWh delivered could be compensated for by higher tariffs. However, the ongoing discussion in Germany is about how to *decrease* network tariffs, and DNOs may fear that the regulation office will follow that route and not accept further increases in tariffs. DNOs may also worry that losses of revenue due to on-site power generation are larger than the expected cost reductions for the electricity grid.

These fears appear reasonable, particularly in the short term. For small consumers, grid tariffs are usually charged per kilowatt hour, while the costs for distributing power are, in the short term, primarily fixed costs. Consequently, in the short term micro cogeneration plants reduce DNO revenues without decreasing costs significantly. Even on a longer term basis, it is uncertain whether cost savings due to deferral of upgrades or capacity reductions will compensate losses of revenue. In the short to medium term, DNOs are burdened with maintaining the balancing power and reserve capacities needed for reliable network operation. The related costs, again, remain with the grid owner, and are in the end paid by the remaining "conventional" electricity customers.

Moreover, the DNOs face supplementary administrative costs related to the German feed-in and compensation system for electricity from CHP (and from renewable energies). They are responsible for paying the CHP-law-bonus to the owner of the micro cogeneration unit and then applying for compensation from the respective balance area.

Altogether, to date, the incentives for DNOs to support micro cogeneration are comparatively small. The issue of economic incentives and disincentives for micro cogeneration should therefore be taken into account in the regulation of DNOs (Leprich 2004; Leprich and Thiele

2004). For example, an appropriate redesign of grid charges that rewards the connecting of decentralized generation units may allow for the elimination of existing disincentives.

7.3 Incentives and Disincentives for Micro Cogeneration

In a world of perfect competition, the dissemination of micro cogeneration would be a function of its cost effectiveness and time. Over time, due to learning effects and mass production, the investment costs decrease and micro cogeneration could become more and more competitive. In the long term, decentralized co-generation has a significant potential, as the expected economies of scale of modular, small-scale, and increasingly standardized systems will make it attractive to an increasing number of potential operators.

To a certain extent, however, the expected economies of scale are likely to be offset by higher transaction costs per unit of capacity installed (Madlener and Schmid 2003). This is even more true for the short term perspective. Altogether, the incentives for implementing and disseminating micro cogeneration are mostly cancelled out by the prevailing disincentives. Table 7.4 summarizes the stimulating and discouraging factors for the actors assessed above.

Most of the disincentives originate from the institutional setting. In the literature on appropriate institutions for systems of innovation, three fundamental functions of institutions are usually distinguished: they ought to reduce uncertainty, manage conflicts and cooperation, and provide incentives (Preissl and Solimene 2003, p. 27). In other words, good institutions reduce the existing transaction costs and risks of the real world context. They should help to *overcome informational barriers* to the implementation of novelties, *manage contractual issues*, and *stimulate the respective markets,* thus facilitating market introduction by setting the right incentives.

As our case study for micro cogeneration shows, the institutional setting in Germany does not yet perform these functions. A number of institutional and regulatory aspects would have to change for micro cogeneration to gain a significant market share.

Table 7.4. Micro cogeneration: incentives and disincentives for actors

Actors	Incentives	Disincentives
Technology developers	Expected market growth Perception of a trend towards decentralization	High transaction costs Dismissive attitude of DNOs
Gas industry	Increase in gas sales Customer retention	Often ownership by electricity companies Existing cogeneration/local heating grids
Large electricity companies	Low investment risks Short lead-time Customer retention	Risk of stranded assets Existing cogeneration/local heating grids
Local energy companies	Customer retention Low investment risk Short lead-time Increased sales volume/self generation	High transaction costs Losses in electricity sales vs. gains in gas sales
ESCOs	Business opportunities	High transaction costs/small margin
DNOs	Disburdening of local network Reduced demand peaks	Ownership structures (vertically) Loss of revenue (reduced transmission)
Customers	Electricity generation at home Environmental benefits Large customers: economic benefits	High transaction costs Small or no economic advantages

Firstly, uncertainty in terms of implementation procedures and economic risks involved with investing in micro cogeneration are still high. The German energy system supports micro cogeneration plants with bonuses paid according to the cogeneration law and the obligation on DNOs to accept electricity fed into its grid. However, the large number of contentious issues – many with respect to technical details of grid access and administrative details – create a significant degree of uncertainty for small cogeneration plants. The related transaction costs, also partly unknown to the operator, are an effective barrier to implementing micro cogeneration, even more so as they further reduce the comparatively small margin to be gained as compared to the separate generation and purchase of electricity and heat.

In order to reduce these transaction costs, the administrative setting would need to be standardized. Network connection standards and procedures will equally allow the minimization of additional administrative burdens, as would a standardization of systems and implementation procedures, including full-service packages. Moreover, to reduce information and search cost, micro cogeneration would need authoritative and independent information about products and systems to be widely disseminated.

Secondly, conflicts are not yet sufficiently managed. Negotiations are bilateral and, thus, little transparent, and court decisions are an expensive way to regulate conflicts. This specifically concerns the relationship of potential micro cogeneration operators and DNOs. On the one hand, this is the result of lack of a body for regulating grid access rules. Even with the new regulatory body RegTP, it is not certain whether DNOs will have a sufficient incentive to foster distributed generation. Yet, a carefully designed incentive-based regulation may stimulate investments in micro cogeneration.

On the other hand, even with a well-functioning regulator, the prevailing ownership structures to which all DNOs belong remain unfavorable for micro cogeneration. The large established energy companies are likely to make use of their market power, while small independent generators do not have the financial backing to survive long legal disputes. Here, *complete* unbundling and the establishment of an independent system operator for the transmission grid may be the best solution. This should be combined with a regulation of grid use tariffs and procedures that gives sufficient incentives to the DNOs to connect decentralized generation (Leprich and Bauknecht 2003; Leprich 2004).

Thirdly, the existing economic incentives are too small for the actors to actually invest in micro cogeneration. Here, economic viability and creating a level playing field for the different sustainable technologies are the issue. Economic viability has not yet been achieved, but micro cogeneration is considered an environmentally friendly technology when applied in the right circumstance. For this reason, public support schemes are reasonable, analogous to those for renewable energy technologies and electricity fed into the grid. Of course, environmentally just feed-in tariffs need to take into account the environmental attributes of micro cogeneration, which are inferior to renewable energies as long as micro cogeneration is operated on fossil fuels. Here, a careful re-evaluation of the advantages of micro cogeneration as compared to central electricity generation seems appropriate (see Sect. 9.1.2). Moreover, the two other components of remuneration for electricity from micro cogeneration fed into the grid – i.e. avoided generation cost and avoided network use – may need revision. All in all, the incentive structure appears to be counterintuitive with respect to the comparative efficiency and environmental benefits attributed to cogeneration, with the consequence that feed-in of high-value electricity from micro cogeneration is currently avoided where possible.

To conclude, the institutional setting is currently the predominant factor influencing actor motivation to invest in decentralized generation. In particular, the potential operators – individuals, contractors, or energy companies – and DNOs do not yet have sufficient incentives to substantially foster micro cogeneration. For a larger-scale introduction it would need strong partners in the electricity generation industry itself and especially among the DNOs. Unfortunately, financial and structural disincentives for these actors are currently dominating the picture, and distributed generation is not part of their strategies. Micro cogeneration is therefore likely to remain for some time in its market niche. In other words, the potentially emerging market for micro cogeneration is still in its very early stages.

8 Institutional Framework and Innovation Policy for Micro Cogeneration in Germany

Martin Cames, Katja Schumacher, Jan-Peter Voß, Katherina Grashof

Aside from the question of technological potential, private and social costs, corporate strategies and consumer acceptance, the innovation path of micro cogeneration is importantly influenced by institutional structures in the field of implementation. Institutions guide the strategies of different actors, producers, consumers and regulators in choosing or not choosing micro cogeneration systems. Institutions are defined here as the social rules that enable or constrain action by setting incentives, providing orientation, or prescribing or forbidding specific behavior (Esser 2000; Voß et al. 2001). Explicit rules of law and public regulation fall under this definition, but also more informal regimes such as policy networks, innovation networks or cartel agreements as well. Important institutions for innovation also include the effective organization of financing, installation, and maintenance services. Indirectly, institutions have already been dealt with under the headings of economic profitability, corporate investment strategies, and consumer acceptance, since these factors are influenced by the setting of rules that guide business operations and consumer activity.

In addition, the institutional environment is closely related to innovation policy, which shapes institutional context conditions so that they are favorable for innovation. Recently, innovation policy has taken on an active role in promoting sustainable production and consumption patterns by shaping institutional conditions so as to bring about innovations which not only spur economic competitiveness and growth, but also feature specific qualities that contain potential for, e.g., increased eco-efficiency or social participation (Kemp 1996; Meyer-Krahmer et al. 1998; Vergragt et al. 2000). Also, the anticipation of innovation effects through prospective technology assessments plays a role in this context (Rip et al. 1995; Simonis 2001).

Section 8.1 focuses on the regulatory framework and institutional arrangements that guide interactions between businesses, consumers, and public policy. It gives an overview of institutional settings, which are currently in place in Germany and have influence on micro cogeneration development. Thereafter, we summarize how micro cogeneration is currently being addressed by innovation policy (Sect. 8.2).

8.1 Institutional Structures

The institutional environment with importance for micro cogeneration in Germany can be broken down into different categories. The most general of these is the overall organization of the *market for electricity,* which influences prices and strategic constellations between actors (e.g., utilities and independent micro cogeneration investors). Grid connection is the core technical interface for micro cogeneration to enter the electricity market. *Institutions for electricity network access* are therefore critical. Economic revenues are strongly influenced by *taxes on fuels and electricity*. General *environmental regulations* play an important role for energy. In Germany, specific regulations are in place for the support of CHP, as it is a strategic element of the national *climate protection strategy*.[1] Finally, the general institutional embedding of practices related to micro cogeneration *investment, installation and operation* greatly influences the readiness of involved actors to adopt the innovation. These aspects will be considered in the following sections.

8.1.1 Liberalization of Electricity Markets and Grid Access

In April 1998, an amendment to the existing energy industry law (*Energiewirtschaftsgesetz*, EnWG) was introduced (Deutscher Bundestag 1998). This amendment ended a several decades long period of regional monopolies in the German electricity industry and introduced competition between electric companies. It established negotiated – not regulated – Third Party Access to the electricity grid and obliged utilities to unbundle their network activities from their generation and sales operations through separate accounts, not, however, to implement either legal unbundling or the unbundling of ownership (Leprich and Bauknecht 2003b: 7).

This amendment substantially changed the conditions for distributed generation, such as for CHP installations. Electricity prices for industrial consumers dropped due to competition by up to 35% between 1998 and 2000, and were in October 2004 still 5% below the 1998 level (VDEW 2004, see also Sect. 7.1.1). Average electricity prices for private

[1] For details on subsidies and support schemes, see Sect. 8.2.3.

households declined in parallel, although to a lesser extent (minus 10 to 15%). As avoided electricity purchases are the most important "revenues" from CHP installations, the decline in electricity prices, as induced by liberalization, first deteriorated the competitiveness of CHP installations significantly. However, prices soon started to rise again, so that the conditions for micro cogeneration improved again.

Apart from the effect on prices, liberalization changed the relations between independent power producers and distribution network operators, and created conditions which Leprich and Bauknecht (2003a, p.3) describe as "double light-handed regulation": a pattern that combines the self-regulation of grid access by industry with a weak approach to unbundling between distribution and supply, on the one hand, and supply and generation on the other. Since the 2003 EU directive to increase competition in the electricity sector came into force in mid-2004, network operators – except for distribution grid operators with less than 100,000 customers – now have time until mid-2007 to legally separate their network activities from generation and sales. In Germany, the formal implementation of these requirements has not yet taken place, but most large network operators have already implemented legal unbundling.

However, because there is no unbundling of ownership (in most of the cases a common holding company exists for both the network and the electricity generation and supply businesses), network operators have a strong incentive to inhibit new suppliers from alienating their customers. Since the four large supra-regional utilities which operate the transmission network account for about 80% of generation, one third of the distribution business, and 50% of retail supply to end-users (without taking into account their minority stakes in regional and local utilities, see Sect. 7.1.2), they continue to have a strong incentive to prevent distributed generation from independent power producers, because of their interest in selling their own electricity (Leprich and Bauknecht 2003a, p.3; Monopolkommission 2004, p.1141). The lack of strong regulation, together with unfavorable incentives from weak unbundling, has led to a number of contentious issues, protracted legal processes and, thus, to competition barriers and legal uncertainty for independent power producers.

The technical connection details are regulated by a number of standards, particularly by VDEW (1996) and the distribution codes of the grid operators (VDN 2003). With regard to grid use, relevant parameters for independent power producers, such as operators of CHP plants, generally include connection costs, grid charges, and compensation for avoided network use due to distributed electricity feed-in.

Grid charges and connections costs. As grid charges in Germany are paid for by those taking power off the grid, they do not directly affect micro cogeneration operators.

Connection costs can be distinguished between deep and shallow costs. Under a regime of shallow connection costs, electricity generators are required to pay only for the local assets specifically required to connect them to the grid. The costs of reinforcing the system beyond the connection assets are recovered through grid charges. Deep connection charges would make many CHP projects uneconomical by making generators pay for all grid-related costs to provide connections between themselves and users. In Germany, the concept of shallow charges is applied, which is basically advantageous for CHP operators. In practice, however, the differentiation between costs of grid connection and network extension is often disputed between distribution network operators and independent power producers.

Some distribution network operators have argued that grid extension only covers strengthening the existing grid to carry the additional load, whereas the cost of grid connection often includes, where necessary, building a new line to the existing grid, putting a high cost burden on some independent power producers (Leprich and Bauknecht 2003b, p.18). Several court rulings require distribution network operators to provide the necessary grid extension up to the last point of a low- or medium-voltage network. But many distribution network operators regard these rulings as case specific and therefore not relevant for their networks. This has led to an unfavorable degree of legal uncertainty concerning connection costs for independent power producers with larger generation capacities. In the case of micro cogeneration, however, this problem should be of minor importance, as potential users are already connected to the grid and their additional capacity will be so small that grid extension will not be necessary.

However, some have criticized that current legislation even requires operators of micro cogeneration installations to apply for permission before connecting their plants to the grid. It seems that network operators have used this opportunity and have considered some applications only with deliberate delay. The German Association for Combined Heat and Power (*Bundesverband Kraft-Wärme-Kopplung*, BKWK) therefore considers it sufficient to inform network operators of the planned connection[2] of plants with less than 150 kW, as this would provide an

[2] Insofar as the house connection capacity is high enough, it is argued that this procedure should not entail technical disadvantages, with the plant itself needing to be installed by a qualified technician.

incentive for grid operators to react quickly and increase planning reliability for micro cogeneration users (BKWK 2004, p.2).

Compensation for avoided network use. Compensation for avoided network use is currently based on the agreement of several industry associations, the so-called *Verbändevereinbarung II Plus* or VVII+ for short (BDI et al. 2001). Compensation for avoided network costs is split into a capacity and an energy component. Plants below and above 30 kW are treated differently. Small plants are not required to install a power meter. On the one hand, this simplifies the process; but, on the other hand, small plants only receive a capacity payment if their feed-in load factor exceeds 2,500 h/a. If the feed-in load factor is below 2,500 h/a, it is assumed to be unlikely that such plants can reduce the peak load on the grid and, correspondingly, only the energy payment is granted. As there is no power meter involved, the distribution network operator assumes a standard profile and covers the costs of imbalance between this profile and the actual load shape of the plant. VVII+ recommends a compensation of 0.25 €ct/kWh; but it is up to the grid operator to determine the actual rate. Compensation for avoided grid use ranges from 0.15 to 0.55 €ct/kWh (Meixner 2003). Mühlstein (2003) criticizes that the calculation of this compensation is not adequate, and discriminates in particular against independent power producers, such as operators of small CHP plants, which feed into the low- and medium-voltage grid. He estimates that the current compensation is two-thirds lower than that which is justified. In September 2004, the Federal Association of New Energy Suppliers (*Bundesverband Neuer Energieanbieter*, BNE) backed this analysis and argued for a fair and legally binding calculation procedure for compensation of avoided network use (BNE 2004, p.4). However, compensation for avoided network use currently only represents about 5 to 10% of the feed-in revenue. Therefore, only major changes in its method of calculation would affect the overall economic efficiency of micro cogeneration (Hohmann 2005).

Leprich and Bauknecht (2003a, p.3) characterize the situation for distributed generation in Germany as two-sided. On the one hand, there is strong support for renewable energies and CHP plants, while, on the other hand, weak regulation exists, with only negotiated Third Party Access, high transaction costs for distributed generation, and a low level of unbundling. Thus, one can conclude that the outwardly strong support for distributed generation is – at least partly – counteracted by weak regulation of the electricity markets and falling electricity prices after liberalization.

However, the legal situation will change in the near future. Due to the European directive (2003/54/EC) to increase competition in the energy sector, which explicitly mentions distributed generation as an important means for network stabilization, Germany has to abolish its negotiated Third Party Access and to set up an energy market regulation authority. This authority has to approve tariffs for network use in advance and, especially important for small CHP, shall ensure fair compensation for avoided network use. The transposition of this directive was to be accomplished by 1 July 2004. Germany did not meet this deadline, but will most likely implement the requirements of the directive to introduce electricity – and gas – market regulation by mid-2005.

Concerning compensation for avoided network use, the German government's current draft energy industry law only contains a discretionary provision. The accompanying draft decree confirms distributed generation's claim to compensation, but has been criticized for neglecting several cases where micro cogeneration would indeed avoid network use but would still not receive any, or too little, compensation (BKWK 2005). Equally, the Federal Council of Germany (the upper house) insisted on a binding obligation to grant compensation for avoided network use in the energy industry law itself (Deutscher Bundestag 2004, p.86). BKWK and other micro cogeneration stakeholders are nevertheless not very optimistic about enhancing chances for micro cogeneration in this respect, the issues in question being highly technical and difficult to explain in the public debate (Hohmann 2005). As mentioned above, the directive also requires the legal unbundling of network from generation and sales activities for large vertically integrated utilities. Currently, this does not require major changes in most integrated utilities, but rather represents a step towards ownership unbundling, the implementation of which is increasingly being discussed at the European level. This would in fact constitute a change for micro cogeneration, eliminating an incentive for network operators to discriminate against distributed generation. But, basically, it can be expected that stronger electricity market regulation will rather improve than deteriorate the conditions for micro cogeneration.

Another issue related to the future chances for micro cogeneration is the development of gas market liberalization. As with the electricity sector, a recent EU directive (2003/55/EG) calls for full competition within gas markets by 2007 and regulated network access; the national implementation of these requirements in Germany is included in the current amendment of the energy industry law. Efficient gas-to-gas competition will presumably reduce gas prices (possibly decoupled from oil prices), improving the position of micro cogeneration operators for choosing their gas supplier. But larger heating plants, being competitors of micro cogeneration, would also benefit from falling gas prices – probably even more so, due to their higher fuel demand. This in turn can alter the relative competitiveness of micro cogeneration. However, the growing horizontal integration of power and gas supply, especially following the merger of E.ON (Germany's second largest electricity producer) and Ruhrgas (Germany's largest grid gas utility, with a wholesale market share of nearly 90%), causes concerns, as it enables E.ON to discriminate against competitors in the electricity generation and retail sectors by offering disadvantageous terms for gas transportation and supply (Mez and Matthes 2002). This would not only affect utilities running, for instance, modern combined cycle gas turbine plants, but also potential micro cogeneration operators.

8.1.2 Energy Taxation

Energy taxation in Germany consists of mineral oil taxes and electricity taxes aimed at reducing energy-related emissions. In 1999, Germany introduced an Ecological Tax Reform (ETR), which increases taxes on energy in a complex way. On the one hand, the ETR raises existing taxes on petroleum products (gasoline, diesel fuel, heating oil, and natural gas); on the other hand, it introduces, and provides for the increase of a tax on electricity. Eco-taxes are levied on final energy consumption (Kohlhaas 2003; Kohlhaas and Mayer 2004).

A significant feature of the ETR is that coal use is generally exempt from taxation, while gas input to electricity production is still taxed via the pre-existing mineral oil tax. This makes for an imbalance within fossil fuel use. In particular, it presents a disadvantage for natural gas use, which is less carbon intensive than coal. This imbalance will be alleviated soon, due to a recent EU Directive on Energy Taxation (EC 2003) that requires fuel inputs to electricity production to be generally exempt from energy taxation. The required exemption of gas inputs to electricity production has yet to be put into national force in Germany. Special provisions, e.g. lower tax rates or tax exemptions, are given to energy intensive industries and electricity producers, including micro cogeneration plants, in order to enhance the cost-effectiveness of such systems.

With respect to *mineral oil taxes*, CHP plants with an overall annual total conversion efficiency (electricity plus heat production) of at least 60% are exempt from the additional mineral oil tax that is induced by the eco-tax (BMU 2004). In addition, fuels used in CHP plants with an annual utilization rate of 70% or more receive a full rebate of all energy taxes levied on their inputs.[3] This is in contrast to other forms of power generation, which are only eligible for a rebate of the new energy tax. Thus, the ETR explicitly favors cogeneration compared to other forms of power generation.

Currently, there are a number of bureaucratic impediments that make it difficult to appropriate the tax savings (Meixner and Stein 2002). The tax is deducted automatically with the purchase of fuels; thus reimbursement of the amount overpaid has to be applied for at the customs office in charge. Such an application involves several steps, including giving detailed information about the technology used, appointing a mineral oil tax representative and deputy, proofing the actual conversion efficiency reached by the plant, etc.; these conditions may be difficult for individual or collective small-scale users to meet.

With respect to *electricity taxes,* "own use" electricity produced in CHP plants with a capacity of 2 MW or less is exempt from electricity taxation. The "own use" requirement has been a subject of controversy, as it requires a spatial connection to the plant rather than ownership (Meixner and Stein 2002). The term "spatial connection" is not clearly defined. A recent supreme tax-court judgment (*Bundesfinanzhof*), however, supported a local judgment (Finanzgericht Düsseldorf, 14 May 2003) that electricity produced in plants with a capacity of 2 MW or less is exempt from the electricity tax if it is distributed and used within a local distribution grid of low or medium voltage. A spatial connection, thus, requires a local grid, low or medium voltage (not high voltage!), and no voltage transformation. In addition, the tax exemption evidently applies to plants operated under third party financing concepts (BMU 2004).

8.1.3 Combined Heat and Power Law

The liberalization of electricity markets in 1998 led to a substantial drop in electricity prices. On the one hand, this was expected and intended; but, on the other hand, it also reduced the competitiveness of CHP plants, which was not intended. Several industrial and public operators of CHP installations announced that they were going to stop the operation of these plants. Some plants were immediately shut down. In order to halt the dismantling of CHP capacities in Germany, a law for the protection of CHP electricity generation (*Kraft-Wärme-Kopplungsgesetz,* CHP law) was

---

[3] There is an option of choosing between annual or monthly utilization rates; this benefits providers who can reach high utilization rates only for specific months in the year (e.g., through increased heat production during wintertime).

introduced in 2000 (Deutscher Bundestag 2000a). The law was mainly intended to protect existing CHP installations of municipal utilities in the short term, and was thus limited to the end of the year 2004. Operators of such CHP installations were granted a bonus of 1.5 €ct/kWh electricity generated in the first year. This bonus was reduced by 0.25 €ct/year thereafter.

The CHP law, however, did not provide a long-term perspective for the expansion of CHP capacity. This was to be achieved through a succeeding regulation that was already announced in the first CHP law (§ 7, Deutscher Bundestag 2000, p.704). Correspondingly, an intensive debate about the design of the expected CHP expansion bill began in 2001. Fixed feed-in tariffs, reverse auctioning of subsidies, and a quota model were the most intensely discussed options. The quota model was developed and favored by the Ministry of the Environment. The basic idea of this market-based instrument is that all electricity suppliers should provide an increasing quota of their total sales from CHP production facilities. To comply with this rule they have to surrender CHP certificates. These certificates would be issued for every kWh produced from CHP, and would be freely tradable on the market. The approach was supported by several German states (*Länder*) and was also included in the German climate-change program that was adopted by the cabinet in October 2000. The industry, however, was opposed to a quick CHP extension, and lobbied heavily against this approach (Meixner and Stein 2002, p.13). Eventually, the quota approach was abandoned and replaced by a combination of i) a voluntary agreement between the German government and industry on the reduction of CO_2 emissions and the promotion of CHP (Bundesregierung and BDI 2001) and ii) a bonus model. Finally, in March 2002 a law on the protection, modernization and extension of CHP – called the second *Kraft-Wärme-Kopplungsgesetz* – was adopted (Deutscher Bundestag 2002a).

In general, the new law obliges grid operators to connect all CHP installations to their grid and to buy the electricity provided by these installations. The electricity price, however, is not fixed, but negotiated between CHP and grid operators according to a "usual price". The term "usual price" was not further specified and became highly contentious. As of August 2004, a revision of the law was put in force with a specific definition of the term. The usual price, thus, is at least the previous quarter's average base load electricity price traded at the European Electricity Exchange. The modification is binding for all parties, with a dynamic component of quarterly changing prices.

In addition to this reimbursement, CHP operators receive a bonus payment per kWh CHP electricity sold. These bonus payments are financed by a mark-up on the grid-use charges. The level of bonus payments depends on the categorization of the CHP installation. There is a differentiation made between several types of existing and modernized installations, on the one hand, and new installations on the other. Whereas all incumbent and modernized installations will receive, over time, slightly decreasing bonus payments up to the year 2010, this is not the case for new installations. Only small new installations, with an electrical capacity below 2 MW, and fuel cells can receive bonus payments up to the year 2010. New micro cogeneration plants fall into the latter categories.

Fuel cells will receive a bonus payment of 5.11 €ct/kWh fed into the grid. As fuel cells are still in the stage of product development and pilot projects, and far from being economically competitive, this bonus will not be sufficient to lift them over the break-even point and – at best – facilitate their market introduction. CHP installations with an electrical capacity of 50 kW or lower will also receive 5.11 €ct/kWh fed into the grid for 10 years from start of operation, if they are commissioned before 1 January 2009. Until recently, investors faced uncertainty regarding the CHP policy after 2005, since the CHP bonus for micro cogeneration was only granted to plants commissioned before 2006. However, in a new amendment to the CHP law this period was extended by three years, giving manufacturers, energy utilities and consumers clear mid-term perspectives for their decision-making.

Meixner and Stein (2002, p.14) criticize that this law will not promote the extension of small CHP installations, mainly because the bonus is only paid for electricity fed into the grid. The profitability of such small-scale micro cogeneration, however, is primarily based on avoided electricity demand. Installations that are designed to cover most of the operator's electricity demand usually feed only 10 to 20% of their generation into the grid. Bonus payments will hardly increase the profitability of such installations. Since the reimbursement for electricity fed into the grid has declined, from usually around 4 €ct/kWh before liberalization to a range from less than 1 to 2.5 €ct/kWh[4], a bonus of 5.11 €ct/kWh will not be sufficient to cover total generation costs, which range from 8 to 12.5 €ct/kWh (Meixner 2003; see also Chapter 4 on the economic performance and perspectives of micro cogeneration plants).

In addition, these small reimbursements do not take into consideration the real value of micro cogeneration electricity. In thermally led systems, electricity is produced at times of high heat demand, which, in many cases,

---

[4] With the new, and binding, definition of the "usual price" paid for electricity fed into the grid, the reimbursement is likely to be higher and more flexible with respect to market changes. From the first quarter of 2003 to the second quarter of 2005, it ranged from 2.5 to 4.2 €ct/kWh.

coincides with times of high wholesale prices (i.e. winter morning or afternoon) (Harrison 2001). Thus, peak load could be reduced if a statistically significant number of micro cogeneration systems were operating. Additionally, the systems are located in proximity to the customer, thus avoiding long distance transport costs. The real value of micro cogeneration could be acknowledged by, for instance, clever settlement procedures or intelligent meters.

An analysis of Traube (2003) confirms Meixner's expectations. In the first year since adoption of the law, only 38 MW of small CHP capacity below 2 MW was installed. A more recent survey by Horn et al. (2004) estimates a CHP law-induced capacity increase of 54 and 57 MW for 2002 and 2003, respectively.[5] If these figures are compared to the average increase in small CHP capacity during the years 1994 to 1996, about 275 MW per year, one can conclude that the new law does not achieve the targets it was designed for.

8.1.4 European CHP Directive

In July 2002, the European Commission submitted a draft directive on the promotion of cogeneration, based on useful heat demand in the internal energy market (COM (2002) 416 final). This was almost five years after a community strategy to promote CHP and to dismantle barriers to its development (COM (1997) 514 final) was presented in October 1997. In this strategy, the commission set the goal of doubling the share of CHP electricity from 9 to 18% by the year 2010. In the directive, however, this figure is mentioned as a non-binding benchmark, but – different from the approach in the renewable energy directive – not as a quantitative target. After almost one and a half years of discussion between the European Commission, Parliament, and Council, a compromise was agreed upon in late December 2003, with the directive being formally adopted on 11 February 2004 (EC 2004).

The CHP directive does not contain any requirements to increase the share of CHP electricity. According to Art. 7, Member States are only obliged to ensure that support for cogeneration is based on useful heat demand. Furthermore, the directive contains a definition of CHP electricity that is similar to the approach agreed upon in the German CHP law

[5] The total capacity increase was substantially larger (145 MW in 2002 and 204 MW in 2003). However, the greater share of this new capacity consists of installations that were commissioned with respect to the renewable energy law, not to the CHP law (Horn et al. 2004, p.25).

(Traube 2002, p.3) and a definition of high-efficiency CHP electricity that might deserve specific support. In addition, the directive contains the following obligations for the Member States:

- to identify national potentials for, and barriers to, the extension of CHP,
- to examine the conditions for grid access, feed-in, and reserve electricity and to facilitate grid access, in particular for smaller plants,
- to assess and simplify administrative procedures in order to make them more transparent, and to reduce barriers to CHP extension,
- to evaluate the progress of promotion schemes for the extension of CHP, and
- to report 2 years after adoption of the directive on all of these issues.

At latest 4 years after adoption of the directive, the commissions will evaluate these reports and prepare an EU-wide interim report, eventually submitting amendments for the directive.

The rapporteur of the European Parliament, Norbert Glante (2003), basically welcomed the compromise agreed upon in December 2003. He highlighted that CHP had been lifted onto the European political agenda and that the Member States must now scrutinize and report on the potential of CHP, with the Commission preparing action plans for the development of CHP if necessary. He admits, however, that he had desired a more ambitious directive, in particular with regard to a quantitative target and a more harmonized definition of CHP electricity.

The treatment of micro cogeneration in the directive was improved only in the very last negotiations. In particular, COGEN Europe had been lobbying intensively for better consideration of micro cogeneration in the directive. They had been arguing that 80% of the market potential for micro cogeneration is geared toward "distressed purchases" due to the breakdown of an existing boiler. Since a household in this situation has no heating and/or hot water, a short-term replacement is required. COGEN EUROPE concluded that these circumstances mean that the whole process for replacing a boiler with a micro cogeneration must be both simple and quick. Customers will not choose micro cogeneration if there is any scope for delay, additional network costs or any additional paperwork. COGEN Europe, therefore, came to the conclusion that it is essential that procedures adopt a "plug and play" approach. (COGEN Europe 2003, p.2)

As the impact of micro cogeneration on distribution networks is minute, it is not necessary to notify distribution network operator prior to the installation of a micro cogeneration plant. Such a simplified procedure would facilitate the penetration of micro cogeneration, as would a type-certified approach instead of the individual certification and testing of each customer's installation by the distribution network operator, which is neither possible nor necessary. Finally, COGEN Europe demanded that the standard for the electrical interface between a micro cogeneration unit and a low-voltage network should be harmonized EU-wide. Otherwise there is a danger that each country will develop its own standard, which would create market barriers and distortions between Member States, limiting the size of the market. The compromise agreed upon by the Parliament did not include COGEN Europe's suggestions literally. Simon Minett (2004), director of COGEN Europe, nevertheless appreciated the compromise because the substance of their suggestions was taken into account.

For the time being, it is too early to determine what kind of impact the European CHP directive will have on the German market for micro cogeneration, since it depends on the way this directive is transposed to German law. The amendments regarding micro cogeneration will, however, facilitate the penetration of micro cogeneration into Germany and, in the medium term, allow competing micro cogeneration technologies, which are already being applied in other countries to enter the German market.

8.1.5 Energy-Saving Decree

In November 2001, the energy-saving decree (*Energieeinsparverordnung*, EnEV) was adopted and set into force from 1 February 2002 onwards (Deutscher Bundestag 2002b). It replaces the insulation decree of 1995 and the decree on heating installations of 1994.

The basic concept of the EnEV is completely new, because it does not simply focus on the insulation of the building *or* the heating equipment, but rather on both features at the same time. The new decree covers the total energy consumption in new buildings, including energy use for space-heating, air-conditioning, and warm water. The permissible energy consumption of new buildings is reduced by about 30% (Hegner 2002, p.1). Downstream energy losses outside the building, the use of auxiliary electricity for heating installations, and the application of renewable energies for heating and warm water are taken into account for their induced primary energy consumption. Due to this approach, developers and engineers can look for integrated solutions and optimize between improving the building and improving its equipment: a weaker insulation standard might be compensated for by very efficient heating equipment and vice versa. The decree is often criticized because it includes a built-in trade-off between efficient heating systems and building insulation rather than promoting both at the same time and thereby reaching higher efficiency gains.

CHP installations are in general very efficient with regard to their induced primary energy consumption. The EnEV, therefore, improves the conditions for application of CHP in buildings. Compared to other heating technologies, the primary energy consumption benchmarks set by the EnEV can be achieved with less effort spent on insulation of a building when CHP is installed. The expenses saved on insulation can then be used to finance the investment into CHP. Buildings where at least 70% of the heat demand is met by CHP plants are, in addition, exempted from the maximum standard for primary energy consumption. All in all, one can conclude that the EnEV promotes very efficient heating installations such as CHP or micro cogeneration.

8.1.6 Emissions Trading

In October 2003, the EU adopted Directive 2003/87/EG, establishing a scheme for GHG emission-allowance trading within the Community: "This directive aims to contribute to fulfilling the commitments of the European Community and its Member States to reduce greenhouse gas emissions more effectively, through an efficient European market in allowances, with the least possible diminution of economic development and employment" (EC 2003, p.32). Basically, the directive controls all GHG of the Kyoto Protocol, although in the first three-year period from 2005 to 2007, only CO_2 will be regarded. It covers installations with a rated thermal input of more than 20 MW in energy activities (power plants, boilers, ovens, refineries, etc.), in the production and processing of ferrous metals, in the mineral industry (cement, glass, bricks, tiles, etc.), and other activities (pulp and paper, lime, etc.) (EC 2003, p.42).

As CHP in general is a very efficient method for energy conversion, it is widely expected that the use of CHP technology will benefit from emissions-trading schemes. However, Stronzik and Cames (2002, p.28ff) pointed out that the current EU emissions-trading scheme might give disincentives to CHP installations, which provide heat for consumers who are not covered by the directive. In these sectors, large CHP plants that are subject to the emission-trading directive have to compete with separate heat and power production units that are not part of the emission-trading scheme. This is particularly the case with CHP plants greater than 20 MW, which provide district heating networks for households or the service sector.

As the thermal-input threshold for installations covered by the emissions-trading scheme is much higher than the thermal capacity of micro cogeneration installations, such installations will not be affected directly by the emissions-trading scheme, because they are usually applied in sectors which are not covered by the scheme: private households and the service sector, in particular. Therefore, they compete mainly with conventional atmospheric or condensing boilers and electricity demand from the grid. However, emissions trading might affect micro cogeneration indirectly through changes in electricity prices. Basically, it is assumed that electricity prices will increase due to emissions trading, which would improve the competitiveness of micro cogeneration, due to the increased value of avoided electricity demand.

The expected price of CO_2 allowances is still very uncertain. Estimations range from 5 to 30 €/t CO_2 (Matthes et al. 2003, p.108f). In Germany, allowances will be distributed free of charge to the covered installations up to the year 2012. Additional costs will, however, emerge due to investment in mitigation measures in order to fulfill the targets. Gilles et al. (2003) expect that large utilities in Germany will – at least partly – pass on such and other costs to their customers. They estimate that wholesale electricity prices might rise in Germany by 16% between 2003 and 2005 due to emissions trading alone (Gilles et al. 2003, p.1). ECON (2004) estimates that wholesale electricity prices will increase by 17% in the short run. In the somewhat longer run, until 2012, ECON expects that the price increase, compared to the reference scenario, will be smaller, about 7% (ECON 2004, pp.31-46). Both studies assume that the marginal costs of emissions trading are passed on to the customer. However, since allowances are allocated free of charge, some utilities might decide – at least in a competitive market environment – to pass on only the additional costs of emissions trading. If this is the case, the resulting price increase will be much smaller.

To sum up, emissions trading will not promote or impede micro cogeneration directly, because the capacity range of micro cogeneration is not covered by the European emissions trading directive. Electricity prices, however, might increase due to emissions trading, improving in turn the competitiveness of micro cogeneration installations, due to the increased value of avoided electricity demand. Correspondingly, emissions trading might promote micro cogeneration indirectly, although it is still uncertain to what extent.

8.1.7 Local Air-Emissions Standards

In Germany, micro cogeneration systems have to be "state-of-the-art" in terms of pollutant emissions. State-of-the-art is regulated by the administrative instruction *TA-Luft*, which has recently (1 October 2002) been revised and now demands NO_X emissions from gas-fired internal combustion engines to be below 500 mg /m³ flue gas for lean engines and below 250 mg/m³ flue gas for all other engines. Carbon monoxide emissions have to be below 300 mg/m³ flue gas for all engines. These values will be easily fulfilled by all micro cogeneration systems; reciprocating engines are typically operated with three-way catalysts or lean combustion.

However, in some other European regions, emissions legislation is much stricter than in Germany with respect to small CHP. For instance, in some Swiss cantons, very strict NO_x emission levels are required (e.g. 50 mg/m³ in the city of Zurich and 80 mg/m³ in the canton of Zurich, 70 mg/m³ in Basel Landschaft). According to the Guidelines for Air Emissions Regulation, Danish CHP systems above 120 kW have to emit less than 65 mg/m³ NO_x and less than 75 mg/m³ CO. Some CHP manufacturers fear that these emissions targets will be transferred to Germany. This would particularly hurt reciprocating engines, whereas fuel cells and Stirling engines (of low-emissions design) could easily comply with these levels.

8.1.8 Institutional Barriers

Apart from the above mentioned regulatory and legal aspects, there are several institutional aspects that may affect the market penetration of micro cogeneration plants. The following outlines a few examples of support or barriers that owners or users may face in the process of installing a micro cogeneration unit; these barriers can cause substantial transaction cost and thus deteriorate the economic competitiveness of micro cogeneration (see Sect. 7.1.4).

Building license. According to the Energy Industry Law (EnWG, Deutscher Bundestag 1998), there is no license needed for either building or operating a CHP unit that produces electricity for own use or for feeding into the public grid. According to the German Building Law, an appropriate building license, however, may be needed. It is the responsibility of the German states *(Länder*) to establish guidelines and directives for the issuance of, or exemption from, licenses.

Authorization charges. The Federal Office of Economics and Export Control (*Bundesamt für Wirtschaft und Ausfuhrkontrolle,* BAFA) charges fees for the authorization of CHP plants which is necessary for receiving the bonus payments guaranteed by the CHP law (see Sect. 8.1.3). The fees apply according to plant capacity: for CHP units up to 2 MW_{el}, the charge amounts to 75 €, large CHP plants, for example of 50 or 100 MW_{el}, are charged around 600 €. This moderately progressive fee puts small-scale CHP plants at a disadvantage. While the fee is an insignificant amount compared to investment cost and expected benefits from the CHP law for large scale, it is a substantial amount for micro cogeneration units. For plants the size of 5 kW_{el} and less, the benefits may hardly balance the fees, therefore, and impede investment decisions. In addition, the authorization procedure at the Federal Office – similarly to the application process for reimbursement of the mineral oil tax – is rather bureaucratic and may discourage potential small-scale (private) investors.

Technical licenses. There are a number of technical factors micro cogeneration plants are required to meet. They include standards with regard to electrical installation; size, location, use, and building materials of the room the unit is to be built in; emissions and pipelines; chimneys and chimney sweeper's approval; and so on. As some items are not clearly regulated, for instance which parts of a micro cogeneration plant require a chimney sweeper's approval, sometimes inadequate security demands regarding the flue gas system of the plants are made. Higher qualifications and a clarification of the duties of chimney sweepers with respect to these small plants would be helpful. In addition to legal requirements, the German association of the electricity supply industry (*Verband der Elektrizitätswirtschaft,* VDEW), has recommended certain security measures for synchronized grid connection. Many local electricity utilities ask for and verify the compliance of a plant with these rules, and sometimes even demand that further measures be taken. Even if the plant owner estimates these requirements to be unreasonable, he still has the burden of proof. If the micro cogeneration plant uses natural gas, the qualifications of the company installing the plant has to be approved by the local gas utility, which also has to certify the proper installation of the plant itself.

Preferential treatment of renewable energy. Germany supports the development of renewable electricity through fixed feed-in tariffs. In addition, grid operators are obliged to buy all electricity produced from renewable sources. Unfortunately, such preferential treatment was not established for CHP electricity, although Art. 3 (7) and 11 (3) of Directive 2004/54/EC (electricity directive) allow for the extension of such treatment to electricity from CHP as well (VKU 2004, pp.2-4). This may lead to a situation where CHP owners, particularly in rural areas, will have to stop generation if grid operators refuse their electricity due to increased generation by renewable sources, such as wind power plants. Up to now, only a few of such cases are known, all of them being industrial cogeneration rather than micro cogeneration installations. Nevertheless, this situation may change with increasing generation using renewable sources. Moreover, it hinders investment in CHP technologies when investors fear that this situation may arise in their area. It remains to be stated that, naturally, all of these factors present barriers to individual small-scale investors who are neither experienced nor trained in these areas. They incur information and transaction costs that may be too high to balance the potential economic benefits of micro cogeneration plants.

8.1.9 Institutional Framework: Discrepancy Between Formal Rules and Actual Effects

In conclusion, it appears that the institutional setting for micro cogeneration in Germany shows ambivalent patterns. For one, many regulations are in place that give direct incentives to micro cogeneration or contain favorable conditions for distributed generation, efficient energy use, or CHP in general, which have indirect positive implications for micro cogeneration. These favorable conditions include shallow connection fees, compensation for avoided network costs, tax exemptions, energy efficiency standards for buildings, and specific bonuses for CHP electricity fed into the grid (Fig. 8.1). Yet, many of these formal regulations are not effective, because they are not properly enforced or because they are overshadowed by more fragmented and informal institutional settings that encroach upon the overall institutional environment for potential micro cogeneration investors, operators, and users. An important point in this regard is the weak regulation of competition in the German electricity market. Incumbent utilities that own the electricity network have a strong interest in hindering the uptake of generation activities by their customers; and they have many different means to do so. This is due to a lack of rigid unbundling guidelines as well as independent and reliable enforcement of network access regulations. The establishment of an independent regulatory authority in 2005 may be a step forward in this respect.

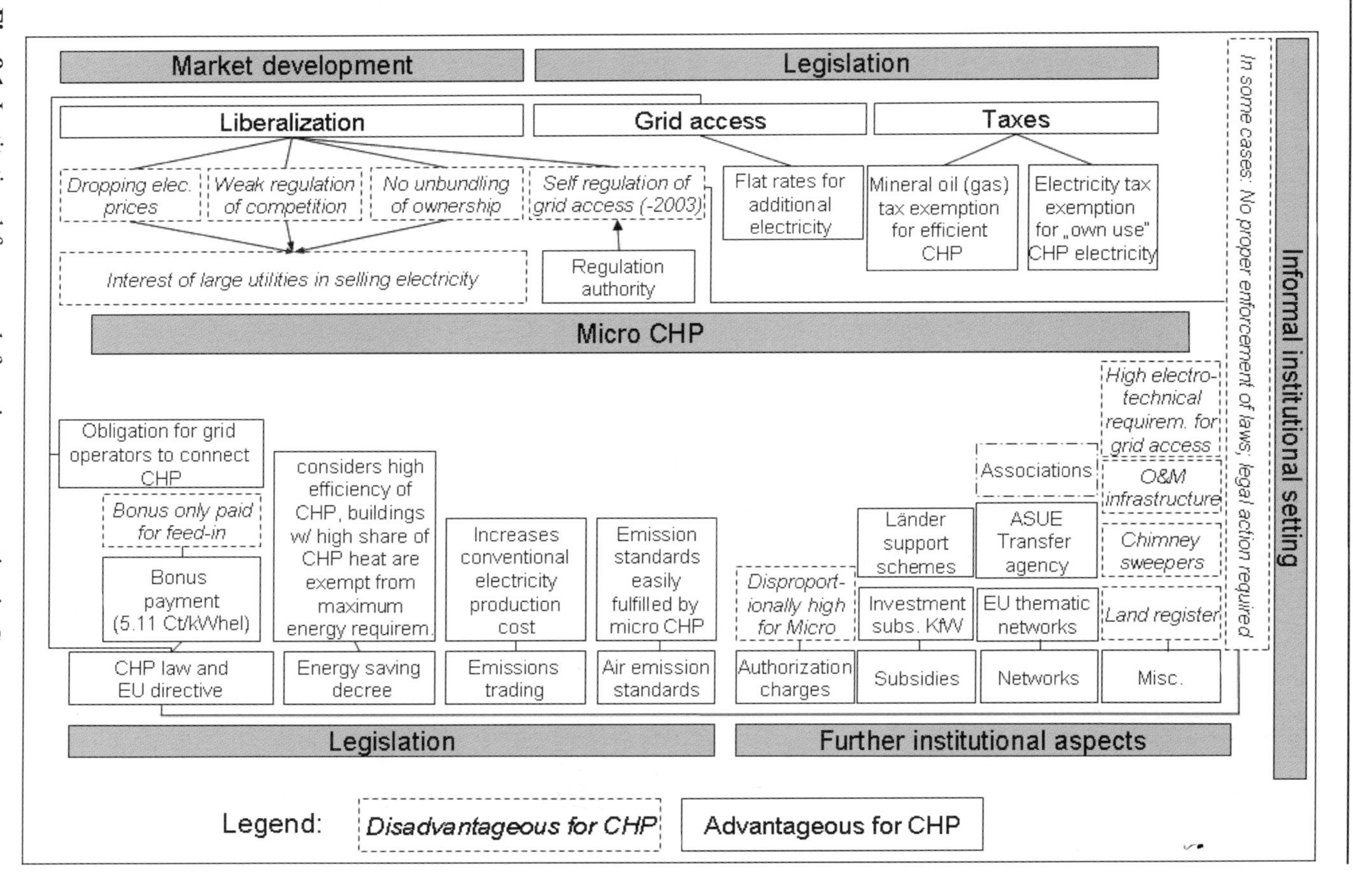

Fig. 8.1. Institutional framework for micro cogeneration in Germany

Implementation failures also became evident under the CHP law. The promotional effect of the feed-in bonus was undermined by a complementary reduction of the non-regulated general feed-in tariff for distributed generation. The general effect of the combined institutional changes since liberalization of electricity markets in 1998 is that newly built capacity of small CHP plants below 2 MW has decreased to less than 15% of the pre-liberalization yearly capacity increase (see Sect. 8.1.3). This flaw has been corrected recently by a precise definition of the "usual price". However, it is too early to assess whether this improvement will revive the market for small CHP.

Of great relevance are many small impediments to micro cogeneration. These do not appear to be influential by themselves, but add up to a considerable increase of transaction costs and uncertainty for investors, so that they effectively compensate for some of the benefits introduced through formal regulation. Among these are disproportionally high registration fees for small micro cogeneration installations, complicated licensing procedures, and incompatible maintenance and service arrangements.

In general, institutional structures for further development and diffusion of micro cogeneration in Germany are favorable in terms of the formal regulations in place. Nonetheless, consideration of the informal institutions that guide the actual implementation of regulations and a diverse array of inconspicuous details within the institutional environment for micro cogeneration activities reveal considerable barriers that can help to explain sharply decreasing growth rates for new installations over past few years.

8.2 Innovation Policy

Against the background of the specific institutional conditions for micro cogeneration in Germany, it is interesting to look at the status of micro cogeneration in innovation policy and the specific activities that are in place to support or shape future development of the innovation process. The following sections give an overview of the programmatic role of micro cogeneration in German policy, relevant actors and their activities concerning the support of micro cogeneration, as well as existing promotional networks.

8.2.1 Status of Micro Cogeneration in Policy Strategies

Micro cogeneration is not an explicit target of German innovation policy. However, it is indirectly addressed from two sides: as a part of CHP in general and as an application of fuel cells. CHP and fuel cells play a substantial role in German climate policy and sustainable development strategies. Moreover, fuel cells are also a target of technology policy.

In late 2000, the German government presented its first national program for climate protection. It contains the policy goals regarding the development of CHP in Germany: primarily, an increase of the share of electricity produced in CHP plants, resulting in a CO_2 emissions decrease of ten million tons by 2005 and 23 million tons by 2010. The national strategy for sustainable development – adopted in 2002 – confirms this goal of conservation, modernization and expansion of CHP, the result of which is hoped to be a reduction of at least 20 million tons of CO_2 by 2010 (base year is 1998). Micro cogeneration, however, is not a specific issue here or in other political statements of the federal government. On the contrary, CHP is always spoken of in bigger terms, as a means of local and district heating, not as a technology suitable for individual buildings. Nevertheless, the strategy for sustainable development states the necessity for supporting the development of a specific technology suitable for micro cogeneration: fuel cells being a "promising future technology of energy supply and use". Both programs and many other government statements generally underline the importance of increased energy efficiency in households.

On the federal level, the Ministry of Economics and Labor (*Bundesministerium für Wirtschaft und Arbeit,* BMWA), currently led by the Social-Democratic Party with strong ties to traditional energy industries, has the main responsibility for CHP issues. The Ministry for the Environment, Nature Conservation and Nuclear Safety (*Bundesministerium für Umwelt, Naturschutz und Reaktorsicherheit,* BMU), led by the Green Party, has only small influence on federal policies towards CHP. This departmentalization of micro cogeneration policy-making, with competences spread over ministries, may inhibit effective micro cogeneration policies to come into place.

8.2.2 Research and Development

Currently, the largest support for micro-CHP-related Research and Development activities (R&D) comes from the federal government. The fourth "Program for Energy Research & Energy Technology"

(Energieforschungsprogramm) funds research in non-nuclear areas, with the main goal of developing climate-friendly technologies. Fuel cells are a central focus of the program. Other micro cogeneration technologies are currently not funded (Kalkutschki 2003; Malinowski 2003). The "Program for Future Investments" *(Zukunftsinvestitionsprogramm)* contains some supplementary funding for fuel cell research including applications of fuel cells, for example, in the form of micro cogeneration (BMWA 2003; Malinowski 2003). Several pilot projects for virtual power plants using fuel cell technologies have been started as part of the national strategy for sustainable development, funded by the programs named above. The Environmental Ministry has set up a technology assessment project to go along with the pilot projects within the *Zukunftsinvestitionsprogramm*, in order to gather further information on potential technical, economic, and ecological impacts of fuel cell applications. From 1993 to 2002, the Federal German Environmental Foundation (*Deutsche Bundesstiftung Umwelt)* funded a field test with small CHP plants using Stirling technology with over € 600,000 (DBU 2003).

On the European level, there is some funding for micro cogeneration R&D as well. Again, unlike other micro cogeneration technologies, fuel cells are explicitly mentioned in the chapter "Energy, Environment and Sustainable Development" (overall budget: € 2,125 million) of the "Fifth Framework Programme" (1998-2002). Projects continuing from the fifth program with German participation are, for instance: i) a study of the design and operation of Micro Grids for increased penetration of micro sources, including fuel cells, micro-turbines, and CHP and ii) a Thematic Network on Combined Heat and Power, which mentions micro cogeneration as one main work package. The European-level fuel cell projects focus on technical improvements in the area of the technology itself, as well as on implementations within virtual power plants (Peisker 2003).

On the national as well as on the European level, R&D funding is mainly devoted to the development of fuel cell technologies and, to a lesser extent, to virtual power plants. Micro cogeneration as such, or technological variations such as reciprocating or Stirling engines or micro turbines, are not a funding area for public R&D support.

8.2.3 Investment Subsidies for Energy Efficiency

As there are no commercial micro cogeneration installations on a fuel cell basis available yet, market introduction schemes practically do not apply. Other micro cogeneration technologies, mainly reciprocating engines (as the most advanced technology) however, could benefit from political support for market introduction.

Through the Reconstruction Loan Corporation (*Kreditanstalt für Wiederaufbau)*, the federal government has set up support programs to encourage efficient energy production and use in private households (*KfW-Programm zur* CO_2*-Minderung, KfW-*CO_2*-Gebäudesanierungsprogramm*). The programs provide credits at a preferential interest rate for renovation, modernization, or construction of buildings. Nevertheless, to date there is practically no demand for these funds for micro cogeneration applications. The program administrator at KfW explains this by pointing to the still unfavorable economic performance of micro cogeneration installations in households, due to low heat demand during the summer season (Götze 2003).

Many Länder have their own programs to support efficient energy production and use. As these programs vary from each other and change with public budgets, they are not discussed in detail; Länder which are most active in this field are Baden-Württemberg, Bavaria, and North Rhine-Westphalia.

8.2.4 Promotional Networks

Currently, there are a number of professional networks in Germany concerned with fuel cells, micro cogeneration, and related technologies. Some of them take a very active part in promoting these technologies and are briefly described in this section.

The aforementioned Thematic Network on Combined Heat and Power, initiated and funded by the European Commission within the *Fifth Framework Programme*, represents a European network to create synergies between many different research, technological development, and demonstration activities all over Europe.

In Germany, the "Working Group for Economical and Environmentally Friendly Energy Use" *(Arbeitsgemeinschaft für sparsamen und umweltfreundlichen Energieverbrauch,* ASUE) – an initiative of the German gas industry – founded the "Transfer Center for New Products" *(Transferstelle neue Produkte)* as a contact point and coordination center for innovative gas applications, micro cogeneration being one of the main technologies supported. The *Transferstelle* is privately funded and looks out for products in other countries that seem suitable for the German market, contacts the manufacturer, and provides necessary information about the German market and regulatory framework. It also works as a lobby organization, for example in the discussion concerning a European directive on CHP. As the focus is on products, which have already been or about to be introduced, reciprocating and Stirling engines as well as steam cells are the main technology focus (Telges 2003).

The "Fuel Cell Expertise Network of North Rhine-Westphalia" (*Kompetenz-Netzwerk Brennstoffzelle NRW)* , founded and funded by the state government, aims at organizing communication between research activities directly or indirectly concerned with fuel cells. The network also initiates cooperation projects between research institutions and industry, while encouraging investments by manufacturers of fuel cell systems and components. It has funded research projects with a total volume of over € 40 million.

A different approach is pursued by the "Fuel Cell Initiative" *(Initiative Brennstoffzelle)*, which was founded by the German gas industry together with some of the most important manufacturers of heating systems. The initiative is currently conducting a large, fuel-cell-based virtual power plant field test in cooperation with several German power and gas companies and partners from other European countries. The project receives financial support from the EU. Other activities of the initiative comprise information services for customers and, in the future, the education and training of electricians and heating installers (Corsten 2003).

8.2.5 Informational Measures

Many actors concerned with micro cogeneration innovation believe that a major barrier for increased market entrance of micro cogeneration systems is a lack of publicity and information. The product and its advantages are not well known among retailers, service providers or potential users. On the other hand, the above mentioned *Transferstelle* (Transfer Center for New Products) has been hesitating to launch a campaign for consumers and for the installation and maintenance businesses, because it considers the availability of different products to be as yet too restricted to give the consumer a real choice (Telges 2003). Several institutions, such as the "Center for Further Education on Fuel Cells Ulm" *(Weiterbildungszentrum Brennstoffzelle Ulm,* WBZU), were recently founded to inform craftsmen on the status and development of fuel cells, including micro cogeneration applications.

Other energy information services – organized by the "German Energy Agency" *(Deutsche Energie-Agentur)*, the "Consumer Advice Center" *(Verbraucherzentrale)*, the Länder and other institutions – provide advice to consumers on the modernization of heating systems. They have a potential role for spreading information on "Electricity producing heating installations". According to current practices, however, they recommend advanced condensing boilers – either because the staff in charge is not informed about micro cogeneration or because economic costs are considered prohibitive (Kafke 2003; Schlösser 2003).

8.2.6 Innovation Policy: Overshadowed by Large CHP, Outshone by Fuel Cells

Altogether, a comprehensive and consistent federal policy towards micro cogeneration does not exist. Although the programmatic orientation of federal policy gives great weight to the expansion of CHP in general, micro cogeneration is not explicitly considered part of a future vision of climate-friendly energy provision in Germany. Larger plants for industrial or district heating applications are generally of higher political interest. Sometimes this is explicitly declared on the grounds of economic arguments; sometimes it can be related to the political influence of municipal CHP plant operators on German federal policy. Market introduction measures, as described in context with the institutional structures above (CHP law, taxes, etc.), mainly target CHP in general with some extra provisions for small-scale plants.

With respect to technology policy, there is a large difference in the amount attention paid to fuel cells, on the one hand, and other micro cogeneration technologies, on the other. Innovation policy follows a clear technology focus in this respect, not a problem-driven approach that starts from differentiated application contexts and respective energy-saving potentials. This one-sided orientation of innovation policy also poses difficulties for communicating the general concepts of micro cogeneration to a large public, because public attention has been too narrowly focused on fuel cells. Support for fuel cells comes from different levels (Europe, German federal government, Länder, and private initiatives) and through different approaches (R&D financing, information management, or network support). R&D financing for other micro cogeneration technologies is negligible. This makes it difficult for products based on other technologies (reciprocating engines, micro-gas turbines, Stirling engines) to compete on the same ground with fuel cells, although they are to an extent much further developed with respect to technological reliability and economic competitiveness.

Review of the standing of micro cogeneration in innovation policy shows that it sits on the fence between large CHP, on one side, and fuel cells, on the other. Large CHP is a major focus of climate-oriented innovation policy through measures for market introduction (mainly financial support). Fuel cells are a recent darling of technology policy because of certain features that make them appear a particularly intelligent technology. Between these two, the specific demands of small CHP installations for private households or the service sector are overlooked; they are currently neither explicitly mentioned on the programmatic level of energy policy, nor are necessary complementary conditions for successful diffusion strategically addressed.

9 Embedding Micro Cogeneration in the Energy Supply System

Lambert Schneider, Martin Pehnt

An important prerequisite for the large-scale introduction of micro cogeneration is its compatibility with the existing and future energy system. This involves different aspects, which we will investigate in this chapter, with the potential impact of micro cogeneration on supply security being a particular important one. Firstly we provide an overview of challenges to supply security (Sect. 9.1) and then go on to describe the impacts of micro cogeneration on both the security of fuel supply (Sect. 9.2) and on the reliability of the electricity network (Sect. 9.3). Furthermore, in the long term, the energy system has to be transformed into a largely renewable system. It is therefore important to assess whether micro cogeneration impedes or supports the use of renewable energies (Sect. 9.4). Finally, we draw conclusions (Sect. 9.5).

9.1 Challenges to Security of Energy Supply

During the past few years, security of energy supply has become an increasingly important issue. Several blackouts in the United States and recently in the summer of 2003 in Europe have indicated that uninterrupted electricity supply is not guaranteed in liberalized markets. In the coming years, the security of both electricity and fuel supply will be confronted with new challenges:

- **High costs of power blackouts in digital economies.** In the emerging digital economy, costs of electricity interruptions are significant. According to U.S. Department of Energy's (DOE) Office of Distributed Energy Resources, power fluctuations and outages cost U.S. business about $50 billion a year (DER 2003). Momentary power disruptions can cost some businesses millions of dollars: semiconductor manufacturers, credit card operations, or brokerage firms for example face 2 to 6.5 million US$ typical cost per hour of interruption (GM 2003).

- **Growing demand for highly reliable electricity.** Estimates by the Electric Power Research Institute (EPRI) indicate that the proportion of U.S. electricity requiring reliable, high-quality power will grow from 0.6 % of current consumption to nearly 10 % by 2020, and that the proportion of enhanced reliability electricity will grow from about 8-10 % to nearly 60 % (CEIDS 2002). On-site production with CHP such as fuel cells could help in achieving highly reliable electricity supply (i.e. 99.9999 %), sometimes referred to as premium power.
- **Integration of wind power.** The proportion of wind power in Germany has rapidly increased from a few pilot plants in 1990 to 16,629 MW electric capacity by the end of 2004. While research is being carried out to improve the quality of wind forecasts, network operators will nevertheless have to deal with short-term changes in wind power supply. Recently in western Denmark, power supply from CHP plants with back-pressure turbines[1] and from wind power plants has already exceeded the power demand. In some cases, wind power turbines had to be closed down to avoid critical power overflows. The possible power overflow is expected to amount to up to 2,900 MW by 2005, whereas peak demand in western Denmark is only about 3,650 MW (Hilger 2002; Jensen 2002; Varming and Nielson 2004). In addition, wind power plants are often built in remote areas with low electricity demand and weak electricity grids. Network operators may have to reinforce the grid to transmit and distribute wind power to consumers. Growing wind power capacities on- and offshore in northern Germany will require investment in additional transmission lines to large demand centers in western Germany. Transmission of electricity from coastal areas to consumers could also get more difficult, since new coal-fired power plants are economically more attractive at coastal sites.
- **Decommission of power plants.** In Germany, power plant operators reduced large capacities after the liberalization of electricity markets, leading to a reduction in the capacity surplus from about 20 % in 2000 to only about 7 % in 2003 (Gilles et al. 2003). While the reduction of the large surplus is economically attractive, it reduces the flexibility of the system in providing reserve power.
- **With distributed generation on the increase, the number of central regulating power plants is decreasing.** In present power systems, large steam power plants regulate the frequency and voltage of the electricity system. With distributed generation on the increase, the number of

[1] In CHP plants with backpressure turbines, electricity and heat generation are directly coupled. Consequently, electricity generation may only be decoupled from heat generation where heat storage facilities are available.

regulating steam power plants is decreasing, while the need for network control is increasing, particularly with increasing wind capacities. As a consequence, the responsibility for network control will have to be shared and managed differently. Several studies addressing this problem suggest decentralizing parts of the operational control of the network with distribution network operators (DNOs) taking a more active role. Van Overbeeke and Roberts (2002) have the vision of "active networks" comprising of many cells, each with its own power control system.

- **Climate change.** Climate change is expected to adversely affect electricity systems. Extreme weather events occurring more frequently may lead to an increase in transmission disruptions and to more abrupt changes in power supply from wind power plants. Rising summer temperatures increase the demand for air conditioning and decrease the supply of cooling water for power plants, as happened in Europe in 2003.
- **Increasing share of natural gas imports.** In the coming decades, power and heat supply is expected to depend to a larger extent on natural gas. While about 44 % of natural gas is already imported in Europe in 2005, this share is expected to grow in 2015 to about 70 % (Boehlert and Bodendieck 2004). Long-term investments in natural gas exploration and transmission infrastructure, particularly in Russia, are required to maintain a high level of supply security for Europe (Öko-Institut 2002). Furthermore, imports of liquefied natural gas (LNG) could gain in importance.
- **Volatile prices for natural gas and oil.** Natural gas prices will continue to depend mainly on oil prices. Oil demand is growing worldwide and oil prices may be volatile due to political instability in the Middle East. Therefore, the management of price fluctuation risks is one of the most important challenges.

As a result of these challenges, the security of energy supply is back on the political agenda. In 2001, the European Commission presented a Green Paper on the security of energy supply, followed by a two-year discussion on how to maintain a high level of supply security in Europe (European Commission 2001). In December 2003, the European Commission proposed a new legislative package to promote investment in the European energy sector, to both strengthen competition and help prevent the reoccurrence of the blackouts that took place in the summer of 2003.

In recent discussions on security of energy supply, distributed generation has played an important role. Distributed generation, particularly in the form of combined heat and power plants, is regarded as a way forward in the mitigation of the risks to energy supply security (Casten 2003). However, embedding distributed power plants in the electricity network is also a challenge, since they affect loads and currents in the grid. In the following, the potential impact of micro cogeneration plants on security of energy supply will be discussed in this context.

9.2 Potential Impacts of Micro Cogeneration on Fuel Supply Security

A broad diffusion of micro cogeneration plants will affect fuel supply security, to which there are several dimensions (Öko-Institut 2002):

- the *long-term physical disruption* of fuel supply due to the depletion of resources,
- the *short term physical disruption* of fuel supply due to technical, political, ecological, or other incidents, such as terrorism, or
- the risk of *sharply increasing prices* due to supply bottlenecks.

Regarding long-term fuel supply, the majority of analysts do not assume that fossil fuels will be in short supply in the next two to three decades, with Russia and Asia being attractive future supply regions (Rempel 2000; Schlemmermeier 2003). As a consequence of increasing imports from Russia and Asia, the transport distance of the average European natural gas mix will rise and, concomitantly, so will the cost and environmental impacts of that gas (BCG 2003).

A minority of analysts expect that the exploitation of worldwide fossil fuel will reach an apex within the first decade of this century. Since energy carriers may be converted (e.g. coal to liquids) and since the exploitation of additional large fossil fuel resources becomes economically feasible with higher fuel prices, the long-term supply restrictions are probably only a minor risk compared to the risks of climate change resulting from combustion of these fuels.

The risk of short-term physical disruption in fuel supply has generally diminished in the last decade, as globalization has integrated economies and reduced the probability of politically motivated supply disruptions. Also the stockpiling policy of the International Energy Agency (IEA) has significantly reduced the risk of short-term supply disruptions. Nevertheless, dependence on natural gas from Russia and oil from the Middle East will further increase, since these regions hold many resources. Particularly the dependence on oil from the Middle East involves the risk of abruptly increasing prices which would have considerable effects on economies worldwide. Increasing oil prices would also directly affect natural gas prices, since the latter are connected to oil prices in long-term supply contracts, while the international coal market would be less significantly influenced. Consequently, the most important aspect of fuel supply security today is the risk that prices for oil and natural gas will abruptly increase.

In the long term, the risk to fuel supply security is best mitigated by enhancing energy efficiency and increasing the proportion of renewable energy carriers. A broader use of coal and lignite is another option but it appears to be less appropriate, since it hampers the reduction of GHG emissions and delays the transformation of energy systems.

Micro cogeneration plants which use natural gas increase energy efficiency. The substitution effects are different for the two products (heat and electricity). Heat supply in Germany comes mainly from boilers which use natural gas and, to a lesser extent, diesel oil. Where natural gas infrastructure is available, most new installations contain natural-gas-fired boilers (see Sect. 3.2). In single-family houses with low energy demand, electric heat pumps may become a more important alternative in the future. Thus, heat generation in micro cogeneration plants mainly substitutes, at present, natural gas. In the future, also electricity may be replaced but to a lesser extent.

On the electricity side, different fuels are displaced. It can be expected that new power plants in Germany, as in other European countries, will be mainly natural-gas-fired combined cycle power plants as well as several coal-fired power plants. Lignite power plants currently have a significant share, but, with the European Emissions Trading Scheme starting in 2005, may become economically less attractive in the long term. The operation of nuclear power plants is being phased out in Germany. Consequently, electricity generated by micro cogeneration plants would replace the construction of combined cycle natural gas and, to a certain extent, coal-fired power plants.

In summary, micro cogeneration plants contribute to fuel supply security by enhancing energy efficiency in using natural gas – a fuel that is vulnerable to abrupt price changes and that is largely imported. However, this effect is partly offset by the fact that micro cogeneration plants shift electricity generation further away from coal and towards natural gas.

9.3 Embedding Micro Cogeneration in the Electricity Network

Historically, distribution and transmission networks are operated to transport power from central power plants to dispersed consumers. Broad diffusion of distributed generation would change this concept essentially. Electricity generation in distributed power plants implies a change in loads and currents in the electricity network, which may affect system reliability, power quality, and congestion in the network. Such a fundamental transformation of electricity generation would require extensive technical and institutional changes.

Table 9.1. Micro cogeneration interaction with the electricity network (Source: Mertens 2004)

Aspect	Problem	Relevance	Technical improvements of
Capacity of equipment (transformers, cables, etc.)	Overload	critical	grid
Voltage variations	Low power quality (grid)	very critical	system & grid
Short-circuit currents	Overload	uncritical	system & grid
Flicker	Low power quality (customer)	uncritical	system
Harmonic distortions	Grid loss, overload	uncritical	system
Asymmetric currents	Non-homogeneous load of phase conductors	uncritical	system
Protection equipment	Malfunction reg. system and grid protection	critical	system & grid
Ripple control system	Malfunction of ripple control systems	uncritical	system

In contrast to large centralized power plants, micro cogeneration plants have the important advantage of being on-site; thus the electricity does not need to be transported from centralized power plants to consumers. In principle, a broad diffusion of micro cogeneration would have beneficial effects on the system, since currents in transformers and transmission and distribution lines would be lowered. As a consequence, transmission and distribution losses are reduced and investments in the transmission or distribution system might be deferred or suspended. Single components, such as transformers, may be downsized when they need to be replaced. Micro cogeneration plants could also help DNOs to overcome local bottlenecks in the distribution system, which would save costs for expensive upgrades (IEA 2002).

With respect to their location and size, micro cogeneration plants are also advantageous compared with other innovative power technologies such as wind power plants. Wind power plants are mostly installed in remote areas with low electricity demand, or even off-shore, often involving reinforcements in the electricity network.

However, increased market penetration of micro cogeneration could also raise several challenges for network operators. The capacity of equipment (transformers, power lines, fuses, switches, etc.) may not be suitable due to changed or reversed power flows. With many dispersed generators, the voltage could vary beyond established limits. In some cases, protection of the distribution network may become more difficult with distributed generators. Distributed generators may, depending on the circumstances, either increase or decrease the power quality of the distribution network through, for example, transient voltage variations or harmonic voltage distortions (Jenkins et al. 2000). These restrictions, as illustrated in Table 9.1, can in some instances be overcome by improvements to the power plants, such as filters, current limiters and, in the case of fuel cells, smart AC/DC converters. In other cases, the restrictions may necessitate modifications of the grid.

In the following, we will analyze some of these issues in more detail, with the assumption that micro cogeneration is becoming more broadly diffused.

9.3.1 Equipment Capacity

Each element of the distribution infrastructure has a limited current-carrying capacity, also referred to as its "thermal rating". If loaded above this limit for an extended period of time, it will overheat, leading to permanent damage (Jarret et al. 2004). Distributed power plants change the current flows in the system, leading to higher or lower current levels. If currents might exceed the thermal ratings of cables, lines, or transformers, the system needs to be reinforced.

In the low voltage (LV) network, the most limiting factor for distributed generation is usually the medium to low voltage (MV/LV) transformer. If power generation is greater than the load of consumers in the circuit, surplus power is fed through the transformer back to the MV level. Such reverse power flows are a technical problem for some transformers. Generally, assurance is needed that reverse power flows will not exceed thermal ratings of transformers.

For micro cogeneration plants, this problem is rather limited, as the capacity of micro cogeneration plants is small. In Germany, LV/MV transformers typically have thermal ratings from 250 to 650 kVA. This would correspondingly imply that 250 to 650 micro cogeneration plants of 1 kVA or 50 to 130 plants of 5 kVA could be connected to a LV circuit, assuming the extreme case that there is no load in the circuit and that all micro cogeneration plants operate at full load (Krewitt et al. 2004).

Today, actual experience and information on network impacts of micro cogeneration are scarce. However, some information is available from power flow calculations that were undertaken for real electricity networks.

In a power flow calculation for a real low voltage grid, Pitz et al. (2003) calculated network impacts assuming that the full technical potential of micro cogeneration in the grid area would be realized. In an extreme scenario, this study assumed that all generation units feed in electricity without any load in the circuit. Even under these extreme conditions, the equipment (cables, transformers, etc.) would not be overloaded, the voltage would remain within the acceptable bandwidth, and the feed-in into the medium voltage grid would be manageable. Only in summer, when the CHP units operate for approximately one hour per day only to cover the warm water heat demand, the thermal rating of the transformer could be exceeded if all CHP units operate at the same hour of the day – which is very unlikely.

Arndt et al. (2004) have undertaken a detailed power flow analysis for an actual settlement in southern Germany. Heat and electricity load profiles are calculated for different types of buildings in the settlement and for representative reference days, differentiating between winter, summer and autumn/spring as well as between weekdays, Saturday, and Sunday. It is assumed that each building is equipped with a micro cogeneration plant, with a settlement's capacity totaling 131.5 kW. The power flow simulation shows that reverse power due to micro cogeneration plants is well below the maximum power flow of 200 kW without micro cogeneration. The most extreme situation occurs in the mornings of winter weekends, where most micro cogeneration engines are operating and the electric load amounts to only 30 kW. In this situation, the reverse power flow to the MV grid sums up to 91 kW.

These calculations show that micro cogeneration plants will generally not cause reverse power flows in transformers that exceed thermal ratings. The total capacity of micro cogeneration plants in LV networks will be rather limited and can be expected to be smaller than the thermal rating of MV/LV transformers, mainly for two reasons: Firstly, the potential for micro cogeneration plants is limited (see Sect. 3.3); thus it is rather unlikely that all buildings in a settlement will be supplied by micro cogeneration. Secondly, the economic assessment in Chap. 4 shows that the electric capacity of micro cogeneration plants has to be relatively small in relation to the electricity demand in order to operate micro cogeneration plants in an economical manner.

However, restrictions regarding the installation of micro cogeneration may occur in cases where other distributed power plants, such as PV or larger CHP plants, are connected to the same LV network. In such cases, an (expensive) reinforcement of the transformer or cables may be required. Also a large share of wind power at the MV network may limit the capacity for micro cogeneration plants. During cold and windy winter days, micro cogeneration and wind power generation could together lead to an overload in the MV network or the MV/high voltage (HV) transformer.

9.3.2 Voltage Variations

Distribution network operators (DNOs) have the obligation to supply their costumers at all points in the network at a voltage within specified limits. In Germany, for instance, the DIN EN 50160 requests low voltage levels to be within +6 % and -10 % of the default voltage at all points in the grid. Low voltage feeders have to be designed in such a manner so as to provide even the most remote consumers with appropriate voltage levels at both maximum and minimum loads. In practice, the size of low voltage circuits is limited by voltage drops due to the resistance of the cables: the larger the circuit, the higher voltage drops along the feeders are.

Distributed power plants raise the voltage, changing the voltage profile along the feeder. Voltage changes depend on the short-circuit level (fault level) in the feeder and the power generation capacity of the plant. Connection of plants is easier near transformers or in grids with high load density, where short-circuit levels are higher.

The German code for the connection of power generators at low voltage level stipulates that rises in voltage be limited to 2 % at the most disadvantageous point of the network (VDEW 2001). As a simple rule, this requirement is assumed to be fulfilled if the rating of the power generator is 50 times lower than the short-circuit level of the feeder. However, this simplified rule can be regarded as rather conservative. Large wind farms have been successfully operated on distribution networks with a ratio of short-circuit level to rated capacity as low as 6 with no difficulties (Jenkins 2000).

A few power flow simulations have been undertaken to assess the effect of micro cogeneration plants on voltage variations. These analyses come to the conclusion that even a broad diffusion of micro cogeneration will not infringe the voltage limits established in DIN EN 50160. To the contrary, micro cogeneration plants may actually stabilize voltage at levels closer to the default value. This is illustrated in Fig. 9.1, depicting simulation results from Arndt et al. (2004) at the end of a radial feeder for a real settlement in Germany (see Sect. 9.3.1). The simulation further revealed that maximal voltage differences along the feeders are higher with fuel cells operating. The maximum voltage difference in the network increases from 6.5 V to 6.7 V if all buildings are equipped with fuel cells. However, in the scenario with fuel cells operating, the voltage is generally closer to the required level.

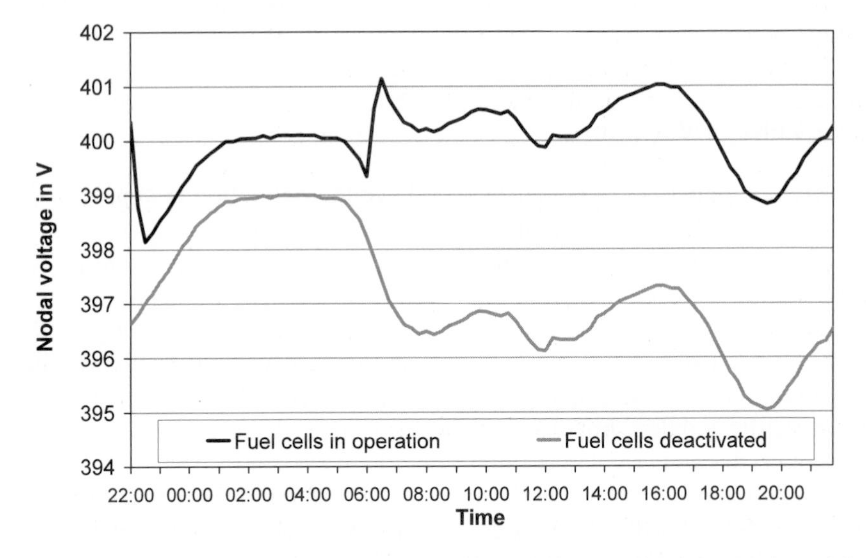

Fig. 9.1. Voltage at the end of a radial feeder with and without operation of fuel cells (Source: Arndt et al. 2004)

A German research project by the Technical College of Darmstadt *(Technische Hochschule Darmstadt)* and the energy utilities RWE Net and Bewag comes to similar conclusions (cited in Krewitt 2004). According to this analysis, a micro cogeneration capacity of 1.5 kW_{el} could be installed at each household (or other consumer site) without causing problems regarding voltage variations.

9.3.3 Protection Issues

Different protection issues can be identified for distributed power generation (Jenkins et al. 2000): Protection of a distributed generator from internal faults, protection of a defective distributed network from fault currents supplied by a distributed generator, and protection from islanding (see below). Furthermore, many distributed power generators may affect the functioning of the protection concept for a distribution network.

Protection of a distributed generator from internal faults is usually straightforward. However, protection of a network becomes more sophisticated with a broader diffusion of distributed power generators. Many distributed power generators may cause flows of fault current that were not expected when the protection system was originally designed. A major problem may be that large quantities of distributed power generation may trigger relays without any network fault, leading to a disconnection of a part of the network. To avoid such effects, adjustments of network protection schemes may be required (Pitz et al. 2003).

If a part of a distribution network is disconnected due to faults, micro cogeneration plants could potentially continue to provide electricity to loads within this part of the network ("islanding"). This may happen if the output of the distributed power plants matches the load in the disconnected part of the network and, consequently, frequency and voltage are not changing and cannot trigger disconnection of the distributed power plant. Such unplanned islanding is unsafe, as personnel performing maintenance and repair measures may not be aware of distributed power plants that are continuing to supply the disconnected part of the network.

9.3.4 Voltage and Current Transients

The connection or disconnection of distributed power plants may result in voltage and current transients. If micro cogeneration plants use synchronous generators, automatic synchronization usually ensures that voltage increases and transient currents are very limited when the plant is connected to the network. Prior to disconnection, the power output can be driven close to zero.

In the case of fuel cells, voltage variations from the cells are adjusted with a DC/DC (direct current / direct current) converter to a constant DC voltage level. A DC/AC (direct current / alternating current) converter generates alternating current and ensures automatic synchronization with the grid. Transient loads can be reduced with the help of resistive dummy loads (Krewitt et al. 2004).

9.3.5 Transmission and Distribution Losses

Micro cogeneration plants generally reduce transmission and distribution losses, since the generated electricity is consumed directly at the site of generation. However, the beneficial effects are difficult to quantify, since distribution losses may vary from 1 % to up to 20 %, depending on the location of the consumer and the grid characteristics.

Mendez et al. (2002) analyzed the distribution losses for different levels of dispersed power penetration. Generally, losses decrease with the penetration level of distributed generation. However, losses increase again with very high penetration levels, leading to significant currents from LV to MV level. Minimum losses are achieved with a high penetration level, if plants are sufficiently dispersed, as would be the case for micro cogeneration. Furthermore, distribution losses can be significantly reduced if reactive power generation or consumption of the plants is controlled.

Arndt et al. (2004) have quantified the savings of distribution losses from micro cogeneration with a power flow simulation, undertaken for a new settlement of detached and semi-detached houses as well as apartment buildings in a small town in southern Germany. If all buildings were to be equipped with fuel cells, distribution losses in the low voltage level would be reduced by more than 80 % in absolute terms. This high reduction occurs because about two thirds of the electricity is generated by the fuel cells on-site, while only one third has to be supplied through the grid. In addition, the distribution lines are less congested when electricity is fed through the grid. According to this analysis, for electricity supplied through the grid, losses are 50 % less with micro cogeneration compared to the scenario without micro cogeneration.

9.3.6 Ancillary Services and Impacts on Peak Loads

In current power systems, central steam power plants have the responsibility of providing ancillary services that are required to maintain high power quality. Ancillary services include, for example, the provision of reserve power and the control of voltage and reactive power.

At present, most power plants in the electricity system are large steam power plants that can provide ancillary services in an efficient manner. However, in the long term, the electricity system may be transformed towards a more decentralized power supply with less predictable generation from wind and solar sources. This has technical and institutional implications. To maintain high power quality and system reliability in an electricity system with a large share of distributed generation, distributed power technologies may need to provide at least some ancillary services to the grid. For example, DC/AC converters from fuel cells could provide voltage control and compensation for reactive power. Synchronous generators allow the control of reactive power, which may be reduced or consumed, depending on the requirements of the network.

However, this potential ancillary service of micro cogeneration plants is not yet utilized. In most countries, DNOs do not yet provide economic incentives for distributed power generators to actively compensate for reactive power in the network. In Germany, the power factor of consumers with distributed generation at low voltage should be between 0.9 capacitive and 0.8 inductive (VDEW 2001).

One of the most important issues for grid compatibility in the future is the provision of reserve power and flexibility in daily power generation schedules. In the following, we analyze the ability of micro cogeneration plants to adapt power generation and to provide power control services in more detail.

In electric power systems, power generation has to be adjusted continuously to the loads of consumers. Variations in load or power supply – e.g., due to tripping of power plants or changes in wind supply – cause deviations in system frequency.

System frequency is controlled by primary control, secondary control, and minutes reserve power. Primary control is the balancing of short-term frequency variations that occur in a matter of seconds and consists of automatic adjustment of power generation in large power plants. In addition, frequency is partly self-regulated by loads that vary with the frequency of the system. In the European electricity network – the UCTE – primary control is conducted jointly by all Transmission System Operators (TSOs). For this purpose, a capacity of 3,000 MW is continuously provided.

Secondary control aims at balancing major quasi-steady-state frequency deviations and controlling power flows between control areas. The TSO of the control area is responsible for contracting and controlling the provision of secondary control power by technical units. For this purpose, all units have to be included in an on-line control loop of the respective TSO. In Germany, the power control capacity of a technical unit must encompass at least ± 30 MW and it must be able to be activated within 5 minutes if necessary.

Complementing secondary power control, the German Transmission Code 2003 also provides for a minutes reserve (VDN 2003). Minutes reserve should restore secondary control and be available in case of the loss of power generating units. The minutes reserve must be activated within 15 minutes after a request from a TSO and must encompass at least ± 30 MW.

Next to these power control services, power supply companies have to establish daily schedules for power generation (injection) and loads (withdrawals). More flexible power generation and load shaping help utilities in balancing power generation and loads. Shifting power generation towards peak hours or shifting loads towards off-peak hours saves costs due to the differences in electricity prices between high-peak and off-peak hours.

For micro cogeneration plants, heat demand is the most limiting factor in providing power control services, as power generation is coupled to heat demand at the facility site. However, hot water storage tanks can help to decouple power generation from heat demand to a certain extent. They are regularly installed with micro cogeneration plants, since they allow micro cogeneration plants to be operated more continuously. If operated and controlled appropriately, hot water storage tanks enable micro cogeneration plants, for a limited time span, to increase or decrease their power generation upon request.

The ability of micro cogeneration plants to provide power control services or to shift daily generation patterns depends on the type and configuration of the plant and on the season. Micro cogeneration plants in the region of 5 kW_{el} are usually equipped with heat storage tanks and a supplementary boiler, while smaller plants of 1 kW may also be operated without a supplementary boiler and heat storage tank. In the latter case, flexibility to adjust power generation is rather limited.

Currently, the overall annual peak demand of the electricity system in Germany is usually during cold and dark winter weekdays, typically in December or January. During this period, there is a large heat demand, and consequently, micro cogeneration plants supply electricity in a relatively continuous manner. This refers in particular to larger plants with a supplementary boiler. Thus, most micro cogeneration plants will provide electricity at full load during winter peak demand periods. However, power generation cannot really be shifted or increased in the short term. The ability to provide power control services is therefore rather limited. In cases of excess power generation, power generation by micro cogeneration plants can be reduced with heat being supplied by the supplementary boiler. However, this reduces the quantity of electricity supplied through cogeneration.

In summer, electricity generation is limited by the demand for hot water, with micro cogeneration plants only operating one or very few hours per day. Plants with heat storage tanks could be set to generate power only during peak load hours, thereby increasing the value of power generation. In this regard, micro cogeneration could be used to a limited extent to mitigate summer peaks. Due to climate change, summer peak loads may become more important in the future, since during hot and dry summers the European power supply is more limited, while power demand increases, e.g., due to the increasing demand for cooling. During the past two years, peak prices at the German stock exchange rose several times for more than one hour per day well above 50 EUR/MWh.

The flexibility of adjusting power generation is greatest in spring and autumn when plants are typically operated for several hours per day. This is the best period for micro cogeneration plants to shift power generation to demand peaks and to also provide power control services. During these periods the demand of peak load can be reduced significantly (Pitz et al. 2003).

In order to shift power generation or to provide power control services, operators of micro cogeneration plants need to have economic incentives. Currently, such incentives are only provided in a limited fashion. As the refund for electricity fed into the grid is considerably lower than electricity purchase prices, micro cogeneration plant operators already have an incentive to install hot water storage tanks and to shift power generation as far as possible to peak demand hours of the facility. In addition, heat storage tanks allow plants to operate more continuously. Some micro cogeneration plant manufacturers are developing self-learning control software that will control daily engine generation, taking into account past experiences regarding the heat and electricity demand characteristics of the facility.

Peak demand hours in a single facility are in most cases different from overall peak loads of electricity grids. For example, peak demand in households occurs typically in the evening, while overall peak loads in electricity grids occur usually around noon. The daily operation schedule of micro cogeneration plants could in principle also be adapted to peak loads of the electricity grids. To this end, the distribution network operator would need to provide additional incentives for operators of micro cogeneration plants to change their generation pattern (e.g., by refunding electricity fed into the grid during peak hours with tariffs that are higher than electricity purchase costs for micro cogeneration plant operators). This type of load shaping could also improve system reliability as a whole, since micro cogeneration plants would, as a result, potentially reduce the demand for reserve power.

Furthermore, operators of micro cogeneration plants would need to receive on-line information on actual control requirements. For this purpose, micro cogeneration plants could be interconnected to a *virtual power plant* (see also Sect. 1.3). The virtual power plant operator could control power generation of all micro cogeneration plants and in this way may provide power control services, within limits related to existing heat demand. In principle, this could include the provision of secondary power control and minutes reserve.

Reciprocating and Stirling engines can, in general, be started quickly and power output can be easily varied. Thus, they can principally be used for power control services. The ability of fuel cells to change power output is still a subject of research. High temperature fuel cells should operate in a continuous manner, as thermal cycles have detrimental effects on the fuel cell materials. During normal operation, power output can only be changed in a limited range. By contrast, power output of low temperature fuel cells is more variable. The limiting factor is the generation of hydrogen in the reformer. Hydrogen storage facilities provide the possibility of short-term power changes but will not be economically feasible in micro cogeneration applications.

In summary, compared to large steam and hydro power plants, the provision of ancillary power control services appears more complicated with micro cogeneration plants. In summer and winter, the potential for power adjustments is limited, whereas there is a considerable potential during the transitional periods of spring and autumn. If connected to virtual power plants and equipped with sufficiently large hot water storage facilities, micro cogeneration plants could shift power generation to peak hours and vary generation upon request, also at short notice.

9.4 Perspectives for Using Renewable Fuels

If climate mitigation goals are taken seriously, the use of fossil fuels in micro cogeneration will ultimately be limited in the long term. So far, natural gas is the dominating and obvious fuel for the three main micro cogeneration technology types (Stirling and reciprocating engines and fuel cells). Among today's available fuels, natural gas is prioritized since it can be processed with moderate chemical engineering efforts and has favorable operation characteristics.

The compatibility of micro cogeneration systems with alternative fuels differs between technologies (Table 9.2). *Stirling engines* generally accept all types of heat sources. Since fuel combustion takes place separately from the Stirling process, the optimization of burner and Stirling engine can be carried out more or less independently, thus facilitating the system design. Also heat from concentrated solar radiation can be used, which appears to be a promising perspective for applications in developing countries.

In the case of *reciprocating engines*, it is more difficult to use alternative fuels. Solid biomass, for instance, has to be gasified prior to combustion in reciprocating engines. This causes problems with respect to tar in the synthesis gas. In small plants, gasification is unlikely to be implemented in the medium-term. Gas from biogenic sources may, depending on the gas composition, cause corrosion.

For *fuel cells*, the efforts necessary to produce a suitable gas from different fuels are even greater. This is because fuel cells require a very clean gas at the entry of the fuel cell stack; therefore, the produced fuel gas (e.g. from gasified wood or reformed hydrocarbons) has to be cleaned properly. The decentralized production of a suitable fuel gas at the micro cogeneration site from biomass or other energy feedstocks is particularly challenging because gasification, reforming, and processing become more difficult to handle the smaller the systems are designed. However, the centralized production and subsequent distribution of hydrogen, methanol, or other fuels suitable for fuel cells, is only an option in the long-term because it requires not only an energy- and cost- intensive conversion step but also a distribution system either by pipeline or via road trailers as compressed or liquid hydrogen or methanol.

Table 9.2 illustrates that natural gas is the favorable fuel for micro cogeneration, particularly for reciprocating engines and fuel cells. Renewable fuels are a medium-term option but involve considerably higher investment costs and further research and development activities. This could hamper the long-term transformation of the electricity system towards renewable energy carriers, particularly since the attractiveness of natural gas for micro cogeneration applications may impair the diffusion of other renewable energy technologies, in particular regarding heat generation.

For instance, the combination of micro cogeneration with solar thermal collectors reduces in most cases the annual operation time of the micro cogeneration plant, decreasing its economic viability considerably. As a consequence, both technologies are economically difficult to combine. Broad diffusion of micro cogeneration may impair the application of solar thermal collectors and vice versa.

A combination of micro cogeneration with wood heating systems, e.g. pellet boilers, as peak load heating devices is feasible but would also be more capital intensive, as infrastructure for two fuel systems would have to be provided.

Table 9.2. Future compatibility of micro cogeneration technologies with different fuels

	Reciprocating engines	Stirling engines	Fuel cells
Natural gas	++	++	++
Fuel oil	++	++	+
Biomass			
Wood pellets	- / o Gasification under development	+ Burners under development	- / o Gasification and fuel processing not yet developed
Other solid biomass	- / o Gasification under development	o Burner technology more difficult for less homogeneous fuels	- / o Gasification and fuel processing not yet developed
Liquid biomass, e.g., rapeseed or pyrolysis oil	+	+	- Reformers not yet developed
Biogas	++	++ Burners under development	+ Under development
Direct solar irradiation	--	+ Solar mirrors under development	--
Hydrogen (from various sources)	+	+	++

"--" = very low, "-" = low, "o" = neutral, "+" = high, "++" = very high

The combination of renewable and CHP systems is much easier in the case of district heating systems. In such systems, different heat suppliers, such as large solar thermal collectors, biomass boilers, CHP plants and conventional boilers, can be combined to jointly supply a heat network. District heating systems achieve similar or even higher CO_2 reductions than micro cogeneration (see Chap. 5) and have, in many cases, a similar economic performance (see Chap. 4). District heating is also of particular importance when a certain minimum size for a supply technology is required, as it is the case for combustion of straw, gasification of wood, long-term heat storage, geothermal energy, and the like (Nast 2003). However, in spite of these advantages, district heating systems suffer from other drawbacks including

- low acceptability, because there is a psychological perception that the heating generation system itself cannot be controlled by the house owner or tenant (e.g. whether it functions or not),
- lack of cost transparency, which is a problem for district heating because, in the initial phase of system development, when high capital costs occur, it is not clear how many customers will connect to the system,
- long implementation process,
- heat losses depending mainly on the number of connections per area[2], and
- a more difficult and costly realization in areas with an existing building stock[3].

For a detailed review of barriers to district heating systems, see for instance Nast (2003). With a general trend towards decreasing space heating demand, the realization of large-scale district heating systems will become increasingly difficult since the relative investment costs and heat losses of the distribution system will increase.

In order to avoid negative environmental impacts, the choice of location of micro cogeneration plants should be based on a prior assessment of competing fuel/technology options:

- If large quantities of renewable primary energy resources are available that cannot be used with micro cogeneration, the utilization of the renewable resources should be prioritized. For instance, if wood residues are available but not in a form suitable for micro cogeneration, a wood heat or larger cogeneration wood plant should be favored.
- If the area is suitable for district heating, a district heating system should be favored. Micro cogeneration plants should also not be operated in areas with existing district heating networks.

If these two conditions are not fulfilled, micro cogeneration may be an attractive technological option. Amongst the micro cogeneration technologies, the Stirling engine offers the highest fuel flexibility. Some of the possible Stirling fuels are especially available in developing countries,

---

[2] Whereas the average heat losses in district heating systems in Germany amount to 10 %, countries with higher shares of district heating even in rural areas and lower population density have losses of up to 20 %, such as Denmark, where 58 % of the houses are heated with district heat (Lauersen 2001).

[3] In development areas, where new buildings are being built, the realization of a high number of connections is easier.

such as solar radiation and small biogas applications, and feature the Stirling as a very promising technology.

Politically, it is important to provide framework conditions that prioritize district heat and the exploitation of renewable energy sources where these options are more appropriate.

9.5 Conclusions

Broad diffusion of micro cogeneration will primarily have positive benefits for energy supply security, both in terms of fuel supply security and in terms of the reliability of electricity networks. Micro cogeneration may reduce the dependence on fossil fuels by enhancing energy efficiency. However, not all micro cogeneration technologies increase energy efficiency substantially (see Sect. 5.2.1). Some engines are even less efficient than separate generation in a condensing boiler and a modern combined cycle power plant. Therefore, high total efficiencies are a prerequisite for positive impacts on fuel supply security and the environment.

In addition, it is important to make use of micro cogeneration only where competing fuel or technology options are less favorable. Whenever renewable energy resources can be used, they should be prioritized. Depending on the respective technology, micro cogeneration systems are also capable of using renewable fuels. However, this involves considerably higher investment costs and further research and development activities, as technologies such as biomass gasification, wood pellet burners, or vegetable oil engines do not yet have a sufficient degree of technical maturity in such small applications. In addition, in areas with high heat density, district heating will in most cases turn out to be more environmentally attractive and economically feasible.

The impact of micro cogeneration on the electricity network is mostly beneficial. As generated power is mainly consumed on-site, congestion of the distribution and transmission system is lowered. As a consequence, distribution and transmission losses are decreased and upgrades of the system may be deferred, if micro cogeneration plants are properly sited.

First analyses of the impact of micro cogeneration on the potential overload of equipment in a distribution network and power quality suggest that micro cogeneration plants could be installed broadly without causing major problems. Voltage is generally stabilized and is expected to remain within established limits even with a high share of micro cogeneration. Furthermore, micro cogeneration plants could provide some ancillary services. The provision of minutes reserve in spring and autumn appears possible, if plants are interconnected to virtual power plants and equipped with heat storage tanks. Even without interconnection to virtual power plants, micro cogeneration tends to result in grid relief and peak shaving due to the economic incentive to generate electricity during periods of high loads within a building and due to a certain coincidence of high heat demand and peak load situations in the grid. Thus, it is questionable whether the additional service benefit of connecting micro cogeneration units to virtual power plants is actually worth the additional effort and cost of communication and control.

There are also some limitations to micro cogeneration diffusion: the protection schemes for distribution networks may need to be adapted, with many distributed generators being connected. In addition, the total connection capacity for each LV network is limited – mainly due to LV/MV transformers – and depends also on the share of stochastic generation in the upstream MV network. These restrictions raise a number of questions:

- A limited connection capacity restricts free and non-discriminatory access to the network. While early movers may connect distributed power plants without infringing the connection capacity, the connection may at a later stage be denied or associated with prohibitively high costs for network reinforcement. As a first step in dealing with this problem, DNOs could be requested to regularly publish and update the available connection capacity (Krewitt et al. 2004). This would certainly increase transparency for all players on the market.
- In addition, rules may be developed to better allocate the costs or economic benefits to the operators of distributed power plants. For example, where a distributed power plant removes a local bottleneck, this should be rewarded through accordingly low (or negative) connection charges. However, such rules for economic incentives would need to be fair and transparent and should be harmonized at a national level in order to avoid arbitrary application by DNOs.
- DNOs should also be provided with adequate incentives to design networks in a manner to enable power generation by a larger share of distributed plants, for example with respect to protection concepts. Such incentives are currently not yet provided. In contrast, vertically integrated utilities, as most utilities in Germany are, have significant economic incentives to discourage distributed power generation by third parties (see Chap. 7).
- Micro cogeneration is the most dispersed type of distributed generation next to photovoltaics. In contrast to many other distributed power

technologies, power is consumed mainly on-site and not generated stochastically. This limits grid impacts considerably and should ease the connection of plants. In order to avoid inappropriate transaction costs for plant operators, connection requirements for very small micro cogeneration plants (e.g. 1 kW_{el}) should be simple and harmonized, for example within the European Union.

It is important to note that many of these issues are interpreted against the background of our current network, dominated as it is by a one-directional system layout and large-scale block-type power stations. With a future power system moving more in the direction of bi-directional electricity and distributed generation, both the backup generation structure and the transmission and distribution system have to be adapted to this development (Fischedick and Nitsch 2002) through, for example, the installation of more flexible gas power plants, appropriate system designs, and so on. By taking such measures, some of the impacts outlined above will be significantly reduced.

10 The Micro Cogeneration Operator: A Report from Practical Experience

Sylvia Westermann

10.1 Starting Situation

An owner in the single-family house or apartment building sector, a trader, a small business owner, or an organisation of like-minded members hears about the option of having an electrical supply from the boiler room of one's own. With enthusiasm, optimism, and information from print news, radio, and Internet he inquires at local heating companies, cogeneration manufacturers, and planning offices about installation options for his application. The status of knowledge and experience of the people he asks will pave the way for a possible installation of a combined heat and power plant on his property. In many cases, from a technical and business management point of view a cogeneration installation will be reasonable if those involved are in a position to optimise the facility so that it is practical. The analysis of the application, the selection of the facility components, and the long-term economical operation will, however, demand much more of the operator compared to the operation of a conventional boiler facility with full power supply from the local electricity suppliers. Experiences such as success, joy, disappointment, and optimism are still always closely related in this field.

One can read lots of informative facts about the use of micro cogeneration in relevant reference books and text books, journals, and informational material of the cogeneration manufacturers. On this basis, it should not be a problem for a planner or operator of a cogeneration facility to implement this innovative and environmental technology. The production of small, market-ready cogeneration modules has evolved from the development and research phase (see Chap. 1). Nevertheless, the selection is limited. In the Federal Republic of Germany – despite almost the same political and economic conditions – there are fields with a great deal and others with very little micro cogeneration use. Furthermore, cogeneration use is hardly mentioned and evaluated in the residential and commercial sections in the daily press. Instead, hopes for the boiler room have been placed in fuel cells, although it is not even close to being ready for the market, while cogeneration motors and Stirling installations are already on the market. Based on a personal experience report, this chapter examines the use of small cogeneration in the residential and commercial sectors.

These experiences were collected by cogeneration planners and manufacturers from the regions of Lower Saxony and Saxony-Anhalt in the field of micro combined heat and power generation, as well as with devices with up to 35 kW of electrical output. Over the last 15 years, they all devoted themselves to this technology.

In addition to the more or less economic success that has transpired, there were other motivations for getting involved professionally in micro cogeneration technology:

- Even if it is scarcely noticed, cogeneration operators make an active contribution to preserving the environment. Resource protection and climate protection by means of an efficient energy production is achieved directly at the consumer's site. But this advantage is not as noticeably tangible as using solar energy on the roof, and it therefore requires more instructional work and time.
- Cogeneration operators make a contribution to decentralising the power economy. If energy savings is extolled and even stipulated by law from the political side, the inner willingness of the population also has to be created for this. If electricity does not simply originate from the socket, but rather consumers are able to understand its complex production, then a more responsible attitude is more likely to develop with regard to the valuable energy type of electricity. Thus, operators of micro cogeneration get the feeling that it is better to operate electricity consuming devices (washing machines, dishwashers, and ovens) that are not time-dependent while heat is being produced, in other words while the cogeneration module is running. A new consciousness of energy use emerges, because the production and consumption processes can be understood on site.

Cogeneration operators help create alternatives to nuclear energy and at the same time promote climate protection. The first step in doing this is the rational, decentralized use of primary fossil fuels. The next step has to be devoted to substituting the use of fossil fuels.

However, these are not the arguments of the average population. Currently, there are still other motives for wanting to operate a micro cogeneration. From experiences with about 200 installations (higher than 5 kW of electric output), which have been put into operation, and in which cases consulting and planning of operational concepts took place, the following types of operators can be recognized:

The idealists would like to do something for the environment and against the monopoly of large power supply companies, based on their beliefs. They have initiative and new ideas that are, however, often difficult to realise, because the conditions for using cogeneration are unsuitable: the demand for electricity and heat is too low or their annual load profile unbalanced. Idealism has often been coupled with a low amount of investment. This resulted in "patchwork solutions" in order to cut costs and maintain customers, which, however, later resulted in significant additional costs or led to a standstill of the installation. This can lead to disappointments and to the negative image of cogeneration technology. In the successful case, the idealists are, however, independent, active participants in spreading cogeneration technology.

Ecological investors are operators that have sufficient financial funds. They want a prestigious facility for the environment, even rely on the statement that cogeneration modules are always economical, and are then even satisfied with a mediocre operational life due to unbalanced energy demand. With good planning, these operators are ideal partners for show piece facilities, but also for further developing the technology on site. Technical malfunctions and business management mistakes should not occur in this case, because they are difficult to communicate factually to the operators.

Independent operators are managing directors or owners of a home, trade business, residential building or the like. These operators make decisions independently, correctly based on facts, and conclusively according to the co-ordinated scope of consultation. During preparation and operation, they are the only contacts for technical and financial matters. This simplifies the co-operation. Projects are often implemented more quickly in these cases.

Operator groups – in the form of a non-trading partnership *(Gesellschaft des Bürgerlichen Rechtes,* GbR) or Limited Liability Company *(Gesellschaft mit beschränkter Haftung,* GmbH) the ideal case for apartment and home owners – require a higher effort of consultation and planning in terms of organisation and legal work, which should be carried out by competent and experienced experts. In this case, companies have several operators that have to be accommodated for one facility. The expenditure of time and processing is often greater, since all arguments and ideals of individuals have to be integrated into the project. Cogeneration and the financial and legal conditions associated with it place great demands on the realization of the entire concept. In this case, work costs ensue that are conclusive and easily comprehensible; however, they have to be organised and potentially co-financed. Surprisingly, cogeneration facilities were not scaled back over the last years for technical reasons, but rather due to mistakes in management, rendering of accounts, and the general framing of agreements.

For medium-sized enterprises, companies, and associations, in which several people are involved in the decision-making specifically related to the subject, similar conditions of application apply as they do for operator groups.

In the operation of cogeneration facilities by operator groups and organisations, overall comparably high costs arise, an effect which is primarily caused by the following aspects:

- The operator group becomes electricity and heat supplier over a time period of at least 10 years with all of the rights and obligations to be established in the shareholders' agreement such as accounting, annual balance sheets, heating cost statement, contractual negotiations with suppliers, tenant meetings, contractual relations with tenants, and management of missed payments.
- Sufficient time has to be allocated for increased costs of professional consulting, planning, and presentation.
- Various interest and work groups have to be co-ordinated and managed. An ultimately decisive leading personality or a specifically defined division of labour for the anticipated tasks are to be clarified at the latest upon commencement of operation of the cogeneration module. Too much democracy has already impeded several cogeneration facilities.

Costs for small cogeneration units require at least the same capacity as for cogeneration units of larger electrical output. However, the cost effectiveness in planning and support goes to the benefit of large facilities. The costs could only be minimized with standardizations and the uniform implementation of regulations.

Costs could also be reduced if micro cogeneration modules were operated by heat and energy service companies like for example, *city plants*. Although several heat suppliers advertise superficially with the option of heat supply by means of cogeneration, this type of operation does not often appear economical for customers. However, due to their management of electricity, city plants often have no economic interest in operating micro cogeneration themselves. When they do choose to operate them on occasion, this is usually a result of environmental-related energy policy in the city parliaments or from public pressure due to successful individual initiatives such as the aforementioned operator models.

Since the deregulation of the electricity market, the operation of small cogeneration by *contractors* has become more attractive. Several energy service providers offer joint self-production of electricity with cogeneration modules to various sites where heat is also sold. However, it seldom succeeds. This model is more likely known when such a company advertises actively on site and has previously not been very recognized in the micro cogeneration sector.

10.2 Options of Operating a Micro Cogeneration Unit

The future operator of a micro cogeneration plant has to make the choice of a suitable, initially primary, energy carrier (see Sect. 9.4). He is usually convinced of the technical and economic advantages of using natural gas. In this case, the planner or skilled worker can fall back on a technology that is most developed with the best cost effectiveness. The relevant cogeneration manufacturers can show a number of references and experiences with different cases of use.

The alternative case of a cogeneration module that uses heating oil is justified when the location of use has no natural gas or a heating facility operated with oil already exists. Micro cogeneration units that use oil require more maintenance, particularly with regard to fuel emission limits. The more costly storage and particularly higher investment, servicing, and maintenance costs stand in the way of the generally lower fuel costs. For this reason, cogeneration units that use heating oil are less in demand in inner city areas than in rural areas.

Micro cogeneration that uses other gases like biogas, sewage gas, and landfill gas are the exception for commercial cogeneration operators. They play almost no part in the residential sector. The cogeneration modules achieve their optimal operational capacity depending on the type and composition of the gas, although the fuel has to be supplied first. The additional costs are less of a burden on the cogeneration unit, but rather accrue due to the gas supply since a sufficient quality of fuel is to be produced. In this case, the cost-benefit ratio has to be estimated. The exchange of complete unit parts after only a 3-year period can be more sensible under the present electricity feed-in conditions than a costly gas supply.

In the case of biogenous gases, there is still the case of the wood gas cogeneration. These continue to be in demand particularly by operators in areas rich in woods or from the wood-processing trade and would present an attractive market as far as quantity is concerned. However, in the last 15 years, no wood gas processing installation for micro cogeneration has been able to establish itself on the market for a longer period so that the use of wood gas for electricity production seems to remain a futuristic vision for the time being.

Finally, the fuel cell cogeneration is worth mentioning, which is frequently extolled as the ideal case of decentralized generation. World-wide research reports of product developers, occasional pilot facilities, and in particular the promising publicity campaigns of various boiler manufactures that want to incorporate fuel cells in their product portfolios give reason for hope. Regardless of the fact that market readiness for a series production at consumer friendly prices is not yet foreseeable (see Sect. 1.2.3), it is apparently underestimated that the business management and political conditions of combined heat and power generation will also not even change with the use of such a future technology. The situation is even strongly manifested in the statement: if it is not possible to implement gas motor micro cogeneration economically today and to operate it in the long-term, micro fuel cells will also fail. The following reasons can be given for this:

- The technical links to the electricity and heating network and the administrative technical conditions (agreements between gas distributors, power suppliers, network operator and customers) are the same as in the case of micro cogeneration that is currently available on the market; at best there is a postponement in the performance ratio.
- If a gas micro cogeneration unit is installed in the basement today, it can be replaced with fuel cells tomorrow. The connection change is almost already prepared.
- The operator management for the self consumption and sale of electricity and heat for refinancing the investment costs is the same.
- The owner and landlord of an apartment building who installs decentralized heating thermae with a view to the accounting costs, or the single family home owner who prefers invisible heating, electricity and gas supply that is as government regulated as possible, is not the fuel cell operator of tomorrow.

10.3 Practical Experiences with the Concrete Operation of Micro Cogeneration

The *rendering of accounts and balance sheets* of micro cogeneration units should be managed correctly from planning up until operation. As a rule, the maintenance company can not afford that. The operation of the cogeneration module is to be evaluated at least for the first years. The expert knowledge of employees at accounting firms for cogeneration central heating units is often not sufficient for that. In analysing so-called crash facilities, i.e. facilities that were taken out of operation before their projected life (at least 10 years), this stood out repeatedly. The realisation of profits with a cogeneration module can be achieved due to the simultaneous power and heat generation with correspondingly high primary energy and maintenance costs. For the most part, this is not to be overlooked "instinctively" by the operators and consumers. Unfortunately, experience has demonstrated that cogeneration operators are not generally aware what their unit has earned and on the other hand are "annoyed" by the multiple accounting arrangements even in the case of external support from accounting firms.

In comparison to the boiler facility, the *maintenance and monitoring* of the micro cogeneration unit makes greater demands on efforts and costs. The modules are more comparable to stationary cars that want to be lovingly looked after and maintained. The maintenance for micro cogeneration engines is done about twice a year (after approx. 2,000-3,000 full operation hours). It is organised independently by the maintenance contractor in corresponding maintenance agreement arrangements. Both partial and full maintenance is offered, which is also in part only designated as agreements for services. If the operator of his own expert personnel can demonstrate or is additionally trained by the micro cogeneration manufacturers, he can carry out the maintenance work himself at more cost affordably (e.g. oil change).

The optimal variants for the operator include, however, full maintenance agreement for a period of time of at least 10 years including remote monitoring of the entire facilities. The costs are calculated according to operation hours and electrical work generated and can thus already be taken into consideration in the planning. These agreements contain the servicing, maintenance, and overhaul of the entire module. That means that during the entire term of agreement, a fully adequate cogeneration module is available. For example, motor overhauls (after 3-5 years) or heat exchanger overhauls are carried out independently by the maintenance company without additional organisational and financial expenditure for the operator. At approx. € 200, the annual maintenance price is relatively high for this service but it is understandable. The estimated cost of the maintenance and overhaul of a micro cogeneration module over its lifetime is as much as the original investment in the facility itself. The purchase of a cogeneration module and the conclusion of a full maintenance agreement should thus be considered a "unit".

The remote monitoring – also of micro cogeneration – should be agreed upon as a standard. Malfunction detections are sent directly to the cogeneration manufacturers or, respectively the maintenance company, and can often be remedied by remote control without the operator knowing about it.

Lastly, a top level control of heat is of utmost importance for a rational cogeneration operation. It has to be capable of organising the cogeneration module and peak boiler operation technically and economically as best as possible. For example, boilers should only be switched on when it is absolutely necessary, frequent start stop procedures have to be avoided, and a performance regulation for the cogeneration unit has to be able to function according to heat and power demand. These are often requirements that traditional controls from the boiler sector cannot (yet) fulfil.

In general, the hydraulic integration of the cogeneration unit will facilitate its basic heat load generation. Hydraulics and control therefore have to be conceptualised jointly.

In this situation, one should not be sparing with finances and quality. In order to estimate the expenditure and use, economic viability calculations are always made for combined heat and power generation plants. The entire concept has to be right: micro cogeneration units require a coordinated hydraulic facility with an intelligent cogeneration module and heat control in order to be able to achieve calculated economic viability.

10.4 Practical Experiences with the Regulatory Framework

For the economic operation of cogeneration plants, the operator has to prepare for a series of political, but also free enterprise framework conditions (see Chap. 8).

The *mineral oil tax law* guarantees a compensation of mineral oil taxes of the primary energy carriers used in micro cogeneration. To be eligible for this, the operator has to register the facility accordingly and receive approval. Once a year, the operator reports the compensation to the customs office by filling out a form and turning in the invoice copies of the energy carrier suppliers.

In practice, it often results in problems. For a specific cogeneration plant with an apartment building complex in Saxony-Anhalt with its own power use, an external review on location was additionally conducted. For the onsite inspection, the customs office required copies of the accounting, annual financial statement, invoices to customers, the agreements with the city plants as well as an office space for the external review, and all that each year.

Overall, the facility operates today without profit, since too little power is generated for consumption for itself. Further causes for this are planning mistakes during the agreement negotiations with city plants. In addition, before the coming into force of the new energy industry law (*Energiewirtschaftsgesetz*, see Sect. 8.1.1), the sale of electricity to tenants was not approved vis-à-vis the electricity supervisory authority. The expenditure of time and money necessary for the annual audit is disproportionately costly for the employees of the customs office, the operators and the tax consultant, and is seen sceptically by potential new cogeneration operators. We are faced with the problem of suitability.

The Combined Heat and Power Law *(Kraft-Wärme-Kopplungsgesetz*, see Sect. 8.1.3) guarantees a minimum compensation for electricity fed into the grid. For this purpose, an application with associated costs has to be filed at the Federal Office for Economy and Export Regulation (BAFA). For a well-designed micro cogeneration unit whose electricity is consumed on site and for the most part does not feed into the public network, the processing costs can definitely correspond to the feed-in compensation in one year. The suitability of micro cogeneration is doubtful even in this case. Furthermore, evaluations of economic efficiency have revealed that for a full input of the generated power, the legal CHP compensation for an economic operation is insufficient (Chap. 4). The operator is forced to consume the electricity himself or, otherwise sell it directly to consumers in the vicinity, bearing all consequences such as supply agreements, risk management, missed payments, accounting, tendering of account, etc. His profit margin arises in this case from the comparatively high and continually increasing electricity prises for small consumers. The law did not achieve the aim of structuring and simplifying the operation of micro cogeneration cost effectively with full input by means of a secured compensation. Only a secured minimum compensation in the context of the consumer-related costs for power generation was achieved.

Cooperation with *network operators and power suppliers*, in whose supply territory statistically few micro cogeneration units are installed, has generally become costly and time-intensive. It presupposes that the operator and his partner have expertise for the general framing of power agreements. Additionally, the agreement arrangements of the power supplier reveal significant differences and often too much is expected of the employees in matters concerning the cogeneration module. They see the suitability of micro cogeneration in the residential and commercial sector in relation to professional power producers in the greater cogeneration sector.

An extreme example will clarify the bureaucratic adversities that can arise in operating a micro cogeneration plant: for a facility that has been shutdown in the meantime, excessive input compensation was awarded due to the fact that compensation was paid according to the CHP law for which no authorisation had been applied for at the BAFA. In addition, three phone calls were made with the request of obtaining authorization with associated costs subsequently on a short-term basis. The case in dispute amounted to 30 €. For the same facility, there was no commencement of operation protocol, because the electricity supplier had changed the structures, the operator changed the responsibilities, manufacturer and seller are no longer on the market, and the installation company filed for bankruptcy a couple years ago. The facility was operated several years. It can be proved that the input compensation was paid, but the BAFA refused to authorise the input compensation according to the CHP law, since the commencement of operation protocol was missing. The operator from the lodging and hotel trade still has year-round balanced heat and electricity consumption today, in which a micro cogeneration unit could be operated cost-effectively at least 6,000 h per year. He has lost confidence in cogeneration technology due to the above framework conditions.

The employees of the electricity supplier are first trained accordingly at a certain number of cogeneration units in the respective area, however, also later with a view to limiting electricity producers and electricity consumers of their own by means of targeted "consulting". In addition, the conditions do not change unavoidably if the electricity supplier also supplies gas, as is frequently the case in city plants. The monopoly of suppliers presents a feature that is not to be underestimated in founding companies for power supply by means of cogeneration.

The disadvantage of micro cogeneration in the residential and commercial sector can take on different forms. This is demonstrated with several, specific examples:

- A training facility operates a micro cogeneration module. The power supply agreement for the excess electricity supply was concluded with the arrangement that no electricity can be sold to the grid. In fact, the electricity feed-in will not occur since only the constant load of electricity demand by the micro cogeneration module is covered. However, by expanding the supply to neighbouring buildings using previously decentralised heating supply (wall thermae), a second module would also still appear cost-effective, provided that the subsequent surplus electricity arising with the heat-powered operation would also be taken on and remunerated by the electricity provider. Even if the power supply agreement is currently legally contestable, out of respect for the bureaucratic expenditure of time and cost, the operator has no interest in expanding the facility.

- The planning and already commenced construction of a cogeneration facility for an apartment building complex in a small town has been halted. The preliminary examinations revealed that the facility can only be operated cost-effectively for electricity use of its own, i.e. sale of electricity and heat to all tenants. This was also consistent with the housing society, which was interested in an ecologically-oriented electricity supply. However, the responsible city plants were not interested in the model: due to a relatively expensive bid for additional electricity and gas, the guarantee of a standard local electricity price for the tenants was at risk. Moreover, at this time many providers of alternative electricity advertised in the public with better conditions and city plan employees coincidentally lived in the respective buildings.

10.5 Public Acceptance

Although support schemes like funds for compensation and reimbursements provide the operator temporally limited economic security, they are more likely to harm acceptance of micro cogeneration in society. Since the CHP bonus split is revealed on the invoices of the electricity suppliers, the normal electricity consumer develops the opinion that a lot of money can be earned with cogeneration at the cost of the people.

Micro cogeneration units are power plants for the small consumer and should stand up equally against boiler, heat pump, and solar facility. However, in the everyday information material, in the news and construction magazines, micro cogeneration can hardly be found in comparison to renewable power generation. The information is usually distributed by means of propaganda by word of mouth or journals and trade fairs. If this technology is to have more application in the residential and commercial sector, a good deal of information and education is still necessary.

10.6 Conclusions

The main problems of micro combined heat and power generation can be derived directly from the functional principle of simultaneous generation of electricity and heat for the small electricity consumer on one's own.

Generating electricity on one's own is up against the economic interests of monopolised electricity suppliers. That means that the spread of micro cogeneration is in the hands of cogeneration manufacturers and representatives of ecological and sustainable energy policies.

Micro cogeneration will continue to face these problems for a while. The advocates of this technology have to be aware of this and be able to accept it. However, only long-term decentralisation of electricity industry on the foundation of a electricity generation mix can provide micro cogeneration with a significant upturn.

On a short-term basis, a rapid increase in the electricity price for small consumers can have an advantageous effect on the use of micro cogeneration. However, even then the subsequent hindrances would have a limiting effect on potential operators: First, the investment and maintenance costs are significantly higher than that of conventional heating; they only make themselves profitable with a long-term and a critical approach to the situation. Second, the technical requirements of heating with expanded functions are more extensive, particularly due to conditions of energy policy and open economy. A generally applicable standard solution has yet to emerge.

Greater confidence in micro cogeneration can only be created by its successful application. The multitude of already existing micro cogeneration manufacturers, providers, planners, and operators today gives reason to hope that such a positive development is well on its way.

11 Micro Cogeneration in North America

Jon Slowe

11.1 Micro Cogeneration Status in North America

Micro cogeneration is at a very early stage of development in North America. This chapter explores the framework, potential for and development of micro cogeneration, principally in North America.

In general, the North American environment is more challenging for micro cogeneration than European and Japanese markets, for a number of reasons:

- The predominant heat distribution systems in North American homes are forced-air distribution, rather than hydronic (water) distribution systems. As heat is recovered from micro cogeneration systems in the form of hot water, an added step is needed to transfer heat from this hot water to air in the distribution system. This adds complexity and cost.
- The cost of warm-air heating systems (usually known as a furnace) is typically significantly lower than boilers, meaning that the differential between a furnace and a micro cogeneration system is higher than the differential in Europe and Japan between a boiler and a micro cogeneration system.
- Across much of North America electricity prices are low, making it harder for micro cogeneration systems to deliver any or significant energy cost savings to end users.
- The electricity market is very fragmented in North America, with a large number of utilities and each utility often having their own interconnection requirements, and sometimes imposing standby and other charges that add cost for micro cogeneration users.
- Across much of North America reduction in carbon dioxide emissions are not valued as much as they are in much of Europe and Japan. Emissions of local air pollutants such as nitrogen oxides are typically of more importance than carbon dioxide emissions.

- Homeowners are likely to place significant value on the ability of micro cogeneration systems to continue generating power if the utility grid fails – something that some micro cogeneration systems developed for the European and Japanese markets are not able to do – as, on average, grid interruptions in the U.S. are more common than in Europe or Japan.

There have been a couple of aborted attempts to develop micro cogeneration systems in the U.S. over the last 15 years. However the spotlight has started to shine on this area again in the U.S., with the U.S. Department of Energy taking an active role in stimulating the development of products and markets. In 2003 they published a Micro Cogeneration Road Map, and in 2004 they awarded four grants to companies to develop micro cogeneration systems for the North American market. In 2005 Climate Energy, based in Massachusetts, unveiled a micro-CHP system built around Honda's internal combustion engine, planning to test units and bring them to market in 2006.

11.1.1 Previous Attempts to Develop Micro Cogeneration

In the early 1990s, the Gas Research Institute teamed up with Kohler, a manufacturer of engines and bathroom furniture, to develop a gas engine cogeneration system for households. The system was built around a 5 kW spark-ignition Yanmar engine. A prototype was developed that successfully ran in a National Association of Home Builders test "smart house" just outside of Washington D.C. for a couple of years, but the project stopped there – the developers just couldn't see the economics working to take the system to market.

Later in the 1990s, Intelligen Energy Systems, a small firm based in Massachusetts, developed and commercialized an oil-fired 5 kW internal combustion engine micro cogeneration system. From 1994, over 100 of these units were sold in New England over a four year period. They also worked on developing a natural-gas fired system, but in 1998 Intelligen ran out of money and the company was forced to close its doors. In trying to ramp up sales, the company faced challenges in securing a distribution, installation and servicing infrastructure, as well as challenges in marketing its product. It also ran into interconnection and other regulatory barriers.

Two micro cogeneration product developers secured agreements to install large numbers of micro cogeneration units, but were unable to fulfill these agreements.

Fuel cell developer H-Power agreed a US$ 81 million deal with Energy Co-Opportunity (ECO) in 2001. ECO was set up as an energy services co-operative of more than 300 rural electric co-operative utilities serving more than 18 million customers. This agreement was primarily for the delivery of residential cogeneration units. Subsequently, H-Power was bought by Plug Power and the deal terminated.

Energy technology development company Ocean Power, having acquired Stirling engine micro cogeneration developer Sigma, announced a deal in 2002 with a private real-estate management company. This deal established the conditions for the sale of 10,000 of their 3 kW Home Power Units, with the installation of these units planned to be complete by the end of 2005. Subsequently Ocean Power ran out of money and filed for bankruptcy.

11.1.2 The Market Today

A small number of companies are developing micro cogeneration systems for the residential market in North America, with one company recently having embarked on selling its 5 kW electrical product to large households and small commercial establishments such as apartment buildings, and another planning to bring a 1-kW product to market in 2006.. The installed base of field-tested and commercially sold micro cogeneration products in North America is estimated to be below 100 systems. The current market is further explored later in this chapter.

11.1.3 U.S. Department of Energy Action

Following a meeting of stakeholders in June 2003, the U.S. Department of Energy published the "Micro Cogeneration Technologies Roadmap", which contained the target that

> "by 2010 environmentally friendly, cost-effective, versatile, reliable, fuel flexible micro cogeneration appliances will be commercially viable for the American residential marketplace."

The driver for the development of this Roadmap was micro cogeneration's ability to meet national energy priorities – energy efficiency, environmental emissions, fuel diversity and energy assurance. The details of this Roadmap are described later in this chapter.

11.2 Framework for Micro Cogeneration

11.2.1 Heating and Cooling Markets

Natural-gas-fired furnaces dominate the U.S. space heating market, although in the north-east of the country there is market for boilers fuelled both by natural gas and fuel oil. It is common practice in much of the U.S. to have separate appliances for space heating and domestic hot water.

In 2003, over 3 million natural gas furnaces were sold in the U.S, with oil-fuelled furnaces accounting for an additional 120,000 units. This compares to only 400,000 boilers sold in the same period – of which approximately 60 % were natural-gas-fuelled, and 40 % oil-fuelled.

For new homes, forced-air systems predominate, powered by furnaces and heat pumps. These systems are found in 94 % of the 1.26 million homes built in 2001.

In terms of existing housing stock, the fuel used for heating and type of heating equipment used is shown in Table 11.1. There are approximately 107 million households in the U.S.

Table 11.1. Fuels and heating equipment of the existing U.S. housing stock

Residential primary heating fuel	Main heating equipment	Percentage of households
Natural gas	Central warm air furnace	41
	Steam or hot water system	7
	Floor, wall, or pipeless furnace	3
	Room heater / other	4
	Total natural gas	**55**
Electricity	Central warm air furnace	12
	Heat pump	10
	Built-in electric	6
	Other	2
	Total electricity	**30**
Fuel oil	Steam or hot water system	4
	Central warm air furnace	3
	Other	<1
	Total fuel oil	**8**
LPG	Central warm air furnace	3
	Room heater	1
	Other	1
	Total LPG	**5**
Wood	Stove, fireplace	**2**
Kerosene, other		**1**

Cooling equipment is also commonly found in U.S. households. Central air-conditioning is increasingly common in households, installed in some 43 % of all households, and the vast majority of new homes. 10 % of households use heat pumps to provide cooling. Other households (22 %) have room air conditioners, and 24 % of households have either no cooling equipment or do not use air conditioners that are installed in their homes. Cooling is therefore a very important part of household energy use, and the development of micro cogeneration systems that can also provide cooling will significantly help micro cogeneration to penetrate the U.S. market.

11.2.2 Electricity Industry

The electricity industry in the U.S. is extremely fragmented, with approximately 3,100 of investor-owned, municipal (government)-owned and co-operative utilities across the country. Investor-owned utilities, numbering around 240 (EIA 2000), serve around three-quarters of U.S. households.

Each state has its own regulator, usually known as Public Utility Commissioners or Public Service Commissioners, responsible for regulating the retail and distribution element of the electricity industry. In many cases, municipal and co-operative utilities do not fall under such regulations.

Coal is the major fuel source for electricity generation, with just over 50 % of all power generated from this fuel. Nuclear and natural gas each provide around a fifth of all electricity, with fuel oil, hydro and other renewables providing around a tenth. The total installed generating capacity across the U.S. is 948 GW.

One of the major differences between electricity markets across the U.S. and those in western Europe is the timing of peak electrical demand. In western Europe peak demand is found in the winter months, exactly when heat-led micro cogeneration systems are likely to be generating electricity. In the U.S., the highest peaks mostly occur in summer months due to the demand for air-conditioning in buildings, when heat-led micro cogeneration systems are not likely to be operating, or at least not for many hours.

Another difference between Europe and North America is that the transmission and distribution system across much of the U.S. is extremely stressed and operating at near full capacity across much of the country. With electrical demand growing at some 2 to 3 percent a year, significant investment is required in transmission and distribution infrastructure.

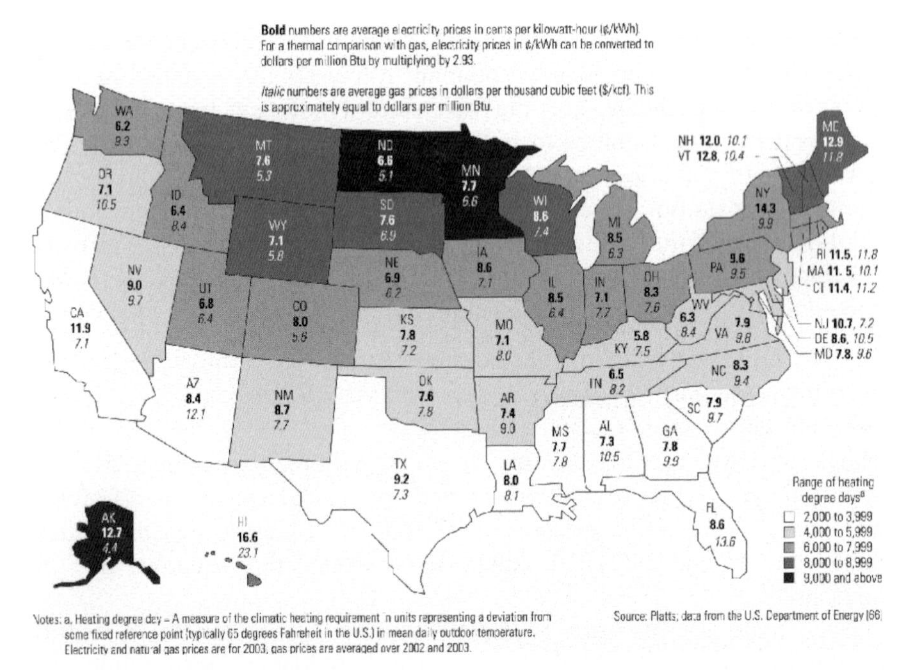

Fig. 11.1. Heat-led micro cogeneration systems have best potential in cold climates with high electricity prices and low gas prices

Retail electricity prices vary significantly across the U.S. Across much of the central and southern parts of the country, they lie between 7 and 8 US ct/ kWh. However in California the figure rises to 12 US ct/ kWh. In New York State it is as high as 14 US ct/ kWh, and above 11 US ct/ kWh for much of the northeast of the country.

11.3 Micro Cogeneration Policy and Incentives

11.3.1 U.S. Department of Energy Micro Cogeneration Policy

The U.S. Department of Energy's Micro Cogeneration Roadmap sets out a target that, by 2007, micro cogeneration systems will need to achieve 5 to 7 year simple paybacks and target costs less than US$ 500/ kW (adjusted to reflect thermal and electrical values) incremental cost for the mass market. Three primary areas are earmarked for action:

- **Define markets.** The Roadmap notes that the U.S. has a very diverse set of regions and types of residential buildings, with many variables affecting the energy needs of residential buildings. These needs will

define the market potential and drive the technical requirements in developing micro cogeneration systems.
- **Develop technologies.** The Roadmap notes that developments in energy storage, cooling, controls and integration technology will enhance the performance and operation of micro cogeneration systems.
- **Accelerate acceptance.** Help to develop standards, execute pilot programs and give utilities, regulators and consumers the information needed to ensure a friendly infrastructure for micro cogeneration systems.

Following the publication of the roadmap, the Department of Energy invited applications for funding to develop micro cogeneration systems. In 2004 four awards were made, as follows:

- boiler and furnace manufacturer ECR International, who is developing a 3 kW Rankine cycle micro cogeneration system,
- TIAX, a research and development company, working on a 2 kW Stirling engine system,
- Advanced Mechanical Technology Inc., working on combining the 5 kW Ecopower internal combustion engine micro cogeneration system (currently being sold in Europe) with a desiccant chiller coupled to a conventional air-conditioning system, and
- United Technologies Research Center, looking at micro cogeneration equipment selection and evaluation for optimized residential systems providing heat, power and cooling.

11.3.2 Solid State Energy Conversion Alliance

The "Solid State Energy Conversion Alliance" (SECA), initiated in 1999 as a unique alliance between government (the U.S. Department of Energy), industry, and the scientific community, seeks to accelerate the commercialization of low-cost solid oxide fuel cells in the size range 3 kW to 10 kW. A number of SECA industry partners, including Siemens Westinghouse, Acumentrics, FuelCell Energy and Cummins Power Generation are leading the development of solid oxide fuel cell systems. The target cost under the program, by 2010, is US$ 400/ kW.

11.3.3 U.S. Cogeneration Policy

The U.S. Department of Energy (DoE) has a policy to increase the amount of cogeneration, aiming to double the 46 GW of cogeneration capacity in

Table 11.2. States with micro cogeneration technologies qualifying net metering (Source: Platts, data from U.S. Department of Energy and DSIRE)

State	Micro cogeneration technologies qualifying for net metering	Size of distributed generation unit and type of customer eligible for net metering	Comments
Arkansas	Fuel cells	Less than 25 kW, residential customers	
Connecticut	Fuel cells	Less than 50 kW, residential customers	Investor-owned utilities must offer net metering
District of Columbia	Fuel cells	Less than 100 kW, all customers	
Georgia	Fuel cells	Less than 10 kW, residential customers	
Idaho	All technologies	Less than 100 kW, all customers	Applies to investor-owned utilities. Residential customers who demand net metering must have $1,000.000 liability insurance.
Louisiana	Fuel cells	Less than 25 kW, residential customers	
Maine	Fuel cells	Less than 100 kW, all customers	Previous regulation included cogeneration systems.
Massachusetts	All technologies	Less than 50 kW, all customers	
Minnesota	All technologies	Less than 40 kW, all customers	
New Mexico	All technologies	Less than 10 kW, all customers	Does not include municipal utilities.
North Dakota	All technologies	Less than 100 kW, all customers	
Ohio	Fuel cells	No maximum output, all customers	Investor-owned utilities only.
Oklahoma	All technologies	Less than 100 kW, all customers	
Oregon	Fuel cells	Less than 25 kW, all customers	
Rhode Island	Fuel cells	Less than 25 kW, all customers	
Utah	Fuel cells	Less than 25 kW, all customers	
Vermont	Fuel cells	Less than 15 kW, all customers	
Washington	Fuel cells	Less than 25 kW, all customers	
Wisconsin	All technologies	Less than 20 kW, all customers	Not compulsory for rural co-operatives, but they usually follow the net metering recommendations voluntarily.

1998 to 92 GW of capacity in 2010. As of September 2004, the DoE reported 80 GW of cogeneration capacity, around 82 % of this installed at industrial facilities. The DoE has not published or sponsored any reports looking at the overall potential specifically for micro cogeneration.

11.3.4 Net Metering

Net metering is a widely used term in the U.S., but the term is sometimes used to define a wider set of arrangements than situations where electricity exported to the grid is paid the same price as electricity imported from the grid. In some situations it is used to refer to arrangements where a fixed price is agreed for electricity exported to the grid, which could be lower than the price paid for electricity imported from the grid. Net metering arrangements pertinent to micro cogeneration for a selection of States are shown in Table 11.2.

What this table does not show is that in many cases the amount of net metering that utilities must accept is capped at very low levels, limiting the use of net metering to stimulate mass market products.

11.3.5 State-level Incentives

Some States have introduced policies to support the development of distributed generation in general. None of these policies are specifically targeted at developing micro cogeneration markets, but some of them are helpful for micro cogeneration. Examples include:

California

In 2001 the California Public Utilities Commission introduced a Self-Generation Incentive Program. This Program provides financial incentives to residences and businesses to generate their own power. Incentives are dependent on the type of technology, and are offered as detailed in Table 11.2.

New York

In July 2004 the New York Public Service Commission approved the introduction of special gas rates for residential distributed generation systems. The Commission Chairman William M. Flynn stated that they wanted to

Table 11.2. Californian Self-Generation Incentive Program

Eligible Technologies	Incentive Offered ($/Watt)	Maximum % of project cost	Minimum System Size	Maximum System Size
Fuel cells operating on non-renewable fuel and running in CHP mode	$ 2.50/watt	40 %	None	1.5 MW
Internal combustion engines operating on non-renewable fuel, running in CHP mode	$ 1.00/watt	30 %	None	1.5 MW

"create opportunities here in New York for new technologies and new ways to effectively maximize energy resources (Flynn 2004)".

Previously, the Commission had directed the major natural gas utilities to offer special gas rates for businesses using distributed generation technologies. The Commission is also gathering data on natural gas usage by residential distributed generation technologies.

Massachusetts

Of all the net-metering laws in the U.S., the law introduced by the Massachusetts Department of Telecommunications and Energy in 1997 is possibly the most meaningful for micro cogeneration, because it applies to all technologies (many net metering laws in the U.S. apply only to fuel cells or renewables), and it specifically prevents utilities from imposing punitive backup charges and liability insurance charges.

It reads as follows: a customer of a distribution company with an on-site generation facility of 60 kilowatts or less in size has the option to run the meter backward and may choose to receive a credit from the distribution company equal to the average monthly market price of generation per kilowatt hour, as determined by the department, in any month during which there was a positive net difference between kilowatt hours generated and consumed. Such credit shall appear on the following month's bill. Distribution companies shall be prohibited from imposing special fees on net metering customers, such as backup charges and demand charges, or additional controls, or liability insurance, as long as the generation facility meets the interconnection standards and all relevant safety and power quality standards. Net metering customers must still pay the minimum charge for distribution service (as shown in an appropriate rate schedule on file with the department) and all other charges for each net kilowatt hour delivered by the distribution company in each billing period.

11.4 Development of Micro Cogeneration Products in the U.S.

Due to the difficulties in developing micro cogeneration markets in North America, there aren't many companies working specifically on micro cogeneration systems for residential applications. Only one company, Vector Cogen, is currently offering a product specifically designed for micro cogeneration applications. They are offering a 5 kW and 15 kW internal combustion engine micro cogeneration system, and have sold around 40 of these systems to large residences and small commercial energy users, primarily in California.

A small number of other companies are developing micro cogeneration systems. Of these, two companies are looking at selecting OEM micro cogeneration technology rather than developing their own technology, a third is using the Ecopower micro cogeneration unit. Only ECR International are developing their own micro cogeneration technology, using a water-based Rankine Cycle. ECR are focusing their efforts through Climate Energy, a joint venture with engineering company Yankee Scientific. In 2005 Climate Energy unveiled a micro cogeneration unit built around Honda's internal combustion engine unit. This provides forced-air heating, and therefore is able to penetrate the U.S. furnace market. They also plan to bring out a hydronic micro cogeneration system to penetrate the smaller boiler market, as well as continuing to develop their Rankine Cycle technology for micro-CHP applications.

Most of the others are working on fuel cells, and these companies are mostly focused on applications other than micro cogeneration. Several fuel cell companies stated, at the end of the 1990s, that they hoped to shortly commercialize fuel cells for residential applications. On the back of this, companies such as Plug Power raised tens of millions of dollars. Today, fuel cells remain several years away from market for residential micro cogeneration applications. Leading fuel cell companies that may target the U.S. micro cogeneration market include:

- **Plug Power.** Around one hundred of their PEM fuel cells have been installed on Long Island, New York. The majority of these operate in electric-only mode at a Long Island Power Authority substation, with others installed running in combined heat and power mode at managed housing (owned by organizations such as hospitals) and commercial property (including a McDonalds restaurant). Plug Power are currently offering a hydrogen-fuelled fuel cell unit to provide standby services to the telecom market, and are working with European boiler manufacturer Vaillant on a combined heat and power unit for residential applications in Europe. In the U.S., they are targeting the off-grid market before targeting the grid-connected residential market. One of their main investors is Detroit Edison, who sell electricity and gas in the Great Lakes region.

- **Acumentrics.** Developing solid oxide fuel cells, one of Acumentrics' strategic investors is Northeast Utilities, who own retail electricity and gas companies in northeastern U.S. Acumentrics are currently testing early-stage models of their fuel cell systems.
- **IdaTech**. Owned by IDACORP, the holding company for electric utility Idaho Power, IdaTech are developing PEM fuel cells, and are working with RWE Fuel Cells to develop a product for the European market. Previously, they had field tested a number of units in Northwestern U.S. and were working with Bonneville Power Authority, but this is no longer the main focus of their work.

Other North American companies developing fuel cells include ReliON, part-owned by Avista Utilities; Ballard, who are focusing on the Japanese market for micro cogeneration through their joint venture with Ebara; Nuvera, testing a 5 kW product in the U.S. and Japan; Fuel Cell Technologies, working on solid oxide fuel cells for micro cogeneration applications; and Versa Power Systems, developing solid oxide fuel cell technology.

11.5 Conclusions

The U.S. and Canada are currently lagging European and Japanese micro cogeneration developments, partly due to differences in heating systems, energy prices and policy frameworks. However, increasing interest from a number of organizations, including the U.S. Department of Energy, is likely to see increasing activity in the next few years. Initially, activity is likely to be focused on niches such as off-grid applications, or in States where there is relatively strong policy and regulatory support for micro cogeneration. Widespread commercialization of micro cogeneration is likely to take several years, as a number of significant barriers still need to be overcome.

12 Micro Cogeneration in Britain

Jeremy Harrison

12.1 Political and Economic Framework

The UK, with its temperate climate, substantial population and number of households connected to a natural gas infrastructure, is ideally suited to the application of existing micro cogeneration technologies. Whilst some other nations are seeking to develop systems with somewhat different characteristics to match their specific needs, the heat to power ratio of Stirling technology aligns well with the electricity and heating demands of typical UK homes and results in relatively long annual operating hours. This, allied to the high price of electricity, both relative to gas and in absolute terms, produces savings more than adequate to recover the initial investment. Furthermore, the additional benefits to the UK as a whole, match well with government aspirations in respect of social, economic and environmental targets.

12.1.1 UK Energy Policy

The UK has been amongst the leading countries committed to combating Climate Change, and committed to achieving a CO_2 reduction of 20 %, more ambitious than the 12.5 % obligation imposed under the Kyoto protocol, although it is likely that only 14 % may be achieved (EEA 2004). Beyond this, following recommendations by the Royal Commission on Environmental Pollution (RCEP 2000), a further commitment to achieve reductions of 60 % by 2050 has been made by the current government. However, the significant early success stories, arising primarily from the abandonment of coal fired generation and the so-called "dash for gas" have largely been swallowed up by organic growth in consumption alongside economic growth and uncontrolled growth in transport emissions.

The publication of the Energy White Paper (2003) proposed extravagant targets for CHP and Renewables, somewhat diluted in the final Energy Act

(2004) which no longer contains specific targets nor obligations for micro cogeneration, although there is an obligation to produce a micro generation strategy by the end of 2005.

Energy Act 2004. The UK Energy White Paper (draft policy document) identified four priority areas:

- security of supply, "to maintain the reliability of energy supplies",
- environment (Carbon abatement), "to put ourselves on a pathway to reduce carbon emissions by 60 % by 2050 with significant progress by 2020",
- fuel poverty, " to ensure that every home is adequately and affordably heated", and
- competition: "to promote competitive markets…sustainable economic growth...improve productivity".

It is clear that there are major conflicts between these goals, particularly the impact of minimizing energy cost on the viability of energy efficiency measures. Although some commentators believe it is possible to reduce energy cost and encourage energy efficiency, it is difficult to see how this can be achieved within a free competitive market. It is for this reason that the UK has developed a number of vehicles to address this issue, such as the Energy Efficiency Commitment (EEC) on energy suppliers. The document also importantly identified the potential contributions of a number of carbon mitigation measures and their relative cost effectiveness. Whilst heavily subsidized technologies such as photovoltaics (PV) had a very high cost (£520-£1250 per ton CO_2 reduction), energy efficiency and micro cogeneration were identified as the most cost effective measures under current conditions with a negative cost of up to £630 per ton (PIU 2002).

As micro cogeneration contributes to all four policy goals, the UK government considers micro cogeneration favorably, although there remain skeptics within some of the departments and support measures so far implemented are meager to say the least. The Act does at least provide an affirmation of support for micro cogeneration and the obligation to produce a micro generation strategy should provide a policy framework for micro cogeneration.

Micro generation strategy. The obligation to develop a micro generation strategy imposed by the Energy Act extends to any generation source below 50 kW_{el} and 45 kW_{th}. It covers CHP, PV, wind and hydro as would be expected, but must additionally consider not only generation of electricity, but also of heat from renewables as well as heat pumps. The strategy must consider the potential contribution to CO_2 and fuel poverty targets, and the potential reduction in demands on the energy networks.

CHP strategy. The recently published CHP strategy (DEFRA 2004) maintains the target of 10 GW_{el} by 2010, set in 1999, when it appeared the earlier target of 5 GW_{el} would be exceeded by 2000. However, unlike renewables, there is no obligation on suppliers to support CHP, and the revised Cambridge Econometrics study (Cambridge Econometrics 2003) shows that even under the most attractive scenario, the 10 GW_{el} target will be missed by a substantial margin. (Interestingly enough, the earlier targets for CHP did not even include micro cogeneration, which is now estimated to provide up to 500 MW_{el} by 2010). Sadly, the UK has so far failed even to reach the earlier 5 GW_{el} target. CHP capacity is stagnant and output is reducing.

It might therefore appear that the negative market environment for CHP generally would be even more severe for micro cogeneration, which faces additional manufacturing and service costs. However, the main obstacles to larger CHP, namely the "spark-spread" between gas and electricity, and the uncertainty of electricity export prices under NETA[1] do not apply to micro cogeneration. Due to the high heat to power ratio of micro cogeneration (typically around 1:8), the additional gas burn required to produce electricity compared with a conventional gas boiler is negligible; even compared with a condensing boiler, it is relatively small. As a result micro cogeneration is virtually cost free in terms of primary fuel input. It is also the case that the spark-spread in the wholesale market has not impacted the domestic ratio where gas is still only around 20 % of electricity prices. In addition, the majority of the value of micro cogeneration to the end user comes from the avoided cost of import, not the value of export; this value is substantially higher in the domestic sector (and is less susceptible to price volatility) than for commercial CHP operators.

The relatively poor market conditions for CHP and the desire to see a contribution in the domestic sector, have led the government to provide a £50million subsidy to Community Energy, although nothing for micro cogeneration.

---

[1] The New Electricity Trading Arrangements (NETA), were intended to overcome perceived abuse of the wholesale market (Pool) by generators. The resulting arrangement exposes buyers and sellers to the risks of "imbalance" on the margin, so that there is value in having certainty in trading volumes. The characteristics of CHP, which is led by heat load, means that CHP export (being uncertain) exposes the buyer to risk, and therefore has a very low value.

12.1.2 Implementation of Policy

The UK suffers from a complex range of government and regulatory bodies, support schemes and accreditation bodies. There is no unitary Energy Ministry. DTI (Department of Trade and Industry) is responsible for various technical and commercial elements of the energy industry, DEFRA (Department of Environment, Food and Rural Affairs) is responsible for environment and sustainable development (including the CHPQA support scheme), ODPM (Office of the Deputy Prime Minister) is responsible for Building Regulations and housing and OFGEM (Office of Gas and Electricity Markets) is the regulatory body charged with promoting competition and environmental objectives in the liberalized UK energy industry. Within this context the level of progress so far achieved is somewhat surprising.

12.1.3 Support Measures

There are two significant support measures for micro cogeneration which may be implemented in the near future. However, both are subject to the outcome of field trials currently being undertaken under the auspices of the Carbon Trust (a QUANGO charged with encouraging carbon mitigating technologies and measures in commercial buildings).[2] Although early results are encouraging, in that they support the anticipated carbon savings achieved in earlier independent trials, it is unlikely that the results will be adequate to trigger the support measures prior to late 2005 for EEC and April 2006 for VAT.

The measures are:

- *Reduction of VAT rate* from 17.5 % (standard rate) to 5 %, the same rate as used for domestic energy and which has also been recently approved for a number of other energy efficiency measures. This is considered simply as redressing an anomaly in the market, that energy and energy saving should be subject to the same VAT rate and as such, can hardly be considered a subsidy. In addition, a recent study showed that this measure would impose a loss of VAT representing less than £50/tonne Carbon, about one quarter the cost of the Renewables Obligation (Ilex 2002).
- *EEC (Energy Efficiency Commitment) approval and enhancement.* EEC is an obligation on energy suppliers to implement energy efficiency

[2] EST (Energy Saving Trust) is the equivalent body with responsibility for the residential sector.

schemes on behalf of their customers. The obligation is measured in TWh annually and the method and cost of achieving the target is up to the supplier (subject to approval of the measures against a standard database). Thus suppliers are constantly seeking more cost effective means of achieving their goals and, as the cheaper options are taken up (roof insulation is now complete for the majority of applicable homes for example), they are forced to seek alternatives. Micro cogeneration is a relatively cost effective measure and, equally importantly, has a high value per installation, thus reducing the management overhead. In order to stimulate the micro cogeneration market, DEFRA has agreed to enhance the accredited value of micro cogeneration (and a number of novel technologies) by a factor of 1.5. In other words, if micro cogeneration actually saves 20,000 kWh (lifetime net present value, NPV) it will be deemed to save 30,000 kWh. This EEC enhancement has been most effective in accelerating the introduction of the most efficient domestic appliances, such that it is now virtually impossible to buy a poorly performing refrigerator in the UK. This support measure is applied only for a short period, until the market matures and the true value of micro cogeneration can be realized through normal commercial channels.

Two key examples of value recovery channels not yet in place, are the trading of exported power from the unit and the attribution of environmental benefits to micro cogeneration. At present the exported power from a typical unit has a market value of around 3p/ kWh, an annual value of £30 and an NPV of around £200. However, it is not possible to trade such small amounts of energy in the wholesale market, so until micro cogeneration customers attain a critical mass and trading systems are modified to trade this power, some form of support is required to recognize the real contribution they are making to the UK power system.

The principle environmental benefit of micro cogeneration is the displaced CO_2 emission from central generating plant. A typical home generating 3000 kWh of electricity from micro cogeneration will displace CO_2 with a lifetime NPV of around €300, based on a value of €30 per tonne and CO_2 displacement in line with the Ilex study, €200 based on average generation mix displacement. Again, there is no mechanism for the householder to acquire this benefit which currently accrues to society as a whole.

12.1.4 Independent Bodies and Lobbying Activity

A great deal of the progress achieved in removing market barriers can be attributed to the efforts of industry alliances and environmental lobby groups. In particular the Micropower council, established to represent the interests of all micro generators (not just micro cogeneration) has successfully lobbied for the introduction of various support measures and legislation and has provided practical support to technical standards committees. Further support is provided by the Green Alliance, SBGI (Society of British Gas Industries) and CHPA (Combined Heat & Power Association).

12.1.5 Current Issues

As it is introduced into the UK market, micro cogeneration has raised a number of issues regarding the anticipated benefits to the community and possible adverse impacts on existing players in the energy industry. In particular, the anticipated carbon mitigation and fuel poverty benefits are coming under scrutiny; indirect impacts on generation mix and distribution networks are also being closely studied.

A number of studies in these areas have recently been completed with results of specific relevance to the UK. The results are surprising, in that the benefits identified generally exceed the earlier claims made by manufacturers. Three studies in particular show that micro cogeneration makes a substantial contribution to fuel poverty, carbon mitigation and diversity of supply targets.

A recent Policy Studies Institute paper shows that micro cogeneration contributes almost as much to fuel poverty as all other measures put together including micro cogeneration (Dresner and Ekins 2004). This paradox is explained by the fact that, if thermal improvements are made to a home, the reduction in thermal load means that the micro cogeneration unit will operate for less hours each year to fulfill the reduced thermal demand. This in turn leads to a reduction in electricity production and as the capital cost is the same regardless of the thermal demand of the home, the payback becomes rather long. In this case, it is unlikely that micro cogeneration would be installed in such homes. In other words the cheap insulation measure has "cherry-picked" the easy savings and made it uneconomic to exploit the full potential for that home. Hence: "Micro cogeneration can do almost as much for fuel poverty as making all possible energy efficiency improvements, including micro cogeneration."

A special report by Ilex, commissioned by Powergen, shows CO_2 displacement for the next 10 years at a level higher than both average generation mix and marginal generation emissions (Ilex 2004). Current government policy is based on 0.44 kg/ kWh which is the average mix, a somewhat arbitrary and in this case, inappropriate measure. Micro cogeneration is shown to displace marginal plant and the study, which matches actual generation profiles for installed WhisperGen units against marginal plant, shows a displacement of 0.54 rising to 0.67 kg/ kWh by 2010. This counter-intuitive result is the consequence of the increasing cost of coal-fired generation which, although it reduces the total amount of coal generation in the overall mix, shifts all coal generation into the margin.

Further indirect benefits accrue to micro cogeneration as it has a profile which supports intermittent wind resources and, by nature of its diversity, reduces the need for back-up capacity. The Environmental Change Institute study based on 20 years of wind and consumption data, concludes that only 400 MW_{el} back-up capacity would be required if micro cogeneration were to support 10 GW_{el} of wind generation (ECI 2004).

The SIAM (System Integration of Additional Microgeneration) study was expected to identify adverse impacts of large-scale implementation of micro cogeneration on Distribution Networks, potentially requiring significant investment in network upgrades (SIAM 2005). In fact, the study showed that in only a few extreme cases would micro cogeneration incur additional short term costs, that in the majority of cases it would have beneficial impacts and the overall benefit to the UK distribution network was substantial; savings in deferred network upgrades and improved operational efficiency were estimated at up to £1.2 billion by 2020 assuming a high penetration level of micro generation.

12.1.6 Legislation and Regulation

The two principal areas of regulation related to micro cogeneration concern the application within the building's energy system, to which the Building Regulations apply, and the connection to the electricity and gas networks.

Recent amendments to the Building Regulations (Part L, April 2002) for the first time apply retrospectively to domestic energy systems. It is now mandatory, when replacing the boiler within a central heating system, not only to replace the boiler with a boiler of specified efficiency, but also to upgrade the entire system to current new-build standards. This efficiency will be raised to 86 % with effect from April 2005 and has a somewhat

confusing result for micro cogeneration. For a micro cogeneration unit with a thermal efficiency of 80 % and an electrical efficiency of 12 %, the overall efficiency is 92 %, well above the threshold. However, the current regulations do not recognize electrical output; indeed quoted boiler efficiencies do not even have to include parasitic (pump and fan) losses in their efficiency ratings. It is therefore difficult to compare micro cogeneration with boilers and work is currently under way to provide a standardized assessment procedure for micro cogeneration units.
The area of electrical connection to the public electricity network has been much simplified by the agreement of Engineering Standard G83/1 (2003), which sets standards for type approved products which can be installed by a competent installer without prior agreement of the DNO (Distribution Network Operator). Work continues to amend regulations for wiring within the home to take account of "double-fed" circuits, and a revision to the IEE Wiring Regulations is expected to be published in 2006 (17th Edition) including advice on connection to radial/ring mains, disconnect times and shock and overload protection. For the time being, micro cogeneration units are being installed using a dedicated circuit to the consumer unit, with its own protection (a motor rated fuse).

There is currently no legislation regarding NO_x emissions from gas appliances, although the current high NO_x emissions from some micro cogeneration (Stirling engines in particular) are likely to be significantly reduced as products evolve to meet increasingly stringent standards.

12.1.7 Energy Prices

Although energy prices are still (31 %) lower than pre-privatization in real terms (DTI), they are no longer regulated. Instead, since early 2002, OFGEM decided that competition would continue to exert downward pressure on prices as effectively as regulation had done until that point. For an initial period, domestic energy prices had been regulated according to a formula based on the Retail Price Index (RPI-X %), and customers became accustomed to annual price reductions. Between the initiation of full competition in 1999 and 2003, prices had fallen in real terms by over 10 %. This situation has now changed dramatically as suppliers have responded to significant wholesale price rises with increases from major suppliers in 2004 of over 18 %.

A typical customer with an electricity consumption of 3300 kWh and gas 18,000 kWh, would expect to pay between £236-£275 for electricity, and £526-£650 in total depending on supplier and method of payment.

12.2 Market Context

The fundamental requirement for economic viability of micro cogeneration is that the marginal investment cost (compared to a conventional boiler) can be recovered from energy savings (primarily generation of electricity) within a reasonable period. This depends on the cost of primary fuel and imported electricity, and the value of export, as well as the amount of each of these. A higher electrical output produces better fuel bill savings, provided it is not achieved at the expense of additional primary fuel input. The ideal market is therefore one where there is a network of low cost primary fuel, relatively high electricity costs, a housing stock which requires space and water heating and a heating season long enough to result in extended running hours. In each of these respects the UK is ideally suited to the characteristics of Stirling engine micro cogeneration.

Under current standards (base case boiler efficiency >78 %) and energy prices as above, the viability of micro cogeneration is limited to homes with a heat loss of at least 12,000 kWh per year in order to give a payback less than 7 years (often considered to be the threshold for consumer investment). For a typical family home (18,000 kWh annual heat loss), the WhisperGen unit would generate 3,000 kWh of electricity of which around 2000 kWh would be used within the home (avoiding import cost of ~7.5p/ kWh) and the remaining 1,000 kWh exported to the network at a value of 3p/ kWh. The total benefit would be £180 and, with a marginal investment cost of £600, a payback of 3-4 years would be achieved.[3]

12.2.1 Housing Stock and Energy Use

Of the 23 million households in the UK, 89 % (20 million) have central heating of some type. The majorities of these were constructed prior to 1970 and consequently have relatively high heat loss of 0.05 kWh/m^2/degree day (compared with Sweden (0.038), Netherlands (0.041), Germany (0.07). Average household floor area is 85 m^2 (Sweden (90), DE (78)). Progressive improvements in Building Regulations since 1970, and the increasing number of smaller households is resulting in

[3] The base case scenario is for a conventional boiler with 78 % thermal efficiency, the current standard. In April 2005 most boilers will need to achieve 86 % so that the micro cogeneration unit will achieve relatively lower savings, about £150 taking account of £25 additional gas cost. However, it is also likely that the higher efficiency boilers will cost more so that the marginal investment also falls and payback is relatively unchanged.

homes with significantly lower space heating demands, but with ever increasing domestic hot water usage.

In contrast to other European countries such as Germany, the UK preference is for owner occupation (68 %), although the tendency to move home on average every 7-8 years does not encourage long term investment in energy efficiency measures, so that short paybacks are essential for successful products. Fortunately, micro cogeneration falls into this category with typical paybacks of 3-4 years. However, for the social sector, investments tend to be made on a longer term basis by the (normally public sector) landlords, both on an economic basis and to fulfill statutory obligations.

A typical UK family home with a gas central heating system requires around 18,000 kWh of space heating annually, together with 5000 kWh of water heating and 3,500 kWh of electricity consumption. The resulting heat to power ratio for the typical home thus aligns well with the characteristics of Stirling engine technology and the relatively low electrical efficiencies are in fact a benefit rather than a disadvantage as was earlier believed. This is significantly different from the U.S. for example, where electrical loads are substantially higher and the case for higher electrical output is justifiable.

12.2.2 Climate

The other major feature of the UK that makes it suitable for micro cogeneration is its climate. As a result of its cool maritime climate, all of the UK has a significant heating season spread over several months. For the majority of the (poorly insulated) housing stock this imposes a demand for space heating of between 2,000 – 4,000 running hours. Although this might appear low compared with the requirements for industrial and commercial installations with anticipated hours of anything from 6,000 – 8,760, the higher unit value of electricity at the domestic level is able to compensate for this.

Interestingly enough, the very severe climates of Scandinavia are not attractive to micro cogeneration. Although the most populous regions have degree-days around 50 % higher than the UK, the heating season is more concentrated and would require the installation of a significant supplementary boiler to meet the extreme peaks. (As with heat pumps, it is neither practical nor economic to size the engine itself to meet peaks). This requirement would adversely affect the economics, even if there were a widespread gas network in Scandinavia.

12.2.3 Heating Systems

Around 18 million homes have gas central heating installed, an additional 1 million have fuel oil or LPG (Liquefied Petroleum Gas) central heating systems and a similar number have coal-fired systems. Electric (non-hydronic) systems, usually incorporating storage heaters, account for around 2.5 million homes.

Hydronic central heating, fired by natural gas is the cheapest in operation for the majority of family homes and electrical systems are generally limited to smaller houses and flats. Indeed, gas heating is almost a status symbol and is assumed to be "best" even to the extent that many will choose LPG off-network in preference to fuel oil in the mistaken belief that, because it is gas, it must be cheaper!

District Heating (DH) is relatively uncommon, with only 1 % market share. This is to some extent a reflection of the high level of owner-occupation and the desire to have independent control, allied to a traditional focus in the UK on low first cost. It is also difficult to see how the conflicting demands of DH (which logically requires all homes in an area to be connected for economic viability) and a competitive market (which demands that all customers may choose their energy supplier and their energy system) can be resolved.

Various studies have identified up to 5 million homes within areas with sufficiently high heat densities to justify district heating (BRE 2003). The areas identified tend to be central urban sites with high rise apartments, unsuitable for micro cogeneration both from a heat loss perspective and due to the physical size and construction of the properties. Micro cogeneration and DH can therefore be considered complementary rather than competing technologies, although, where practical, micro cogeneration is preferable on economic and environmental grounds (Harrison 2002).

12.2.4 Boiler Market

The UK boiler market of 1.5 million sales annually, is weighted heavily towards replacement of existing boilers (1.2 million) with around 100,000 for new systems in existing buildings and 200,000 for installation in new homes.

Of these, an increasing number are wall-mounted, with combi-boilers also becoming increasingly popular, due primarily to space considerations. It might therefore appear that floor-mounted micro cogeneration units would face a challenging market. However, there are still estimated to be

7-8 million old, cast iron floor-mounted boilers in place with very low operating efficiencies. With a replacement rate of 7 %, there is thus a potential annual market of up to 500,000 floor-mounted micro cogeneration units in the UK. Within this sector, the smaller output micro cogeneration units will be ideally suited to homes with annual heat loss above 12,000 kWh and the larger units suited to those above 23,000 kWh (Fig. 12.1).

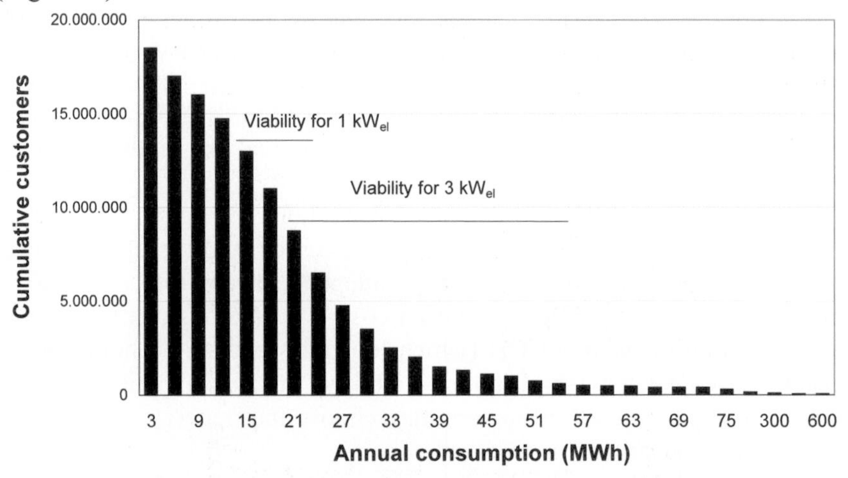

Fig. 12.1. UK domestic gas consumers (dark gray area indicates viability for 1 kW_{el}, gray area for 3 kW_{el})

12.2.5 Technology Developments

As micro cogeneration enters the commercial phase, the products themselves are reaching a level of maturity which primarily requires developments to be of a "design for manufacture" nature rather than addressing fundamental issues of performance and reliability. The products also need to be adapted to the peculiar requirements of each market, and this section outlines technology developments of specific application to the UK market.

At the same time, it is becoming apparent that the implementation of micro cogeneration can be accelerated by the availability of "enabling technologies" such as advanced controls and metering which improve the performance within the home and which allow the true value of micro cogeneration to be realized.

Thermal storage. When micro cogeneration products were first being developed in the UK, it was believed that generation should be closely aligned with demand in order to avoid export and maximize the value of micro cogeneration. Indeed, BG Technology proposed 1m^3 of thermal storage and sophisticated electronics in order to completely avoid export and ensure operation which could be considered grid-isolated. This was to avoid the former complexities of applying for permission to connect to the grid. Although grid connection has now been simplified to the extent where it no longer represents an obstacle, thermal storage still has practical benefits in terms of reduced engine cycling and peak lopping, allowing greater operational flexibility and improving system performance.

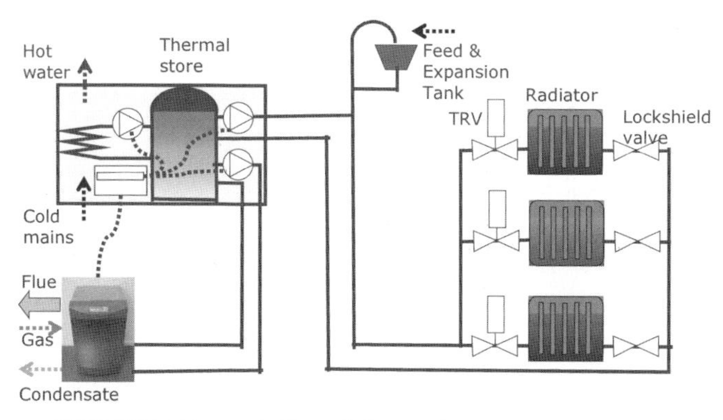

Fig. 12.2. Gledhill thermal store package

The Gledhill thermal store (Fig. 12.2) has been developed to meet these requirements; it is effectively a packaged hot water cylinder in which the stored water is part of the primary circuit. The package incorporates three pumps. The first pump runs whenever the store temperature falls below 55°C and continues to run until the store temperature reaches 75°C, taking heat from the micro cogeneration unit. A second pump is switched by the programmer/thermostat to take water from the tank to the radiators whenever there is a demand for heat. The third pump operates when there is a demand for hot water and passes stored water through a high capacity plate heat exchanger, which instantaneously heats cold mains water. This has the additional benefit in the UK where pressurized hot water systems are not the norm and where the resulting hot water pressure is unsuitable for showers.

So far trials have demonstrated clearly that it is possible to de-couple the production of heat and power on a diurnal basis, eliminating the engine cycling problems (which could adversely impact engine life and maintenance) as well as allowing the engine to run only at its optimum efficiency level and minimizing the parasitic losses during stopping and starting; this is particularly relevant during the marginal heating season (spring/autumn) when only a small thermal output is required, but for extended periods of the day.

In the example illustrated (Fig. 12.3), the blue line represents operation with a conventional heating system and shows 16 cycles on the day in question, several of which never achieve full electrical output before being terminated. The red line shows 4 cycles for the same operating conditions and a significantly higher electrical production.

What has not yet been attempted in practice, although certainly feasible in theory with the correct controls and communications, is to operate the unit with its primary target of electricity production to match periods of peak value, either to the householder by matching own loads, or to the energy supplier by matching peak wholesale market prices.

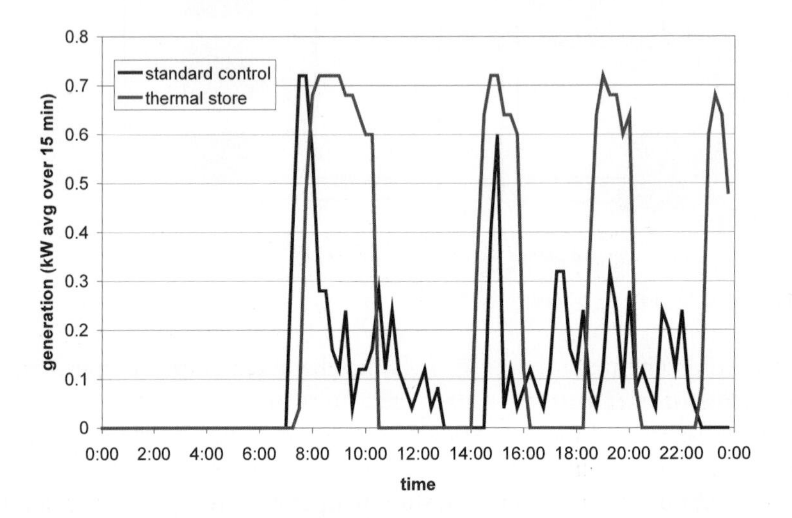

Fig. 12.3. Comparison of cycling performance between conventional system and thermal store

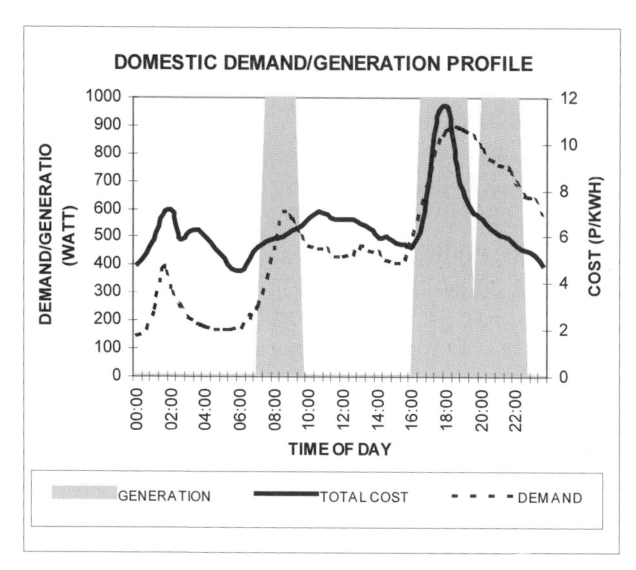

Fig. 12.4. Variation of electricity cost throughout a typical winter's day shows the value of micro cogeneration. Generation coincides substantially with peak supply cost, as does domestic demand. Demand weighted value of micro cogeneration is around 3.4 p/ kWh over the year compared with an average pool price less than 2.8 p/ kWh

Metering & control. From a very early stage, the regulator (OFGEM) imposed a requirement for separate metering of imported and exported electricity. This was deemed essential in order to understand energy flows within the distribution network and as a means of attributing the true value of generation to its correct source. Initially, metering on a half-hourly (HH) basis was envisaged (as with the commercial market), but the excessive administrative costs (around £800 annually per site) meant that it would not have been viable.

A solution for micro generators was agreed and now generators up to 30 kW_{el} may register exports on a non-half-hourly (NHH) basis (P81 standard), normally simply showing kWh totals on a cumulative basis as for standard meters. However, industry infrastructure limitations have led to the use of export sub metering in the short term and exports are not traded in the wholesale market. Considerable effort is currently being applied to developing agreed metering, settlement and trading standards for micro cogeneration.

In the short term however, it is still possible to operate micro cogeneration, it is just that the export has no tradable value. Even before the advent of micro cogeneration, the metering industry was facing considerable technical and operational challenges. The competitive market had raised innumerable issues relating to meter reading accuracy particularly regarding change of supplier (which can occur every 28 days), change of tenancy and change of tariff. The use of pre-payment meters for debt recovery and for fuel poor also raised difficulties in ensuring that these groups were not denied access to energy in emergencies. This, combined with the recent separation of metering as a business entity, has forced suppliers to review the type of metering, their ownership of metering assets and operation procedures. Currently the meter is owned by the DNO (Distribution Network Operator) and maintained by a MOP (Meter Operator) on behalf of an energy supplier who pays a daily charge for the use of the meter. This means that the meter can only be changed on the instruction of a supplier, a situation favoring micro cogeneration manufacturers who partner with an energy supplier.

As the NHH meter only registers total kWh consumed and the wholesale market is traded on a HH basis, it is necessary to apply agreed (assumed) demand profiles to all customers. As a short term solution the import profile for micro cogeneration is assumed to be the same as for standard customers, and a simplified export profile has been agreed. Current monitoring work is aimed at identifying actual profiles before micro cogeneration is widespread.

Considerable progress has been made in establishing industry processes and it is hoped that, by the end of 2005, suppliers will be able to install import/export meters and trade both in the existing settlement process, against agreed profiles.

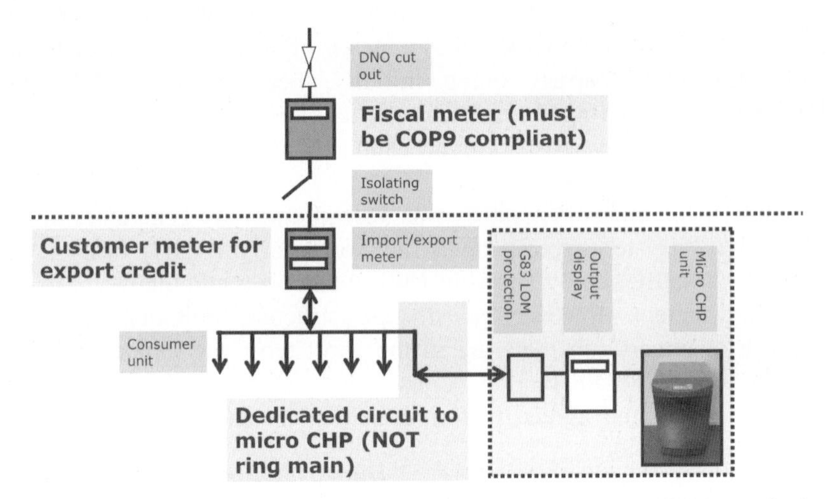

Fig. 12.5. Current connection arrangement to comply with P81 (metering) and G83/1 (network connection)

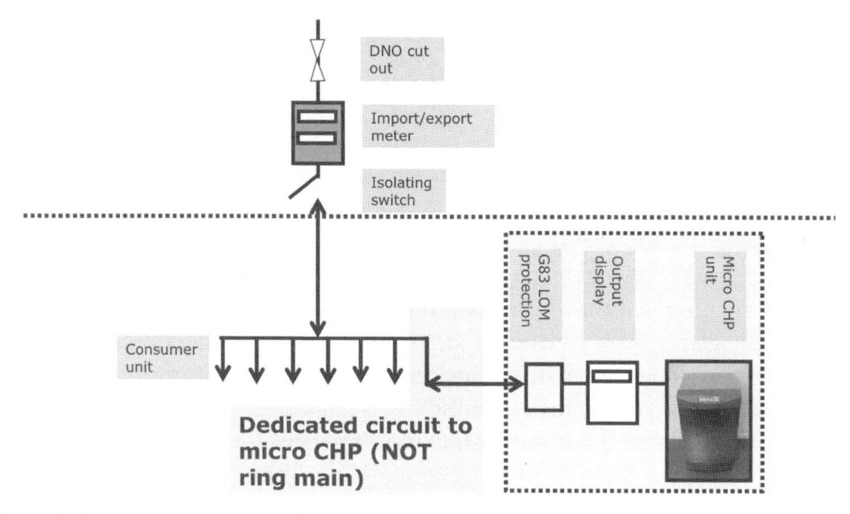

Fig. 12.6. Proposed connection arrangement with import/export meter registered for settlement

12.3 Major Commercial Activities and Actors in the Country

12.3.1 UK Energy Industry Structure

Privatization of the UK energy market began in 1990 and is now fully competitive for all customers. At the retail level this allows customers to change supplier every 28 days, although the practicalities involved often lead to much longer change times. However, suppliers are constantly aware of the choice their customers face and strive to offer the most competitive tariff or service.

The transmission grid remains a monopoly operated by NGT and central plant, operated by a number of major companies, is connected to that grid. All system users are subject to the same use of system (TUoS) charges. The distribution network (33kV and less) is operated by regulated distribution companies (DNO), vestiges of the original local electricity boards prior to privatization. Again use of system charges (DUoS) are regulated and the same for all users. Supply licenses exist for 13 regions in Great Britain, aligned with the DNO franchise areas. However, these licenses are owned by an increasingly small number of energy supply companies. These companies supply direct to individual homes and it is generally considered that a critical mass of 5 million customers is required to justify the infrastructure costs. This is quite different from other markets, such as Germany, where the major suppliers' customers tend to be municipalities or smaller local suppliers who own the end-user relationship.

It should be clear from this that the only margin in the value chain, which is truly competitive is the suppliers' margin; wholesale energy is a market open to all, use of system is the same for all. In this environment, micro cogeneration offers a considerable competitive advantage to any supplier owning it, or with customers using it. This is primarily due to the beneficial generation profile (Fig. 12.6) of micro cogeneration which aligns well with peak wholesale prices, which is itself strongly influenced by domestic demand, in turn responding to peak heating periods.

Although regulated and separated, most energy companies are becoming vertically integrated to the extent that they have matched generation access and supply to mitigate risk. Also, after a decade or more of aggressive acquisition and competition, the supply market appears to be stabilizing with recognition that it is unprofitable to constantly acquire new customers only to lose existing customers, a process known as "churn". In the UK, this can be as much as 25 % annually, compared with the German market with almost insignificant churn rates below 5 %.

12.3.2 Routes to Market

The two major energy suppliers, Powergen and British Gas (Centrica) have both made arrangements to commercialize micro cogeneration to their energy customers. British Gas, the former monopoly gas supplier, has an established business offering service and installation of gas boilers, so has a ready made network.

Recent developments in the wholesale energy market, which have caused both players to raise prices to end-users, have had a serious impact on British Gas, which has lost several hundred thousand customers to its rivals.

Micro cogeneration offers companies such as these an opportunity to improve the profitability of their energy trading portfolio (micro cogeneration reduces the amount of high cost electricity a supplier needs to obtain on behalf of its customer, improving margin), to offer very competitive tariffs and to offer an energy saving technology which reduces the number of kWh purchased (further saving customer bills) and maintains total margins for the supplier.

The only product commercially available in the UK for individual homes is the WhisperGen unit from Powergen. This is being offered to individual householders on a fully installed basis using Powergen's approved installer network, and by direct sale from a dedicated team. This strategy is aimed at overcoming the major obstacles which prevented the widespread introduction of high efficiency boilers to the UK market.

The UK faces a chronic skill shortage in the heating industry and Powergen has established training courses to ensure installations are of an appropriate standard and that service support is available. The lack of service support had been identified as one of the key obstacles to the uptake of condensing boilers, and all involved are anxious to ensure that micro cogeneration is implemented smoothly and as quickly as practicable.

The other obstacle faced by condensing boilers was the structure of the boiler distribution and installation market. Distribution through major national merchants disconnected the manufacturer from the installer and products were often selected only on the basis of cost or availability, rather than appropriateness for the application. The installers effectively became the specifier, providing advice, again on less than objective grounds, and were naturally reluctant to upgrade their skills or take on the challenges of condensing boiler technology.

British Gas are expected to use a similar route to market, making use of their existing network of installers and service engineers, although it is possible that BG MicroGen (their supplier) may make products available through the merchant route as well.

The other potential player in the UK is a boiler manufacturer, Baxi, who have acquired the German Dachs unit as well as European Fuel Cells (developing a 1 kW_{el} fuel cell) and the rights to the Inergen (1 kW_{el}) organic-Rankine cycle unit. This latter development, which may reach market by 2007/2008, promises low cost manufacture and compatibility as an add-on module to Baxi's range of boilers. Baxi seem committed to the existing delivery chain and it will be interesting to see how they manage installation (which can legally only be carried out by G83 accredited installers) and after sales support with a relatively complex product.

12.3.3 Market Status

There are a number of products currently undergoing various stages of development and trials. MicroGen expect to launch their 1 kW_{el} unit in 2007, following their current field trial, Baxi Inergen and Disenco (a 3 kW_{el} Stirling engine based on the Sigma unit) are both under development and a number of early stage prototypes are being tested within the Carbon Trust sponsored trial program. This is aimed at validating manufacturers' performance claims prior to confirmation of government support schemes and as a means of accreditation for compliance with Building Regulations.

Powergen are in market test phase and are offering WhisperGen units installed for £3000 (including VAT) for individual homeowners and for £1350 (plus VAT) to house builders. House builders have proved surprisingly keen to install micro cogeneration given their reluctance to embrace any new technology, particularly with a relatively high initial cost. However, the prospect of increased building standards and the potential to "trade-off" insulation against heating system enhancements may contribute to this enthusiasm.

Fig. 12.7. The first of Powergen's 80,000 micro cogeneration customers, with a WhisperGen unit installed in an existing 1930's home in Nottingham

A condition of the sale is that customers must also purchase their energy (gas and electricity) from Powergen as well, although they are at liberty to subsequently change to any other supplier 28 days later. This is an indication of Powergen's motivation to enter this market. The energy package includes gas and electricity at exactly the same price as for standard customers, but with the addition of an export tariff of approximately 3p/ kWh (about 50 % of import tariff). This is in line with the actual value of export which will be obtained once the requisite infrastructure is in place.

Powergen recently announced their intention to install a minimum of 80,000 units by 2010 as a result of the positive market response. However, the exact timing will be subject to the implementation of agreed industry and regulatory changes referred to earlier. In particular, the metering and settlement issues must be resolved in order to obtain the full value of micro cogeneration.

There are also numerous issues regarding commercial agreements between DNO and supplier (connection terms) which may impact the timing of mass market launch. However, overall the UK market seems an ideal environment for existing micro cogeneration technologies, which may be further exploited by fuel cell and other technologies as they become available.

13 Micro Cogeneration in Japan

Yasushi Santo

13.1 The Energy Situation of Japan

To thoroughly understand development of micro cogeneration in Japan, it is relevant to briefly describe the overall energy situation in Japan. Geographically said, Japan consists of four major islands located closely together – from north to south, Hokkaido, Mainland, Shikoku, and Kyushu – and one farther south, Okinawa. Japan's total land area is quite similar to Germany's, and its population is about 128 million. Seventy per cent of the land is covered by forests and mountains, where people have difficulty living, resulting in the fact that most of the major cities are located near the coastline. Its climate is rather mild, except for the Hokkaido area. Major cities such as Tokyo, Osaka, Nagoya, or Fukuoka belong to that temperate zone. Consequently, heating demand is not high, while cooling demand is rather high compared to Europe, because in summer, Japan experiences high humidity combined with high temperatures, often above 30° C.

13.1.1 Primary Energy Supply and Power Generation

The primary energy supply for Japan (604 million kl crude oil equivalent) consists mainly of imported fossil fuels. At the end of 2000, 52 % was from oil, 18 % from coal, 13 % from liquefied natural gas (LNG), 12 % from nuclear energy, and 3 % from hydro power stations. In order to reduce the risks associated with heavy reliance on imported energy resources, Japan's energy policy has been targeted toward reducing the import of oil, especially from the Middle East. Japan's level of dependence on oil, its largest source of energy, has decreased drastically since the first oil crisis in 1973/74; its share was reduced from 77 % in 1973 to 52 % in 2000. On the other hand, during this period, nuclear energy and natural gas have increased their shares in the supply of primary energy.

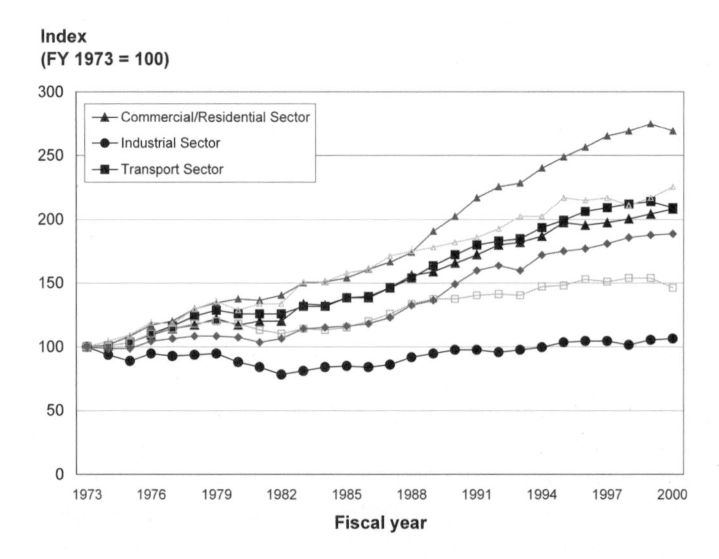

Fig. 13.1. Development of energy demand in key sectors

With the exception of the two major oil crises, the demand for energy in Japan has been consistently increasing, with, for example, significant growth (16.2 %) of energy demand in the commercial/residential and transport sectors from 1990 to 2000 (Fig. 13.1).

In contrast, energy demand in the industrial sector has been relatively stable as a result of the introduction of highly efficient processes in response to the overall requirement of reducing reliance on imported energy and in order to maintain industrial competitiveness in the global market. In light of such circumstances, more efficient means of energy utilization need to be introduced to smaller-sized businesses and the residential energy market.

Regarding electric power generation, as of March 2002, Japan had a generation capacity of 230 GW in total, out of which 22 % used oil as fuel, 13 % coal, 26 % LNG, 20 % nuclear energy, and 20 % hydro power. In order to reduce the heavy reliance on imported oil, nuclear power capacity has been expanded since the start of commercial operation of a nuclear power plant in 1966. But it is becoming significantly difficult to find new sites, due to resistance from people in the candidate regions and also because of decreased trust in the safe operation of nuclear power plants. This skepticism was caused in part by false inspection reports that Tokyo Electric submitted to the government in 2002 and by a high-pressure steampipe explosion which occurred in August 2004 at the Mihama nuclear power site, owned by Kansai Electric, killing 5 people – though without leakage of radioactive materials.

The electricity generated in fiscal year 2001 was 924 billion kWh in total. Oil thermal plants generated 8 %, coal thermal 21 %, liquefied natural gas 27 %, nuclear power 35 %, hydro power10 %, and new energy sources, including renewables, 0.3 %.

13.1.2 Electric and Gas Utilities

There are nine electric utilities on the four main islands and one in Okinawa. They are, from north to south, the Hokkaido, Tohoku, Tokyo, Chubu, Hokuriku, Kansai, Chugoku, Shikoku, Kyushu, and Okinawa Electric Power companies; all are investor owned. Except for Okinawa Electric, they are interconnected; but because of the electricity network's history of development, Hokkaido, Tohoku, and Tokyo Electric send 50 Hz electricity and the rest 60 Hz.

The total number of gas utilities in Japan is a little less than 230, among which the major ones are Tokyo Gas, Osaka Gas, and Toho Gas in Nagoya; these account for more than 75 % of the total gas sales in Japan. Almost none of the gas utilities in Japan are interconnected by national pipelines; rather, each has its own transmission and distribution network, which is quite different from Europe or the U.S. The necessity is apparent for Japan to install national high-pressure gas pipeline networks; but due to the enormous costs entailed, no private entities have shown interest in the project. In the areas where no gas networks are available, many households use liquefied petroleum gas (LPG). The number of LPG users is about the same as the city gas users. Some gas utilities distribute LPG-air (LPG-air is a gas with lower Btu than LPG as the result of injected air) or produced gas. The Japanese government is encouraging the conversion to natural gas.

13.1.3 Energy Market Deregulation

The electricity market is open to new for to customers of above 500 kW currently; and the ceiling will be lowered to 50 kW in 2005. All-out fights have erupted between electric utilities and non-electric utilities, including gas utilities subsidiaries, attempting to capture new customers. Many of the large-scale industrial users are enjoying decreased electricity costs as a consequence of this increased competition.

The gas market is open to competition for customers with a gas consumption more than 500,000 m^3 per year. Electric utilities which import much LNG as a fuel for power generation are coming into this market to sell their gas to large scale gas users, and several large-scale industrial gas users have switched to the gas provided by electric utilities, supplied via dedicated pipelines or via gas utilities' pipelines.

13.2 The Reciprocating Engine Micro Cogeneration Market in Japan

Micro cogeneration systems do occupy a share of the total CHP market. As yet, their market share is still very small, but the number of installations of micro cogeneration units with a capacity of less than 10 kW_{el} is expected to increase significantly, especially since 1 kW micro cogeneration systems were introduced to the Japanese market in early 2003. In the long run, the residential and commercial application of PEMFC will eventually enter the market; but they are still in the field testing stage (see Sect. 1.2.3). The general trend of the CHP market is shown in Fig. 13.2.

As of 2002, total generation capacity of CHPs in Japan had reached approximately 6.5 GW (excluding the capacity at paper mill steam plants) at 4,500 sites. The number of commercial sites has been increasing rapidly. The figure does not chart the category of residential applications, which opened up in late 2002. It is likely that, in the future, these charts will show the category of residential applications starting from the year 2003.

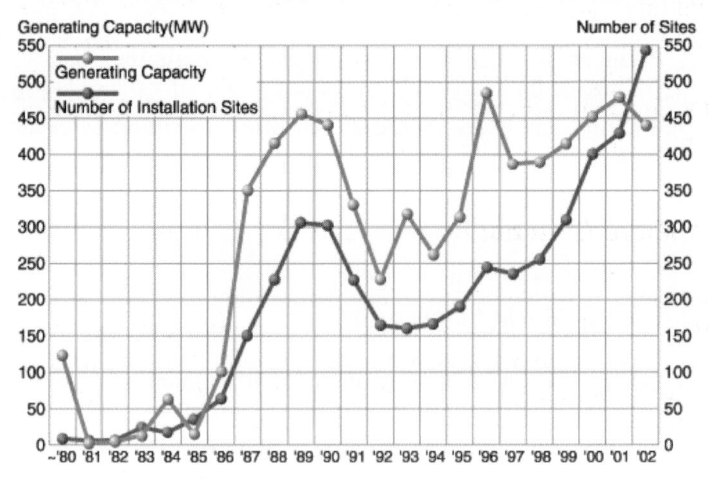

Fig. 13.2. The number and generation capacity of CHPs in Japan for each fiscal year, 1980-2002 (Source: Cogeneration Center 2004)

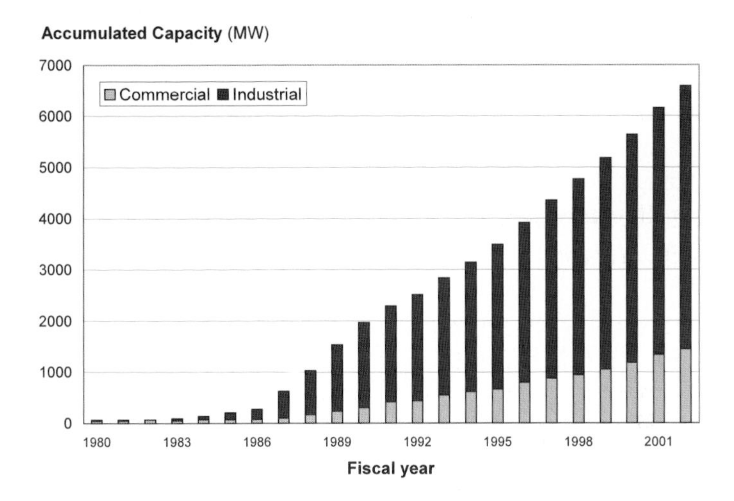

Fig. 13.3. The accumulated capacity for CHPs in each fiscal year, 1980-2002

Natural-gas-fueled CHPs can receive a variety of government subsidies and/or tax incentives and lower-interest loans from the Development Bank of Japan. The government is promoting them in order to achieve higher-efficiency energy utilization, seeking to increase the speed of oil replacement and, thereby, air pollutant reduction. These CHPs, especially the medium- and small-scale models with capacities of less than 500 kW_{el}, are sold through the sales channels of major gas utilities, which are making efforts to increase their gas sales and also to attack the electricity market. Also, it is very common that gas utilities request CHP manufacturers to develop the units which best fit the gas customers' needs. Gas utilities and CHP manufacturers jointly develop detailed specifications.

13.2.1 Micro Cogeneration for Commercial Use

The capacity of micro cogenerations for commercial use range between 5 and 10 kW_{el}. 15 kW units were previously available on the market before; but their sale was terminated a few years ago. The reason for this is that CHPs with capacities of less than 10 kW can be operated by owners themselves, without requiring certified operators. Such a deregulation was introduced as a result of the gas utilities' lobbying the government.

When the units less than 10 kW capacity were introduced to the market, these systems were operated without being connected to the grid, in order to avoid the time-consuming application procedures and high fees demanded by electric utilities for grid connection. The functional operations were achieved by adding load-switching devices for calculating and balancing the customer's whole electricity and heat demand. However, most recent installations are connected to the grid using low-cost inverters[1] incorporated with each CHP unit. Yet, they are not allowed to sell their excess generated electricity to the grid.

According to the data provided by the Cogeneration Center of Japan, the number of micro cogeneration units ranging from 1 kW to 15 kW sold in fiscal year 2003 – ending prior to the end of March 2003 – was approximately 1,100 units at 900 sites, with total capacity having reached 8,800 kW. Actually, at this point there are almost no units above 10 kW on the market, mainly due to the above mentioned regulatory restrictions regarding the installation and operation.

Gas engine micro cogeneration models have been developed jointly by major gas utilities, such as Tokyo Gas, Osaka Gas, and Toho Gas, in cooperation with engine manufacturers. Because gas utilities know (more or less) precisely the heat demand requirements of their customers, micro cogeneration manufacturers need initiatives from utilities even for marketing their products. Most of the units are sold through the marketing channels of gas utilities, with initial maintenance services usually being provided by the utilities.

Micro cogeneration plants of less than 10 kW have been marketed mainly to commercial customers with high heat demand, such as public baths, restaurants like McDonald's and noodle shops, homes for the elderly, hospitals, and healthcare centers. They can receive a government subsidy of around 700,000 yen/unit. Most units for commercial use are grid connected, but do not sell generated electricity to the grid. These units receive maintenance services from gas utilities under service contracts.

13.2.2 Micro Cogeneration for Residential Use

Another interesting market segment are systems for residential use below 1 kW_{el}. At the end of March 2004, approximately 3,000 gas-engine micro cogeneration units were running in 3,000 houses in Japan. The unit capacity of micro cogeneration for residential use is actually a little less than 1 kW_{el} with 3 kW heat recovery. The 1 kW units were test marketed

[1] The engine generator generates AC (alternating current) which will be converted into DC (direct current). The DC is converted again to AC to be used by appliances in the house. This is to electronically isolate a small generating machine from the grid. This inverter is cheaper and more reliable than the device formerly required by electric utilities for grid connection.

in late 2002 and put on the market in March 2003. Their sales have been increasing far faster than anticipated.

The Ecowill was developed through the joint efforts of Honda, Osaka Gas, Toho Gas, Saibu Gas, Chofu-seisakusho (a boiler manufacturer) and Noritsu (a water heater manufacturer); it was put on the market as an ordinary sales commodity in March 2003 with the name "Ecowill". The price of the unit was set at 750,000 yen, to make it competitive with the conventional heating and water heating units if the government subsidy of approximately 200,000 yen is applied. In addition, a lower gas tariff for this application was prepared.

The tariff, designed by Osaka Gas, a leading marketer of the units, is significantly lower than the ordinary one. A press release claims that Ecowill users can satisfy most of their heat demand and 40 % of their electricity demand. Its operation is controlled so as to minimize gas bills by memorizing the electricity- and heat-usage pattern of each user. The information is fed into a control panel which is mounted, for example, on the kitchen wall and shows current figures on how much has been used in a month, how many kWh are being generated currently and in accumulated numbers, and the contribution to the reduction of CO_2 emissions. The average Ecowill users pay 63.79 yen/m^3 compared to about 110 to 120 yen/m^3 for ordinary gas customers.

One thing to note is that Tokyo Gas, the largest gas utility in Japan, is not involved with the development and sale of this 1 kW CHP. Instead, it is pouring its efforts into the development of 1 kW PEMFC systems to be applied to residential CHP.

Table 13.1. The installation of 1 kW gas engine micro cogeneration units, 1999-2003 (Source: Cogeneration Center of Japan)

	~ 1999	2000	2001	2002	2003	Total
Units	0	0	0	522	2419	2941

Regulations and interfaces with electric utilities. The gas-engine micro cogeneration is theoretically a power plant, which means that it has to comply with the regulations covering all generation plants, regardless of their capacity. But the government has made special arrangements for these systems to be installed very close to houses and to be allowed to operate without being attended by certified operators. The distance between houses in Japan is normally very narrow, typically, about 5 meters wall to wall, and houses are usually set 1.5 meters back from the property line, in order to fully utilize the limited space. A 1 kW gas engine CHP should be installed between a half meter to a meter away from the wall of a house to reduce heat loss from the pipes connecting the unit to heating or hot water appliances.

A 1 kW unit will be connected to the grid, and in the case of power outage, will be automatically turned off. Islanding is not allowed. When a unit is installed, an inspection by the electric utility of the area is required in advance, and an installation permit has to be acquired from the utility. After the completion of installation, a utility inspector visits the site to make sure that the unit is properly installed according to the specifications rendered by the electric utility. Such units are designed to generate electricity up to 1 kW, and accidentally over-generated electricity is switched to a heater for warming up water in the water storage tank.

13.2.3 Development Initiatives

Gas utilities in Japan are trying hard to protect their residential market from the attacks by electric utilities and electric appliance dealers. In the past, almost all cooking and heating appliances were fueled by gas or kerosene. But these days, electric heat pump air-conditioners and cooking appliances, especially ones using IH heating systems, and electric heat pump water heaters are raising the level of competition.

At the same time, all electric utilities in Japan are promoting the sale of all-electricity-powered houses (i.e., with no gas appliances), which can enjoy lower rates, including a far-lower night rate, if they are not connected to a gas pipeline. Sales promotions appear daily on television, emphasizing the safety and cleanliness of electric appliances for kitchens. The share of all-electricity-run homes among newly built houses is rising sharply; it is said to be reaching 50 % according to an electric utilities report. The benefit of such homes is a lower rate of 6.36 yen/kWh at night, 19.64 yen/kWh during the evening, and 26.09 yen/kWh during the day, compared to the ordinary flat rate of 23 yen/kWh. These rates are for Kansai Electric Power Company, which has been most aggressive in promoting the expansion of all electricity-powered homes.

Under such pressure, gas utilities need to develop some effective ways to protect their market. One of them is to have a way of providing electricity generated by gas to residential customers. In the future, fuel cell systems could be an ideal solution. But the currently available competitive method is to employ a small-scale gas-engine-driven power generation unit, which should be small enough to meet the electricity demand of a single household.

The residential energy market has been shifting toward being favorable for micro cogeneration. The number of houses, especially newly built ones with floor-heating systems, has been increasing.

Furthermore, the higher total efficiency of micro cogeneration has raised its attractiveness for environmentally conscious people. Also, house builders and developers have been wanting to add something new that raises the quality and value of their products. The government subsidy to promote environmentally friendly energy systems was made available for CHP systems in principle due to their using natural gas as a fuel and demonstrating a high level of efficiency.

13.2.4 Prospects for Gas Engine Micro Cogeneration in Japan

The market for micro cogeneration is estimated to grow steadily for commercial and residential uses. After the Kyoto Protocol becomes effective in February 2005, the Japanese government will surely promote the sale of CHP in general and focus their promotional efforts toward the commercial and residential markets because they are concerned about the higher growth rate of energy consumption in these sectors. The Cogeneration Center of Japan estimates that about 5,000-6,000 1 kW units will be sold by March 31, 2005; and in the case of the CHPs of less than 15 kW, 400-500 units will be sold annually for some years. Some manufacturers are thinking of exporting their products to Europe or the U.S.; but it would be hard to provide attractive benefits to possible users, especially because there are no government incentives and no great value can be added to recovered heat due to the low cost of natural gas, especially in Europe. Furthermore, good partners will have to be found to provide maintenance services to users.

13.3 Fuel Cell Development

Stationary fuel cell systems are strategically important devices for gas utilities trying to increase gas sales and, at the same time, to acquire methods to deliver different and higher quality energy (electricity) to customers. For this reason, major gas utilities in Japan such as Tokyo Gas, Osaka Gas, and Toho Gas have been continuing to devote research investments toward developing fuel cell systems different from Phosphoric Acid Fuel Cells (PAFC). They have their own laboratory and are knowledgeable about technologies for converting natural gas into hydrogen (reformers).

These utilities favor PEMFC (see Sect. 1.2.3) for residential use. Tokyo Gas announced on December 6, 2004 that it will begin leasing 200 PEMFC 1 kW class units for ten years, from March 2005, under a special agreement with users. The units coming onto the market will be manufactured by Matsushita Electric Industries and Ebara-Ballard. The cost of leasing is 1 million yen, covering the ten-year period including maintenance costs. According to a press release, Tokyo Gas expects to be able to reach the stage of massmarketing in 2008. Following the press release, Misawa Home, one of the country's leading housing developers, announced that the PEMFC units will be installed with their highest-ranked residential houses, pre-including the leasing charge in the cost of the houses, in cooperation with Tokyo Gas. It is likely that the temperature of recovered heat of these units will not be high enough to be fed into the ordinary heating systems, such as floor heating. It would be necessary to add a supplementary hot water heater for such purposes.

Osaka Gas and Toho Gas, the second and third largest gas utilities in Japan, respectively, have announced that they would put PEMFC onto the market in early 2006, one year later than Tokyo Gas. But they will be selling their 1 kW gas engine CHPs, whereas Tokyo Gas is leasing. Both gas utilities plan to start marketing 0.75 to 1 kW_{el} PEMFC early in 2006; but the president of Osaka Gas has indicated the possibility of delaying the schedule further. The possible reason for this is that core technologies have not been developed fully enough to maintain the level of reliability needed for sale to ordinary households.

In addition to utilities and gas suppliers, oil refinery operators such as Nippon Oil (ENEOS) and Idemitsu Kosan are developing 1 to 5 kW range PEMFC for residential or small commercial uses, including fuel cells fueled by kerosene or LPG. As previously described, Japan has no natural gas transmission network covering whole islands. The number of LPG users, therefore, equals that of natural gas users. In order to keep or expand their LPG market, Nippon Oil has developed a system to generate hydrogen from LPG and combined it with PEMFC stacks developed by Sanyo Electric. Beginning in March 2005, they will start renting 150 of these units to customers around the Tokyo area for three years, with a rental cost of 60,000 yen per year, excluding the installation cost. Under the rental agreement, the users have to provide the data requested by Nippon Oil.

Japanese companies are also developing Solid Oxide Fuel Cell (SOFC) micro cogeneration systems. Kyocera, TOTO (leading ceramic toilet products manufacturer), and Mitsubishi Materials, together with some gas utilities, are developing designs for cell components and systems stacks for SOFC. They have been trying mainly to reduce the operation temperature of cells in order to extend the life of cell stacks without appreciably sacrificing the high efficiency of SOFC.

The government fund for developing PEMFC in fiscal year 2004 was 4.1 billion yen for a more reliable and less costly cell structure, and 2.37 billion yen for assessing the performance and building the basic analytic systems for the assessment. Another 100 million yen is also earmarked for LPG-driven PEMFC. The same year, 1.52 billion yen was devoted to SOFC systems development and 1.01 billion yen for raising the reliability and performance of SOFC, while reducing its cost.

13.4 Conclusions

Under the Kyoto Protocol, Japan is obliged to reduce its global-warming gas emissions by 6 % of the 1990 level. Yet its CO_2 emissions continue to rise; and it is currently estimated that Japan must now reduce its emissions by 14 % to meet the mandated target. While Japan needs to promote all possible ways to effectively reduce CO_2 emissions, its energy policy is not supporting the expansion of renewable energy. For this reason, the most promising means for reducing CO_2 emissions would be to raise energy utilization efficiency. In this regard, the government will continue to support the marketing of micro cogeneration, especially those models suited for residential applications, such as the 1 kW gas-engine-driven CHP "Ecowill" and 1 kW-range PEMFC. The total government subsidy budget for Ecowill has been doubled for the Fiscal Year 2005. Also, the government is preparing the expansion of its program for the installation of residential fuel cell systems. It is anticipated that the market for 1 kW sized micro cogeneration will show steady growth in Japan.

Still, much additional funding is required to make fuel cells a commercially attractive option. There have been rather optimistic expectations built up regarding the PEMFC systems. But there is the fact that PAFC took almost 30 years to reach the stage of technically feasibility and reliability for practical applications, though its cost per kW has not yet been reduced enough to be able to compete with conventional power systems.

From such a perspective, gas-engine-driven micro cogeneration will acquire wider acceptance. Oil-fueled engine micro cogeneration smaller than 15 kW is not available in Japan. LPG-driven ones are available, but the systems cannot receive advantageous arrangements such as government subsidies or tax benefits. For such reasons, gas engine micro cogeneration for commercial applications will show a steady growth.

Honda is the only supplier of engines for the 1 kW micro cogeneration. Since the growing demand for this sized micro cogeneration is assured for some years to come, new participants might emerge. In order to compete with electric utilities, which are supplying very attractive tariffs for all-electricity-powered homes, gas utilities and LPG distributers have to develop competitive solutions. If the 1 kW Stirling engine micro cogeneration which Powergen in UK have been selling (see Chap. 12) demonstrates good reliability and cost effectiveness, it is possible for major Japanese gas utilities to import the system for Japanese customers. It would, however, be necessary for them to spend some time to modify the system to comply with the requirements of the household energy market in Japan.

The market for gas engine micro cogeneration of less than 10 kW_{el} will show steady growth, mainly in major gas utilities' territory such as Tokyo, Osaka, Nagoya, and Hakata in Kyushu. As the result of energy market liberalization, electric utilities which are large natural gas users could enter this market by delivering gas to commercial customers via the gas network owned by gas utilities. But this possibility looks very slight for the coming few years. Furthermore, no one knows yet what the impact will be when fuel cell micro cogeneration comes into the Japanese energy market in five years.

14 Micro Cogeneration in the Netherlands

Michael Colijn

14.1 Introduction

The Netherlands is gearing up for mass market introduction of micro cogeneration. There is significant demand from the market in general: interest from the installation industry, requests from building societies and demo projects with the energy utilities. The government has included micro cogeneration in its energy transition program, and is looking for micro cogeneration to help develop the path to a more renewable energy future. Micro cogeneration developers – local and international - have identified the merits of the Dutch situation and are actively engaging government, energy companies and the boiler industry.

Before the end of 2005 the micro cogeneration market in the Netherlands will have made a step forward to becoming a part of the installation and heating services industry. By early 2008 the micro cogeneration market should have already established itself firmly as competition to the current high efficiency condensing boilers as the products being brought out can directly replace the current stock of central heating devices.

14.2 Potential

To determine the potential of micro cogeneration in the Netherlands, the boundaries of technology use have to be defined. This includes the generator type, the fuel choice and the capacity.

Micro cogeneration technology is very diverse: not only are different technology types being developed, such as Stirling engines, gas motors, steam cells and fuel cells (see Sect. 1.2), the basic fuel used to drive the technology can differ from natural gas and hydrogen to wood chippings and fuel oil. For the Netherlands, the most common heat fuel for homes and light industrial premises is natural gas. For any mass market application, Dutch micro cogeneration would use natural gas first, and only for niche applications would other fuel types be considered.

The Netherlands has one of the highest penetrations of natural gas in the world. For all practical purposes, virtually all buildings are connected to the natural gas mains network. Of more than 7 million domestic and light industrial sites, more than 95 % have a connection to the gas network. The theoretical potential for micro cogeneration is thereby more than 6.65 million units, and this makes it an attractive potential market for micro cogeneration.

However, a more practical method for analyzing the market is to base it on the number of central heating boilers that are placed each year. Each year some 385,000 central heating boilers are placed in homes in the Netherlands. Just fifty to eighty thousand of these are for new homes or premises; the rest are replacements in existing homes. This means that more than 300,000 boilers are replacements.

For micro cogeneration to be a mass market product it has to be able to replace the existing heating technology of a home. Of the 300,000 or more replacement boilers per year, the current micro cogeneration technology available will not be suitable for all homes. There are two limiting factors that drive the economics of micro cogeneration in the current regulatory framework: the heat demand and the electricity demand. The electricity demand has a winter peak in the Netherlands that mostly coincides with the heat demand. The heating season is substantial, with six to seven months of space heating required; the remaining thermal capacity required for hot water throughout the year.

The average Dutch home require some 1600 m^3 of natural gas per year for hot water and heating purposes. The same home uses on average 3500 kWh of electricity. According to ECN, the instantaneous electricity demand does not come above 1 kW for 95 % of the time over a given year.

Micro cogeneration technology should therefore aim to fit around this average demand for natural gas and electricity. The most promising technologies at present are those that can cover the maximum heat demands of between 15 kW and 24 kW, and cover a substantial part of the electricity requirement – though not more than the annual consumption. When an average heating device runs for 2500 hours per year, a micro cogeneration of 1 kW electricity generating capacity would produce an equivalent number of kWh: 2500. This is equivalent to 2/3 of the annual consumption. If more electricity is produced than is required, the household is faced with electricity export to the grid, and at present there is no adequate system for dealing with exports. This would then make the micro cogeneration uneconomical, and not attractive to the end user.

When taking these factors into consideration, the number of premises suitable for micro cogeneration, given the current technology, drops down to 150,000 per year.

A final requirement is comfort. The average boiler in the Netherlands is light weight (40-60 kg), compact (it hangs on the wall in the attic / under the roof) and silent – it produces less than 42 dBa. Moreover, on the morning of a cold winter the average boiler is able to heat a home within 30 minutes. 85 % of all boilers sold are high efficiency condensing. And over 65 % are combination boilers able to produce hot water and heat directly from the device. Micro cogeneration has to match this level of comfort, now accepted as standard by a population who have been exposed to increasing sophistication and comfort of the central heating system over the last 30 years.

Therefore, when the current micro cogeneration technology is compared with the boilers on offer now, its still higher weight, above average noise level and generally larger space requirements will further reduce the potential uptake. Nonetheless, based on these conditions, micro cogeneration can still directly replace the current boiler in some 50,000 homes per year.

This is a substantial number in a market that has been the most rapid in uptake of new heating technology over the last two decades. When taken at this growth potential, micro cogeneration would have become a significant contributor to the electricity pool in over 5 % of all homes in just 10 years. With further advancements in micro cogeneration technology to allow direct replacements of existing boilers more easily, the potential will only increase.

14.3 Players

There are a number of micro cogeneration developers active on the Dutch market at different stages of development

14.3.1 Micro Cogeneration Developers in the Market

The Dutch market has seen one player already active from some ten years: this is the German manufacturer Senertec (see Sect. 1.2.1). In total Senertec has sold and installed some 100 micro cogeneration units known as the Dachs. The Dachs is a hand built gas motor micro cogeneration, meeting all the strict requirements of heating devices in Germany. However, the Dachs has an electrical output that far exceeds the demand of an average home, and is therefore more suitable for higher demand facilities such as small hotels and hospitals. Furthermore, the 500 kg weight and large size of the Dachs limit its use to facilities with ample space for installation.

The other player selling units commercially in the market is WhisperTech of New Zealand (see Sect. 1.2.2). WhisperTech has brought its Stirling based WhisperGen unit to the Netherlands, after first launching in the UK. The current unit of WhisperTech is ideally suited to the Dutch market in terms of outputs. It delivers 7 kW_{th} and 1 kW_{el} output – a combination well suited to a well-insulated modern home with modest heat demand. The drawbacks of the current model are weight (124 kg) and noise (slightly above the 45 dBa), though both of these will significantly improve in future models. It is floor standing, and only the size of a washing machine, allowing it to be built into a kitchen, or a modest space in the garage. WhisperTech are currently engaged in a 50 unit demonstration project together with Gasunie and the 12 Dutch utilities. When this project proves successful, the WhisperGen will become the first mass market micro cogeneration in the Netherlands.

14.3.2 Micro Cogeneration Developers Trialling

Vaillant are engaged in a trial in the Netherlands with their fuel cell micro cogeneration. The 12 units being tested are part of their wider European trial that involves more than 30 units. The unit is very large, to allow for easy access during the trial, and there is scope for making it more compact. Vaillant have indicated that their current model is not going to appear in the Dutch market commercially before 2010. This means that the end of the current trial will led to a reduced level of activity with the fuel cell activity in the Netherlands. Furthermore, the current unit also has outputs that exceed the demands of an average home, thereby reducing the market to light industrial applications, as for the Dachs.

ENATEC have trialed some 12 units in the Netherlands over the last years. As a Dutch developer of a Stirling-based micro cogeneration, their original plan to launch in 2003 did not pull through. The company is a combination of research institute ECN, electricity utility ENECO and boiler manufacturer ATAG. The outputs of the unit are similar to the WhisperGen, and they would eventually compete in the same segment. In 2004 and 2005 ENATEC have been very quiet, with no further announcements of developments, trials or manufacturing agreements.

14.3.3 Other Developers

MicroGen have been actively engaging with installation and energy companies for the last three years. They have participated in a government feasibility study, and have developed concrete solutions to regulatory barriers still existing in the Dutch system. MicroGen are developing a 1 kW_{el} Stirling engine at their UK developed center. Their wall hung solution to micro cogeneration aims to look like a condensing boiler, making it an easy to fit replacement. However, despite numerous announcements, they have not been able to bring their unit to market in 2004 as planned, and are now aiming for 2007. The main issues seem to be noise and vibration, and lifetime guarantees for the engine.

There are several other developers active in the Netherlands. These include Honda with their gas motor and a Dutch gas turbine developer. Honda does not yet have a product suitable for the Dutch market and will probably be some 3 years away from having a product on the market. The turbine developer does not have a product yet, and it is expected that basic prototype development will require another three years. Hence, the main players for the Dutch market in the near future are Senertec, WhisperTech, MicroGen and ENATEC.

Apart from these technology developers, there are other players of importance in the value chain such as boiler manufacturers and energy companies.

14.3.4 Manufacturers

The Dutch boiler manufacturers have been known for their innovation and acceptance of the condensing boiler. Now that developments of this product are nearing the end of life, a next generation of products could be micro cogeneration for them. The industry, however, has consolidated significantly over the last years, and only few players are independent. This means that scope for developments has been reduced as smaller Dutch players have been absorbed into larger multinational organizations.

ATAG is already involved in the ENATEC development, though recent developments indicate that some changes are afoot in the consortium. Up to now, no concrete numbers have been announced for manufacturing, though the question is whether this is due to the boiler side of the development, or the engine.

NEFIT is now part of Bosch Buderus and is no longer independent. However, the Nefit company will continue to function as an R&D site, and hence, any new developments would be created there, including micro cogeneration. Bosch, of course, are working on their own Stirling based micro cogeneration through their Junkers subsidiary. This development is a 450 W_{el} unit, expected for launch in 2008. However, this is a general development, not specifically for the Dutch market.

REMEHA have been actively participating in the Dutch government's transition program, indicating an interest in new technology, and particularly in micro cogeneration. REMEHA is active in the Netherlands and Germany, and would be well suited to manufacture micro cogeneration as they are independent, and in need of some new technology. To date, however, nothing concrete has been announced.

AGPO are similar to NEFIT in that they have become part of a larger company: Ferroli. Their R&D function makes them of interest to the micro cogeneration developers as the AGPO arm steers developments that impact the whole group. The down side is that they have no manufacturing capacity in the Netherlands, and would require commitment from their parent to engage in the development process towards a new product.

Intergas is an independent Dutch manufacturer. However, their target segment is the bottom end of the market, where micro cogeneration does not naturally fit at this moment in time. Intergas may enter the market in a few years, but is unlikely to be among the first to develop micro cogeneration.

The Magic Boiler Company is a new development in the Netherlands looking to manufacture a range of micro cogeneration technologies. They will manufacture white label micro cogeneration, and thus act as a risk buffer for both developers and boiler manufacturers. The Magic Boiler Company is developing its concept in the north east of the Netherlands, close to the boiler industry, research centers and the German border. The Magic Boiler company has developed a modular manufacturing concept, that allows it to integrate generators of different types, using mostly standard heating technology. Their aim is to come to market early to mid 2007 with the first products for the Dutch and German markets.

14.3.5 Gasunie

Although Gasunie is not a manufacturer of micro cogeneration, nor a mass market distributor, they have been extremely active at promoting micro cogeneration. Gasunie has been testing and developing several technologies over the last 10 years. They engage with developers to bring the product to the next level of market readiness, and ensure that sufficient publicity is given to these developments to open doors to the market.

One of the main concerns of Gasunie is that micro cogeneration is perceived as a risk, both by the end user, as by potential manufacturing partners. The first is needed for the market, the latter to make the product. The installation branch can help convince the end users that the technology is now sufficiently reliable for them to make a purchase decision.

The manufacturing partners, however, are concerned about betting on the wrong horse: what happens to the technology investment if a better technology arrives on the market in a short space of time? Gasunie aims to remove part of this worry by co-offering "ready to manufacture" developments through their cooperation with developers.

14.3.6 Electricity Utilities

The Netherlands has 12 electricity and gas utilities. All of them are engaged in the micro cogeneration discussion to greater or lesser extent. However, there are four that are actively looking for a method to engage and position themselves as leading the micro cogeneration market: Essent, Nuon, Eneco and Delta.

Essent is the Dutch utility that has traditionally covered the south, and part of the north east of the Netherlands. Essent has no position in solar energy, which they believe is economically unfeasible, though they have active engagement in wind industry projects. Through their vast installer network of the "Inhome" organization, they are interested to market micro cogeneration as the innovative solution. With more than several thousand installers, their ability to roll out a mass market product is strong.

Nuon, traditionally active in middle and north west Netherlands, have previously embraced solar and wind energy, as part of their division Nuon International Renewables. They are now more focused on profits than before, and view micro cogeneration with a healthy sceptical optimism: better than solar photovoltaics, but still to prove itself. Like Essent, they want to be first movers and therefore actively engage with developers, the trade association and the government. Again like Essent, a very strong installation network would allow roll out easily once products become available.

Eneco has been less engaged in government projects than the other two. To balance that, this utility of traditionally middle and western Netherlands regions, have had their share in ENATEC, thereby being well ahead of the others in micro cogeneration interest. For the moment, Eneco looks to continue its path of participating with ENATEC, thereby confirming its interest in the technology for the future.

Delta is the smallest utility of the four, and is present in the south west of the country. However, Delta has been active in wind energy, and has engaged both with the trade association Cogen like the others, and in the government energy transition program. Together with Altran consulting they have developed concepts for micro cogeneration, but no actual projects have developed as a result yet.

14.4 Projects

At the beginning of 2005 there is only one project for micro cogeneration planned in the Netherlands. This is the 50 units demonstration project with WhisperTech and Gasunie.

The total number of micro cogenerations in the Netherlands is the sum of 100 Dachs, 50 WhisperGen, 12 Vaillant fuel cells, 12 ENATECs and a hand full of lab units of different technology developers.

The Netherlands is not currently leading in terms of number of installed, commercially available micro cogeneration units. However, the government, utilities, developers and Gasunie are extremely ambitious and determined to succeed in making micro cogeneration a standard technology.

In 2006, Gasunie is hosting the International Gas Union (IGU) conference. For this, Gasunie aims to have 1000 micro cogeneration units installed, preferably in a cascaded virtual power plant set-up. This will be video-linked to all delegates at the conference as an example of what is possible using very small scale generating technology.

To facilitate the uptake of micro cogeneration, however, the government will have to improve the installation climate for this technology as there are a number of structural barriers.

14.5 Government Actions

The government is actively seeking new ways to develop a more renewable energy infrastructure in the Netherlands. To do this, they have engaged industry and research institutes in several programs. The biggest of these is the Energy Transition Program (www.energietransitie.nl).

14.5.1 Energy Transition

The Energy Transition Program is an initiative of the Ministry of Economics. This ministry holds the Dutch energy portfolio and is responsible for security of supply, supervising the energy regulator and developing an energy future. The ministry has set up a project team of civil servants and energy industry experts, and has organized a multitude of workshops together with industry to ensure buy-in from developers, manufacturers, researchers and distributors.

The program consists of four main streams in energy, of which the Team New Gas is one. Micro cogeneration falls under this category, in one of two subclasses called efficient gas use or combined heat and power using gas.

Under this program the ministry has financed feasibility studies for promising developments, with a goal to getting actual projects out of these studies. The projects in turn are able to request part financing so that the novel elements in all projects are covered, thus allowing the learning curve to be as cost neutral as possible. Repeatability will then become easier for suppliers of technology and roll-out in terms of marketing and distribution, and installation.

Unfortunately, only one of the developers managed to supply units for this program: Whispertech. The expectation is that by end of 2005, early 2006 more developers will be able to place a few units in trials or demonstration projects, albeit in limited numbers.

14.5.2 Eco Innovation

The Ministry of Environment is responsible for renewable energy production and carbon dioxide emissions. However, they do not hold the energy portfolio. Nonetheless, they launched an ambitious program entitled Eco Innovation in the Spring of 2004, ahead of the Dutch Presidency of the EU in July 2004.

The Eco Innovation Program aimed to bring forward examples from industry that show economically attractive means for cleaner and greener manufacturing, energy consumption and waste management.

Using several workshops throughout the EU, the Ministry gathered information on all possible methods and examples, and compiled a book of the 18 best examples of eco innovations. This book was presented at the Council of Environment Ministers in July 2004. CHP, and micro cogeneration in particular, are mentioned as the single best solution for achieving carbon dioxide emission reductions.

The Ministry does not have an active financing program like the Ministry of Economics. However, they have shown interest to develop case examples using micro cogeneration over the next year or two. As the Ministry's full name is "Ministry of Health, Planning and Environment" they also hold responsibility for urban planning and housing stock. In that light their support of micro cogeneration is significant as the impact on housing regulations can be very positive for micro cogeneration.

14.6 Regulations

The Dutch situation with regards to regulations is dismal at face value: there is no right to connect to the electricity grid, applications for connection to the electricity grid have to be made in advance and are unclear, there is no feed-in fee for excess kWh, the certification process is not in place for technology, and all home owners who buy a micro cogeneration would potentially be earmarked as entrepreneurs generating a dramatic bureaucratic paper mill.

Although all these points are true, history paints a different picture.

14.6.1 Right to Connect

The Netherlands needs a right to connect to be written in to the electricity law. This will remove any doubt with the future buyer of micro cogeneration that they will have to uninstall the unit once they have bought it.

The lobby effort to get this done is being made primarily by MicroGen's and WhisperTech's Regulations Manager as their products need to compete directly with condensing boilers. Several angles are being used: the European Embedded Generator Norm that is being drafted in Brussels; the Ministry of Economics Energy Transition Team; the Ministry of Environment; the trade association Cogen and the regulator DTe.

Although it is important to secure the right to connect firmly in the law, there have been no difficulties connecting small generators to the grid up to now: solar photovoltaic (PV) panels have been sold over the counter for the last 10 years to home owners who have plugged up to 600 W straight into the nearest socket. No meters have been changed in this process, and mostly, no mention of the installation was made to the local network company. For even the smallest generator, formally the network company should be informed in advance, the meter would have to be changed, and a certified electrician should have checked the wiring prior to installation.

Solar PV, however, is viewed as a niche product. Micro cogeneration is set to become mass market. It is therefore important to secure the right to connect, and specify under which conditions such a right exists. Furthermore, the energy market has fully opened up since July 2004, thereby adding competition to the market. The once more relaxed energy companies may take a more formal stance now that their income is driven purely by market forces.

14.6.2 Fit & Inform

Apart from the right to connect to the grid, it is important that micro cogeneration be allowed installation in the same timeframe as a condensing boiler. If micro cogeneration takes longer to start with installation due to an advance information requirement, its chances of becoming a mass market product are greatly reduced. This is because most condensing boiler sales are distressed purchases: when the boiler breaks down the owner wants a replacement within 24 hours.

At present, any generator installation needs to seek advance permission at least three working days in advance according to the Netcode. This can be changed to a same day information, like the UK's G83 standard has included.

14.6.3 Meter Changes

The current meter market is theoretically liberalized in the Netherlands. Every home owner could choose to have a meter replaced by an independent company, and have a contract with that company for a different meter. However, the process of changing is very costly and time consuming, so that in effect, there is no meter market.

In order to reduce costs for the end consumer, the meter change should be part of the micro cogeneration package: installation performed by an accredited installer could easily include the meter. This would prevent one call-out fee; it would also reduce the time needed for the home owner to be away from work on a separate day from the unit installation.

Furthermore, choice of meter would again reduce the costs of the meter change. Where a meter is currently charged to the end user for amounts upward from € 50, an accredited import-export meter is available from € 18.

Therefore, a liberalized meter change market, with same day meter change, using a meter of choice would reduce costs and time needed for the overall installation, and improve both customer experience and efficiency.

14.6.4 Feed-In Tariff

In the Netherlands, there have been no phenomenal feed in tariffs like in Germany. Renewable energy has received up front support up to 50 % of the investment costs, but no structural support on the electricity side.

Former programs for condensing boilers included up front support as well, yet showed that 60 % of end consumers bought the boiler without seeking subsidies, despite their availability.

The current electricity law, however, allows "renewable domestic generators as well as hybrid generators that make use of fossil fuels" to feed up to 3000 kWh into the electricity grid per connection per year. This allows those kWh to be subtracted from the used kWh and thereby receive a full value of the electricity.

The difficulty with this inclusion from May 2003 is that there are too many uncertainties over the definition of renewable and over hybrid systems. The parliamentary initiator of this amendment mentioned micro cogeneration, yet it is not written into the law. What is needed is to clarify the current amendment and clearly state that micro cogeneration is to be included in the list of permitted types of domestic generators.

14.6.5 Certification

The current technical standards that apply to micro cogeneration are for generators up to 5 MW. This is due to the novel aspect of micro cogeneration where they did not exist before. Clearly the requirements for a 5 MW plant are different than for a 5 kW_{el} micro cogeneration unit, and a simpler version needs to be made for the domestic generator types.

The European embedded generator norm is one document that will be of value. However, that will need to be adopted in national standards, and a national certification procedure. Hence, the requirements for installers and for developers to their current products should be spelt out at national Dutch level, rather than waiting for an EU norm. This will provide adequate feedback into the Brussel's process where the EU norm is still being debated and contested by other national representatives.

14.7 Expectations

The Netherlands is poised to become an extremely active micro cogeneration adopter. There are many developers actively engaged in the market, looking for both a distribution market, as well as potential manufacturing capacity.

The government has earmarked micro cogeneration as a promising emissions abatement technology, and has given it the initial support required to kick start projects.

The utilities and Gasunie have shown their interest in the technology by taking the initiative to develop a project of their own, and by planning for a big demonstration virtual power plant in 2006.

What is now needed are two elements: the government needs to adapt its legislation and standards to allow easy adoption of the new technology in this eager market. And the developers have to stick to their promise of delivering a product to the needs of the market, and within the timeframe required: as soon as possible.

15 Summary and Conclusions

Martin Cames, Corinna Fischer, Martin Pehnt, Barbara Praetorius, Lambert Schneider, Katja Schumacher, Jan-Peter Voß

This book has dealt with micro cogeneration, the combined production of heat and power in small units for individual facilities, such as single- and multi-family houses or small commercial enterprises. Its main characteristic is that power and heat generation takes place immediately at the consumer's site. Micro cogeneration represents a socio-technical configuration that differs from the dominant architecture of contemporary electricity systems based on large-scale power generation with long-distance transmission to end-users. With micro cogeneration, consumers could potentially become operators of power plants. This would affect the roles and interests of a number of players in electricity markets to a more than marginal extent. For instance, electricity distribution and retail structures would be affected if micro cogeneration was broadly diffused.

In this book, we have assessed the potential contribution of micro cogeneration towards a sustainable transformation of the electricity system from various thematic angles linked to specific disciplines and from different national perspectives. The following sections of this final chapter give a summary of the main findings. We then look at the specific setting and potential for micro cogeneration in Germany and indicate starting points for improving the implementation context for this (and other) innovations, with the ultimate objective of contributing toward a sustainable energy system transformation.

15.1 Towards a Decentralized Energy Supply: Summary

Conversion Technology

The core of micro cogeneration is the application of small-scale conversion technologies that concurrently generate electricity and heat. Different technologies may be applied for this purpose. Reciprocating engines are by now the most reliable and cost-effective technology; yet they have can some ecological disadvantages, such as higher NO_x and CH_4 emissions. Another option is the Stirling engine, which operates with exchangeable external heat sources and, therefore, offers high fuel flexibility. Fuel cells directly convert chemical energy into electric energy without combustion, leading to low emissions of air pollutants and noise.

Reciprocating engines are well-developed technologies, with more than 10,000 units having been sold in Germany to date. Stirling engines are currently entering the market, in particular in the UK, while fuel cells are as yet in a development and testing phase. Further technologies are also being developed, including Rankine steam cycles, thermo-photovoltaic and thermo-electric processes.

Micro Cogeneration: A Cluster of Innovations

The innovativeness of "micro cogeneration" goes beyond the conversion unit as a technical artifact. Technologies become innovations only if they are actually used, and for this they need to be embedded in existing socio-technical contexts.

Micro cogeneration may thus rather be conceptualized as a socio-technical **innovation cluster** and not as a technological innovation, even more so since the technologies it builds on have been available for a long time. Within this cluster, the development of conversion technology interacts with other innovations that may promote or inhibit the application of micro cogeneration. Vice versa, these innovations may be promoted or inhibited by developments in the conversion technology. Such a cluster comprises technological innovations, such as new materials or control technologies, but also social and institutional innovations such as the training of installers, user routines, company strategies, or adjustments in the political, regulatory and economic framework conditions. To assess future opportunities for micro cogeneration, it is necessary to take these different perspectives into account.

Currently, micro cogeneration represents a rather insignificant contribution to the worldwide power generation portfolio. Taking Germany as an example, we estimate that, by the end of 2004, a stock of about 60 MW micro cogeneration capacity was installed, producing about 0.04 % of overall electricity generation. Thus, though sales numbers are increasing, micro cogeneration is still in a very small niche market and far from being introduced to a broad market.

In the public perception, high expectations are associated with the introduction of certain technologies, such as fuel cells, while other – more developed – CHP options, such as Stirling and reciprocating engines, are hardly known yet. Correspondingly, perceptions and expectations on the level of government, industry, and science will impact on the level of policy support and incentives granted to micro cogeneration. Implicitly, such expectations are also captured in energy scenarios.

An assessment of available **energy scenarios** of different types (forecasts, technology scenarios, policy scenarios, explorative scenarios) reveals that micro cogeneration is treated differently across scenarios, sometimes even within the same scenario type. Besides these differences, there are also commonalities within scenario types. In forecasts, whose nature is rather to extrapolate from current trends, micro cogeneration does not appear as an important element in future energy systems. In technology foresight studies it is recognized as a possible fuel cell application, with fuel cells being considered a potential break-through technology. What counts, though, is the novel technical principle (the fuel cell), not its application. In policy scenarios, micro cogeneration appears mainly in scenarios with strong climate protection goals as part of a more general CHP strategy, which mostly includes larger CHP plants. In explorative scenarios, micro cogeneration plays an important role in possible developments toward technological decentralization.

Some scenarios point to issues that need to be considered if a coherent micro cogeneration strategy is to be developed. On the one hand, its potential contribution to energy supply depends on demand drivers such as population development, building stock and structure, heat demand, and the structure of the heating systems. The expected decrease in heat demand, for example, may reduce the potential for micro cogeneration applications. On the other hand, micro cogeneration also faces competing – and ecologically attractive – supply options like larger CHP, based, for example, on biomass, solar heat, or electricity import from renewable sources. However, the scenarios show that, despite a decreasing heat energy demand and an increasing share of renewable energy sources, micro cogeneration could contribute to the future energy system. In a sustainability scenario for Germany, micro cogeneration is expected to provide some 3 % of the German electricity demand by 2050. This is a small – yet relevant – building block for an energy policy working toward a sustainable transformation of the electricity system, and forms the starting point for analyzing the current situation as well as the incentives and barriers to exploring its potential.

Economic Performance

Whether or not micro cogeneration can live up to its potential and gain a considerable market share, thus contributing to a sustainable transformation of the electricity sector, will – among other factors –

depend on its economic performance. The analysis shows that micro cogeneration is increasingly becoming an economically interesting option for operators. To date, a number of systems larger than 5 kW_{el} have been successfully commercialized. Smaller systems of about 1 kW_{el} may soon break through. They could fully substitute for boilers in single-family houses. Under current conditions, single-family houses are appearing to become a particularly promising market segment: micro cogeneration plants are only economically attractive if most of the electricity is to be used or sold on site and not primarily fed into the grid. In the case of apartment buildings, tenants may choose to be supplied by an external energy company and, thus, jeopardize the economic viability of the cogeneration unit.

Generally, economic viability of micro cogeneration depends on the attractiveness of competing heat and electricity supply options. In urban areas with high heat densities, district heat may be an economically (and ecologically) more attractive option, whereas in areas with low heat densities micro cogeneration appears more promising.

The economic attractiveness of micro cogeneration for operators builds largely on country-specific regulatory advantages, such as tax exemptions, the payment of a CHP bonus for electricity fed into the grid, and avoided concession levies and grid charges for electricity generated on site. If these incentives are excluded, and as long as the external benefits of the improved environmental performance of micro cogeneration are not being taken into account, none of the micro cogeneration technologies assessed here is economically viable in Germany yet. Furthermore, with state-of-the-art micro cogeneration technologies, a trade-off between environmental and economic performance must be acknowledged. In particular, small Stirling engines designed for single-family houses have a rather good economic performance, but a relatively low electrical efficiency. Reaching the highest electrical and total efficiencies possible is crucial for achieving emission reductions at reasonable costs.

Ecological Performance

With respect to ecological performance, we conducted an environmental life cycle analysis (LCA) of micro cogeneration units, particularly of gas-fuelled ones. Compared to separate production, generating electricity and heat in micro cogeneration units leads to **primary energy savings** and **greenhouse gas** (GHG) **benefits**. Based on natural gas as a fuel, GHG emissions per kWh electricity and heat produced are typically 20 % – and under certain circumstances up to 45 % – lower in comparison to condensing boilers and combined cycle power plants. In some cases, for example of technologies with low electrical and total efficiency, however, only little – if any – GHG mitigation can be achieved, compared to separate heat and power production with state-of-the-art technologies.

Compared to **district heating**, micro cogeneration does not offer significant energy and GHG emissions advantages. District heating systems have the disadvantage of high heat-distribution losses; but these are offset by significantly higher electrical efficiencies compared to micro cogeneration. Therefore, we regard micro cogeneration not as a competing, but rather as a supplementary technology to district heating, meaning that it should optimally be applied in cases where larger district heating is not viable for infrastructural or economic reasons. In rural areas, for instance, building density is often low, causing long transport distances and thus high investment costs and large distribution losses for district heating networks. On the other hand, natural gas (or heating oil) is readily available making individual heating solutions like micro cogeneration attractive.

The comparative performance of CHP technologies with respect to climate and resource protection depends on their electrical and thermal conversion efficiency. To exploit the advantages of micro cogeneration, it is important to optimally integrate the system into the heating system of the house (low return temperatures, use of condensation heat, etc.).

Micro cogeneration plants also cause emission of **air pollutants,** such as nitrogen oxides (NO_x). Here, fuel cells – and to a lesser degree Stirling engines – offer the advantage of low emissions during operation. Reciprocating engines, in contrast, emit significant amounts of NO_x, CO, and hydrocarbons. However, the assessment of local air quality shows that for gas-fuelled systems, even if residential areas were fully equipped with reciprocating engines, pollution load would be below the level of irrelevance as defined by German law. Other emission sources, such as transportation, contribute significantly higher NO_x loads to the atmosphere.

Another relevant parameter for ecological performance is the **fuel** a micro cogeneration system runs on. Besides natural gas, heating oil, LPG, and renewable fuels (such as vegetable oil), biogas, gases produced from solid biomass, and other primary energy sources may also be used. Reciprocating engines based on heating oil are widespread and exhibit similar GHG advantages as natural-gas-based engines when compared to conventional modern oil heating systems. However, the NO_x emissions of these systems are significantly higher than in the case of other micro cogeneration technologies.

The most flexible technology regarding the choice of fuels is the Stirling engine. The use of biomass, e.g. wood, is more complex in micro cogeneration than in larger plants because the required process components (e.g. gasification, gas clean-up) are more difficult to realize on a small scale than in larger systems. Integration of renewable energy carriers into micro cogeneration systems will, therefore, happen later in this market segment.

User Perspectives

For an ecological assessment, it is also important to take into account interactions between technology and user behavior. Analysis of the literature reveals that **"rebound" effects** could be relevant to the overall performance of an innovation. This means that efficiency gains through technology may be overcompensated for by an increase in energy consumption, e.g. through additional energy-consuming appliances or altered consumer behavior. Conversely, efficient and innovative technologies may also sensitize consumers about energy issues and stimulate more energy-conscious behavior. In the case of micro cogeneration in Germany, a rebound effect may occur especially from application of the "Energy Saving Decree", which allows for less efficient building insulation if cogeneration is used for heat supply, potentially leading to an offset of efficiency gains from micro cogeneration. Whether consumers are sensitized depends heavily on the form of feedback given on energy consumption and production as well as on the entire system of incentives, including financial ones.

Pioneer users play an important role in the diffusion of new technologies if testing them and giving feedback to manufacturers and if spreading the word and acting as multipliers. We have explored the characteristics and motives of pioneer users of micro cogeneration: first analyzing the literature on pioneer users of other innovative energy technologies and then conducting a postal survey of applicants for a fuel cell field test as well as holding group discussions both with applicants and actual users. The analysis reveals that pioneers of micro cogeneration come from a well-educated, established, middle-class population with good income. Their lifestyles are rather traditional, they usually have families and live in their own houses in rural areas or small towns. Their education is very often of a technical nature, spurring interest in new technologies. Striking is their relatively high age and the almost complete absence of female pioneers. Pioneers are environmentally conscious and positive that innovative technology will help to solve environmental problems. To make their own contribution to their diffusion, many of them already possess "green" energy technologies like solar heat, heat pumps, or efficient household appliances. These technologies might hence serve as a "door opener" for micro cogeneration.

However, pioneers also wish their home energy system to be cost-effective, reliable, and user-friendly, at least in the long run. In these domains, they spot deficiencies in fuel cell-based micro cogeneration. Helpful tools to overcome uncertainties with respect to cost and performance are reliable information (especially if based on personal experience), contract arrangements, and support schemes. Pioneers are willing to give an immature technology a chance, but strongly point to the necessity of remedying the deficiencies in order to reach a mass market. Other micro cogeneration technologies than fuel cells might perform better in this regard. On the downside, especially reciprocating engines are sometimes perceived as being "dirty". Despite the perceived shortcomings, pioneers tend to be very positive about their technology, willing to promote it, and function as communicators and multipliers.

German Market Structure: Institutional and Policy Framework

If introduced on a larger scale, micro cogeneration would substantially change a number of technological and structural features of the markets for electricity and heat supply. The attitudes of major players in the energy market towards micro cogeneration depend on whether they perceive themselves to be winners or losers from the expected impacts of such a larger-scale introduction.

Market liberalization led to substantial changes in the structure of the German energy industry. The market is now dominated by four vertically integrated electricity companies, which also own major gas companies and shares in all levels of electricity and gas supply. Competition is stagnating and newcomers are rare. The established electricity industry does not focus on small cogeneration units, but rather follows its traditional path of centralized electricity generation. On the other hand, the natural gas industry and local utilities, insofar as they are not owned and dominated by the electricity industry, are interested in micro cogeneration. Yet, a prevalent strategy towards fostering the broad-scale introduction of micro cogeneration could not be found among any of the German gas companies.

Distribution network operators (DNOs) do not currently have much motivation to foster distributed generation. The existing economic (dis)incentives and ownership structures do not provide attractive terms for connecting distributed generators. DNOs may receive new incentives to facilitate the connection of micro cogeneration, depending on the evolution of an incentive-based regulatory scheme by the new German energy regulatory body. This means that, at this point, technology manufacturers themselves and third party financing companies (or energy service companies), together with a handful of gas companies, are the main drivers for micro cogeneration.

Liberalization has led to a sharp decrease in **electricity prices**, which made the operation of existing, and the installation of new, CHP plants more difficult. However, electricity prices, in particular for households, have started to rise again and are expected to mount even further over the next few years. In addition, emissions trading is likely to further increase the cost of conventionally produced electricity and thus improve the competitiveness of micro cogeneration due to the higher value of avoided electricity purchase. Moreover, German legislation has enhanced the economic opportunities for micro cogeneration plants, in particular by exempting micro cogeneration plants from electricity and natural gas taxation, but also by introducing a bonus for electricity fed into the grid.

All in all, the promotion of CHP is an established activity area of German **climate and innovation policy**. The latter has one focus on energy technologies, with explicit political support for decentralized technologies. Thus, after the first shock of liberalization, the general institutional framework for micro cogeneration has improved considerably. However, due to changing governmental priorities and an as yet unclear continuation of CHP policy, the perspectives remain uncertain. Micro cogeneration has not yet taken on a role of its own in German energy policy. It is either subsumed under CHP in general or entrained by the fascination for fuel cells. The specific socio-technical features of small decentralized CHP are not considered elements of a general vision for future energy systems, nor are they systematically cast into policy strategies.

Also, the generally favorable economic conditions for the operation of micro cogeneration plants seem not to be sufficient to effectively promote their diffusion. Grid operators use the currently weak regulation of grid access to hinder the emergence of independent power producers. A sufficient procedure for managing conflicts – between, for example, micro cogeneration operators and distribution network operators – has not yet been established. Due to the lack of regulation, negotiations about grid access and tariffs are bilateral, and disagreement often has to be settled by court decisions – an expensive way of regulating conflicts. The introduction of a German regulatory body in 2005 may change this situation.

An additional barrier is the low level of information on the customer side. Technologies other than fuel cells are little known and elicit little enthusiasm. Because of small plant size and operation by non-professional electricity producers, complicated rules and lack of access to information give rise to relatively high transactions costs and reduce the economic attractiveness of micro cogeneration plants. Acceptance of novel marketing and operation strategies that might reduce these information and administrative cost – such as third-party financing – is comparatively low. General financial support, such as feed-in bonuses for CHP, even with higher rates for small plants, is therefore not sufficient to effectively promote this kind of innovation.

15.2 International Experience

While our analysis has so far focused on Germany, the comparative perspective on a number of other countries nicely shows how differing framework conditions may significantly influence the potential for micro cogeneration.

Micro Cogeneration in the USA

In the USA, infrastructure and economic conditions are much less favorable for micro cogeneration than in Europe. Heating systems are generally based on forced-air technology, that is, furnaces. That fact adds cost and complexity to micro cogeneration systems, which usually generate heat in the form of hot water. Moreover, investment costs for a furnace are much lower than for a boiler, increasing the difference in investment costs between micro cogeneration and "conventional" systems. The generally lower level of electricity prices further reduces the value of avoided energy purchase; and, as peak loads due to air-conditioning usually appear during summer, they do not coincide with the maximum operation hours of heat-driven micro cogeneration. Moreover, a patchwork of different network operators and utilities, each with their own interconnection requirements and tariff systems, makes it more difficult for micro cogenerators to become connected. Conditions for micro cogeneration are most favorable in the north-eastern part of the USA, where electricity prices are high and there is a high space heating demand.

In the USA, policy priorities are more mixed than elsewhere. Climate protection is not as high on the agenda as in Europe, but micro cogeneration fits into policy priorities like security of supply and energy efficiency. Against this background, the Department of Energy has launched a "roadmap" for the development of micro cogeneration that sets targets for costs and defines areas for action: defining of markets, development of technologies, and acceleration of acceptance. Furthermore, programs exist for increasing the share of CHP. Fuel cells are receiving special interest; state support programs and policies, such as the introduction of net metering, focus with few exceptions on fuel cells.

All in all, only a few companies in the USA are developing micro cogeneration for residential applications, and only one of them, Vector Cogen, is currently marketing a reciprocating engine. Climate Energy is set to field test several units based on Honda's reciprocating engine, and hopes to market the product in 2006. Most of the other developers are working on fuel cells, many of which are at the same time designed for applications other than micro cogeneration.

Micro Cogeneration in Great Britain

Great Britain poses almost exactly the contrary picture. High annual heat demand – caused by a prolonged heating season and significant heat losses of buildings –, a good gas infrastructure, and comparatively low gas prices together with high electricity prices are supportive to render residential micro cogeneration economical. In such a context, energy supply systems with a high heat-to-power ratio are particularly attractive. Micro cogeneration would become much less attractive should quality insulation be provided, because operation hours and, thereby, electricity production would decrease.

Micro cogeneration also suits the four main policy goals articulated in the Energy White Paper of 2003, namely, security of supply, carbon abatement, combating fuel poverty (that is, providing affordable heating), and competitiveness. Consequently, the Energy Act of 2004 obliged the government to develop a micro generation strategy. Support measures will possibly soon be introduced, among them a VAT reduction and a favorable treatment of micro cogeneration within the Energy Efficiency Commitment. Other framework conditions, such as building and connection regulations and network access, are also favorable, and there even exists a lobby for micropower. Even though the market is currently restricted to the replacement of floor-mounted boilers, a potential annual market of up to 500,000 units is estimated. A number of technological advancements are currently being developed in order to meet the specific demands of the British market, such as thermal storage, metering technology and standard load profiles.

Table 15.1. Framework conditions for micro cogeneration in different countries

	Germany	USA	UK	NL	Japan
Political and support framework					
Relevant energy / climate policies	Ambitious target for CO_2 reductions through cogeneration; liberalization of electricity market; energy saving policy (i.e. Energy Saving Decree)	Deregulation; no CO_2 emissions target; security of supply policy; energy efficiency policy	Ambitious CO_2 reduction target; Energy White Paper 2003/ Energy Act 2004; building regulations; security of supply policy; fuel poverty policy	Ambitious CO_2 reduction target; elaborated climate policy implementation plan incl. energy efficiency policy; environmental protection targets	CO_2 target to come; ongoing process of liberalization; deregulation and privatization; focus on energy security; CO_2 target to come
Cogeneration policy	Cogeneration law	DoE program to increase cogeneration	£ 50 million subsidy for local CHP	Cogeneration Incentive Programme (1988); operating support via provisions in energy tax	(Minor) part of policy to increase energy supply security
Micro cogeneration policy and support schemes	Subsidy / remuneration scheme for electricity feed-in from small cogeneration	Micro CHP Technologies Road Map (June 2003); financial incentives on state level	Energy Act 2004: micro - cogeneration strategy due by end of 2005, support measures (VAT reduction) to be introduced	Feasibility studies financed under Energy Transition program, business purchasing micro cogeneration is entitled to tax breaks on investment	Government subsidies; tax deductions; low interest loans; lower gas tariffs for gas-fired micro cogeneration; promotion activities
Connection to electricity grid	Obligatory, bilateral negotiation with DNO; little standardization	Different regulations in different states	Standardization in 2003	No regulation of or right to connect; no feed-in fee; in practice, no difficulties in connecting	Grid connection, but no selling of generated electricity to the public grid
General situation on energy market	Horizontal / vertical concentration; high energy prices; regulatory agency to come	Low electricity prices; liberalized markets; diversified supply structure	Comparatively low gas prices; high electricity prices; high supplier change rates	Large domestic natural gas reserves; market liberalization; unbundling; regulated TPA	Liberalized markets; gas suppliers under increasing pressure from electricity companies; high importance of LPG
Market and technology situation					
Technology focus	Reciprocating engines (market leader); fuel cell development	Fuel cell development	Small Stirling engine (1 kW)	Mostly reciprocating engines; small Stirling engine (1 kW) to come	Gas-fired engines; fuel cells
Economic situation of micro cogeneration	High transaction costs; small margins, yet increasing sales numbers (few in residential sector)	Poor commercial viability	Small Stirling expected to be competitive in residential sector	Medium-size reciprocating engines for small business; otherwise no economic advantages	Few incentives; gas-fired engines already occupy (small) market share
Attractiveness for residential sector	District heating prevalent; well-insulated houses; decreasing heat demand; small economic incentives	Mostly warm-air heating (furnaces); few water-based systems; unfavorable infrastructure / climate conditions; no incentives	Favorable climate conditions; little insulation; high level of hydronic central heating; little district heating	Majority with access to natural gas and individual heating systems; little incentives to use micro cogeneration	Mostly gas appliances; little electric heating systems; some incentives for implementation
General setting for micro cogeneration					
Institutional setting	Setting not yet sufficiently supportive	Setting not supportive	Advanced system of support for distributed generation	Little support yet	Supportive in many regards, but not yet successful
Actors and coalitions	Little networking; no advocacy coalitions	Lobbying by SECA (an alliance between government, industry and the scientific community)	Strong advocacy coalition; lobbying by large companies (Powergen, British Gas)	Little advocacy; micro CHP working group (led by Cogen); some distribution companies involved in field trials	Lobbying by major gas utilities (Tokyo Gas, Osaka Gas, Toho Gas)

The two major energy suppliers, Powergen and British Gas (Centrica) have both made arrangements to commercialize micro cogeneration to their energy customers. Powergen is currently planning to install a minimum of 80,000 WhispherGen Stirling units by 2010, with customer retention being an important reason. Furthermore, British Gas is planning to launch a MicroGen Stirling unit in 2007, and a major boiler manufacturer, Baxi, is currently developing cogeneration units based on fuel cells and Rankine cycle engines.

Micro Cogeneration in the Netherlands

The Netherlands are currently on the cusp of becoming an active micro cogeneration adopter, but have not yet adjusted the institutional setting to be sufficiently supportive. The country offers a well developed gas supply infrastructure and the potential to replace about 50,000 traditional boilers by micro cogeneration units per year. If implemented at this rate, about 5 % of all homes would be equipped in about 10 years time. Private households, however, currently obtain few incentives to implement cogeneration. There is no feed-in remuneration for electricity generated from micro cogeneration units and as yet no connection standard.

At the same time, the Netherlands have a liberalized energy market with regulated grid access. They announced an ambitious CO_2 reduction target and an elaborate climate policy implementation plan which also includes cogeneration as a means to increase the efficiency of energy provision. As early as in 1988, a cogeneration incentive program had been launched, and the new Energy Transition Program explicitly fosters micro cogeneration by financing feasibility studies. Moreover, a business that implements micro cogeneration is entitled to tax breaks on related investments.

Despite the above incentive programs, there is little advocacy for micro cogeneration yet. A handful of distribution companies is involved in field trials. Only four out of twelve Dutch electricity and gas utilities are engaged towards micro cogeneration. The most active among them is Gasunie which has been testing micro cogeneration for 10 years already. Recently, a micro cogeneration working group was created, led by the international cogeneration federation, Cogen Netherlands. As for the results, in early 2005, there was only one project for micro cogeneration planned, a 50-unit demonstration project by the gas utility Gasunie and the Stirling engine developer WhisperTech. Technology development is currently focusing on such small Stirling engines, while reciprocating engines are already available on the market. In total, less than 200 units have been implemented so far.

Micro Cogeneration in Japan

Compared to most other countries presented in this book, micro cogeneration is comparatively well developed in Japan. Here, the main proponents are major gas utilities. Japanese gas companies are under increasing pressure from electricity companies who aim at supplementing gas-run appliances by electric ones. Japanese utilities are promoting all-electric houses with bargain tariffs; these arrangements compete with micro cogeneration. Gas companies therefore aim at using micro cogeneration to protect their market share. To this end, they make use of their existing distribution channels and knowledge about their customers' heat demand. Micro cogeneration also benefits from government subsidies, tax deductions and low-interest loans designed to support gas-fired CHP in order to replace oil and reduce air pollution. In the future, abatement of climate gas emissions is likely to become more important and trigger further government activities towards efficient energy technologies.

As a result, micro cogeneration already occupies a – albeit small – market share. The common technology is gas-fired engines, being applied both to commercial and residential uses. In 2003, approximately 1,100 units for commercial use were sold. In the residential sector, the "Ecowill" units of a size smaller than 1 kW_{el}, developed by Honda in co-operation with gas suppliers and boiler and water heater manufacturers, were introduced in 2002. By May 2004, about 3,000 units had already been sold with the help of a government subsidy and a lower gas tariff. Although connection to the grid, was recently made easier and is now the standard mode of operation, excess electricity is usually not fed into the grid but rather used to heat water.

Gas companies complement their current activities concerning gas engines with considerable investment in fuel cell development, supported by government R&D funding. Some companies have announced the introduction of fuel cell systems to the market from 2005 on. But until they reach competitiveness, the market for gas engine-based micro cogeneration will continue to grow. Due to the primary energy structure of Japan, fuels such as kerosene and LPG are considered for micro cogeneration purposes as well.

15.3 Conclusions

So far, our assessment has been illustrating that micro cogeneration could form a small but relevant building block of a more sustainable future energy system in Germany – and also in other countries. As the economic and ecological analyses have shown, the performance of micro cogeneration in both respects varies by technology and strongly depends on the respective implementation context; in many circumstances, other solutions for electricity and heat supply seem preferable. Such competing options include district heating in areas with high heat densities and the use of renewable energy carriers where they are available at reasonable costs. Any strategy to foster the use of micro cogeneration therefore needs to be embedded into a **broader transformation strategy** towards a sustainable energy system. Building on our research for Germany and the experience from other countries, we now develop some proposals how such a strategy or policy should take into account micro cogeneration. We start by considering competing technology options and close the chapter with some thoughts regarding the current institutional and economic setting for micro cogeneration in Germany.

Competing and Complementary Options

Demand-Side Energy Efficiency Measures: Generally, it is fine to improve the efficiency of generating electricity and heat, in order to save energy resources and avoid emissions. However, above all, priority should be given to measures to *reduce* the demand of both, and then to supply the remainder in the most efficient and environmentally sound way.

Sometimes, both approaches are competing, due to sub-optimal policy regulations. For example, the German Energy Saving Decree provides for such a trade-off between building insulation and efficient heat supply systems for new buildings. This is counterproductive, as it allows for less efficient insulation of buildings when CHP is used for heat supply. Such an offset of potential efficiency gains should be avoided, since it is important to implement enhanced insulation standards when buildings are being constructed. Future improvements in insulation would be considerably more expensive, while, by contrast, micro cogeneration could also be implemented at a later stage. In addition, at the end of the life time of the cogeneration unit, it might be replaced by a conventional heating system which would then operate in a less well insulated house.

Similar offsets might occur when consumer behavior changes as a result of the installation of a micro cogeneration system (e.g., higher comfort requirements). Such rebound effects could be reduced through adequate feedback mechanisms and price incentives for potential micro cogeneration users.

Micro Cogeneration and Renewable Energy Resources: From a climate protection and sustainability perspective, renewable energy resources should be prioritized wherever they are available at reasonable costs, as they simultaneously reduce both climate gas emissions and the consumption of fossil resources. Some micro cogeneration systems are also able to use renewable fuels. However, technologies such as biomass gasification, wood pellet burners, or vegetable oil engines do not yet have a sufficient degree of technical maturity in such small applications. They require further research and development activities in order to reduce investment cost and increase reliability. Currently, where local biomass resources are available, small district heating systems or larger commercial applications with biomass based CHP plants may be better suited than micro cogeneration to make use of such potentials.

Also, micro cogeneration may impair the diffusion of thermal solar collectors and vice versa, since a combination of both as a heating source is economically less favorable. As a near zero-emission heating device, solar thermal heating is ecologically superior over micro cogeneration based on fossil fuels. Hence, in order to create a level playing field and encourage optimal decisions from an economic *and* environmental perspective, the framework and support conditions for both technologies need to be adjusted – with the ultimate objective being to adequately internalize the related external cost and benefits.

Micro Cogeneration versus District Heating: Our analysis shows that district heating with cogeneration is an economically and ecologically comparable or even superior competing option, particularly in areas with high heat density. In addition, the use of biomass is technologically easier for district heating systems than for dispersed micro cogeneration plants. For these reasons, in urban areas with high heat density, district heating with cogeneration should be prioritized. Micro cogeneration may be particularly interesting in areas with more dispersed buildings, such as settlements with single-family houses.

Recommendations

From these more general principles, we have derived some more specific recommendations for a strategy to embed micro cogeneration into the electricity and heating market. Particularly with respect to the institutional framework, the concrete shaping of this strategy is highly country-specific. Therefore, the following recommendations focus on Germany.

Scope and Focus of R&D Policy: German research and development policy has so far strongly focused on selected technologies, such as fuel cells. Several other technologies (Stirling engines, steam cells, etc.) appear also quite promising for stationary applications; we hence recommend that

R&D policies broaden their scope. Additionally, R&D policy should focus not only on the technology itself, but comprise the full innovation *cluster*. This may also include "promoting" innovations, in other words: accompanying technologies and institutional and regulatory changes whose development would simultaneously advance micro cogeneration.

Technology Development Needs: The basic concepts of the different micro cogeneration technologies have been known for a long time; and some of them are already sufficiently reliable for commercial uses. However, there is still scope for technological and material improvements in terms of cost and efficiency. Such innovations include:

- Burner technologies for biomass, such as wood pellet burners, could enhance the use of renewable fuels in micro cogeneration Stirling or steam engines. Given the strong need for climate change mitigation and, thus, the broad utilization of renewable energy carriers in the long-term, this seems an important strategy element for making micro cogeneration fit into a long-term perspective.
- Intelligent and cheap grid access, control and metering technology would reduce costs and increase the acceptance of micro cogeneration by distribution network operators. For instance, the benefit from cogeneration could be maximized by prioritizing and deferring loads (demand response) and by employing time-resolved metering of electricity feed-in. Feedback (e.g. displays) on current consumption and generation of electricity and heat would make micro cogeneration more economical. It could also induce secondary effects, such as enhanced awareness of energy issues. Standardized electronic equipment for micro cogeneration plants may help in grouping them to larger packages, ultimately to virtual power plants. Advanced IT applications for micro cogeneration could also be a first step toward smart home energy management.
- Advanced heat storage technologies would make micro cogeneration technologies more flexible and compatible with different operation strategies.
- Combinations of micro cogeneration with other innovative conversion technologies, such as heat pumps and refrigeration engines, in this small capacity regime require further R&D. This, however, would open up new markets such as the provision of cooling, if successfully implemented.

Regulation and Institutions: As it shapes the market and thus the prospects for micro cogeneration, the regulatory and institutional framework plays an important role in its development and diffusion. As described above, the regulatory framework in Germany is characterized today by significant monetary incentives on the one hand and discontinuity and non-transparency in the overall policy towards CHP on the other. In the following, we highlight issues that appear particularly important in the further development of this institutional framework.

- Compulsory and transparent network connection standards and procedures, possibly on a European level, would reduce transaction costs and allow for a further standardization of micro cogeneration systems and installation procedures, such as full-service packages. Moreover, the administrative requirements for the installation of micro cogeneration plants could be simplified. This refers, for example, to the registration fees for micro cogeneration and the procedures to receive tax reimbursements.
- The regulation of grid-use tariffs should provide sufficient incentives to distribution network operators to support the connection of distributed generators. For example, grid regulation could include a certain target for connecting high-efficiency cogeneration plants, which would then be rewarded in the approval of future tariff adjustments. Furthermore, more active network structures, with enhanced interaction between distribution network operators and micro cogeneration operators by means of communication technologies or spatially/temporally resolved feed-in tariffs could help to provide ancillary services, such as relief of grid congestion and peak load shaving.
- An important institutional barrier for micro cogeneration is the prevailing ownership structure of the distribution network operators, which currently all belong to established electricity supply companies. They are likely to make use of their market power, while small independent generators do not have the financial backing to survive long legal disputes. Here, the current legal unbundling is not sufficient; a complete ownership unbundling and the establishment of an independent system operator for the transmission grid would be the best solution to create a level-playing field for distributed generation.
- In certain niche markets, conditions for economically attractive operation of micro cogeneration are fulfilled to a higher degree than in average objects, rendering micro cogeneration more competitive. When considering such early markets, not only the economy, but also additional factors such as visibility, outreach impact, and commitment of operators should be taken into account.

In addition to such institutional aspects, **financial incentives** have been shown to be effective in stimulating the market, for example in the case of renewable energy technologies. Here, both investment subsidies and financial incentives for energy generation have supported their broader diffusion into the market. However, any financial support scheme – also for micro cogeneration – should be designed carefully, in order to avoid counterproductive effects. A number of aspects have to be considered and balanced:

- The economic incentives for micro cogeneration in Germany and many other countries are considerable. In Germany, this includes the exemption from natural gas and electricity taxation, the CHP bonus, and remuneration for decentralized feed-in. Some of these economic incentives (exemption from electricity taxation, saving of concession levies) only apply if electricity is consumed on-site. This is a major barrier for many applications, in particular in case of multi-family houses, where micro cogeneration plants are only economically attractive if consumers in the building are willing to purchase electricity from the operator of the micro cogeneration plant. From a societal perspective, subsidies for micro cogeneration should not depend on who purchases the electricity but should rather be related to the environmental benefits gained.
- Ultimately, to become interesting for the building industry and energy service companies, decentralized cogeneration would need a support scheme that gives sufficient incentives, regardless of the amount of electricity consumed on location and/or fed into the grid. In Germany, the provisions of the cogeneration law are already a significant step in this direction, but do not yet encourage investment in micro cogeneration with electricity being fed into the grid. The recent adjustment of the "usual price", a component of the feed-in tariff, was also an important step toward improving this situation.
- At the same time, any further increases of feed-in remuneration and other support schemes for micro cogeneration should also consider effects on and benefits from alternative technologies for a sustainable energy sector transformation. It would not make sense to foster micro cogeneration at the expense of other, equally or more sustainable energy technologies, such as district heating or thermal solar collectors, when the latter are more suitable from both the economic and ecological perspectives.
- Similarly, current public support schemes provide a number of economic incentives for micro cogeneration, while larger cogeneration systems, as used for district heating, are supported to a lesser extent, despite the related higher level of energy savings. Future German policies towards sustainable energy systems should therefore aim at

providing a consistent framework, in order to supply similar support for technologies with similar environmental benefits.[1] It is remarkable, for instance, that only a very small number of the Senertec Dachs systems sold are low-NO_x systems. Public support schemes could be designed to reward low pollutant emission levels or to demand a certain emission level, such as the emission levels required by the German environmental label "Blauer Engel" (blue angel). In this context, only highly-efficient cogeneration should then benefit from the exemption from natural gas taxation.

The ultimate aim should therefore be to create a level institutional level-playing field for all innovative energy technologies, one which reflects the environmental performance of the respective systems.

Awareness Raising and Innovative Financing Concepts: For broader market diffusion, and in order to reduce information and search costs, authoritative and independent information about products and systems needs to be disseminated more broadly. Currently, only few potential operators and intermediaries, such as boiler installation firms, know about the different micro cogeneration technologies, as they usually focus on heating systems *or* electricity, not on their combined production.

- Probably the most successful way to market micro cogeneration is to sell them as innovative heating systems with an integrated supply of electricity. Thus, building owners who have to replace an old heating system would take cogeneration systems into consideration. For this approach to be feasible, boiler manufacturers, plumbers and heating installers would be important strategic partners for micro cogeneration developers, because they are usually consulted when it comes to replacing the heating system.
- An alternative approach, realized in Great Britain, would be to market micro cogeneration via the electricity market. This has the advantage that large electricity supply companies provide technical and financial back-up and thus guarantee reliability of heat and electricity supply. These approaches are complementary. Which one is more realistic depends heavily on the motives and commitment of the actors in the heat and electricity market in the respective countries. In Germany, however, the established electricity industry is not interested in decentralization yet.

---

[1] We acknowledge that higher financial support may be justified for technologies where assessment suggests that costs will decrease along learning curves, and where the expected environmental benefits are high.

- As a means to bring micro cogeneration (and other environmentally benign technologies and services) to the customer, innovative financing concepts may be helpful. For example, end-users need information on options for co-operation with third party financing organizations such as energy service companies or energy agencies. Energy service companies benefit from synergies in the operation of many plants, such as lower natural gas purchase and maintenance costs. Due to the high up-front investment cost, this is an important aspect for successful diffusion of micro cogeneration.
- In order to spread information and raise awareness about micro cogeneration, campaigns targeted at specific groups, and in particular at pioneer users may be an important step.

To conclude, if implemented at suitable locations, micro cogeneration could form a noteworthy contribution to the overall energy supply. It is now up to policymakers to adjust the framework conditions in order to release its potential. Energy and climate policy should identify a reasonable and well-balanced set of measures to simultaneously increase energy efficiency, the contribution of renewable energy sources and the share of cogeneration in heat and electricity supply using both micro and larger cogeneration systems.

16 References, Links, Authors and Abbreviations

16.1 References

Introduction

Club of Rome (2002) Das Naturkapital unserer globalen Umwelt ist bedrohter denn je. Memorandum des Club of Rome. Frankfurter Rundschau, Frankfurt, Aug 5, 2002

Enquete-Kommission (2002) Nachhaltige Energieversorgung unter den Bedingungen der Globalisierung und der Liberalisierung. Deutscher Bundestag, Berlin. www.bundestag.de/parlament/kommissionen/archiv/ener/index.html. Last accessed: Mar 15, 2005

Rifkin J (2002) Wasser marsch. Frankfurter Allgemeine Zeitung, Frankfurt, Sept 24, 2002

Schumacher EF (1973) Small is Beautiful. Penguin, London

Chapter 1: Micro Cogeneration Technology

Arndt U, Wagner U (2003) Energiewirtschaftliche Auswirkungen eines Virtuellen Brennstoffzellen-Kraftwerks. VDI-Berichte 1752:165-179

ASUE (2001) BHKW-Kenndaten 2001. Module, Anbieter, Kosten. Energiereferat der Stadt Frankfurt, Arbeitsgemeinschaft für sparsamen und umweltfreundlichen Energieverbrauch, Frankfurt Köln

Educogen (2001) Educogen. The European Educational Tool on Cogeneration. Co-financed under the SAVE programme, Cogen Europe. www.cogen.org/projects/educogen.htm. Last accessed: Mar 9, 2005

EU (2004) Directive 2004/8/EC of the European Parliament and of the Council of 11 February 2004 on the promotion of cogeneration based on a useful heat demand in the internal energy market and amending Directive 92/42/EE. European Union, Bruxelles

Feldmann W (2002) Dezentrale Energieversorgung - zukünftige Entwicklungen, technische Anforderungen. Conference: Energie Innovativ 2002. VDI Verlag, Düsseldorf

Jänig C (2002) Perspektive. Lokale Energie - Geschäftsbericht 2001. Stadtwerke Unna GmbH, Unna

Lewald N (2001) Das EDISon Projekt. Forschungsverbund Sonnenenergie. Themenheft 2001: Integration erneuerbarer Energien in Versorgungsstrukturen. Berlin

Pehnt M (2002) Energierevolution Brennstoffzelle? Perspektiven, Fakten, Anwendungen. Wiley VCH, Weinheim

Roon S (2003) Betriebskonzepte und Produkte virtueller Kraftwerke im liberalisierten Energiemarkt. Supervised by Pehnt M and Moers A. Thesis (Diplomarbeit) at the Technische Universität Berlin

Stephanblome T, Bühner V (2002) Virtuelles Kraftwerk: Energiewirtschaftliche Voraussetzungen & Leittechnik-Software. Conference: Energie Innovativ 2002. VDI Verlag, Düsseldorf

WADE (2003) Guide to Decentralized Energy Technologies. World Alliance for Decentralized Energy, Edinburgh

Chapter 2: Dynamics of Socio-Technical Change: Micro Cogeneration in Energy System Transformation Scenarios

Arthur WB (1997) Increasing Returns and Path Dependence in the Economy. University of Michigan Press, Michigan

Axelrod R, Cohen MD (2000) Harnessing Complexity. Organizational Implications of a Scientific Frontier. Free Press, New York

Berkhout F, Smith A, Stirling A (2004) Socio-Technological Regimes and Transition Contexts. In: Elzen B, Geels FW, Green K (eds) System Innovation and the Transition to Sustainability. Edward Elgar, Cheltenham, pp 48-57

Bijker WE, Hughes TP, Pinch TJ (1987) The Social Construction of Technological Systems. MIT Press, Cambridge

Canzler W, Dierkes M (2001) Informationelle Techniksteuerung: öffentliche Diskurse und Leitbildentwicklungen. In: Simonis G, Martinsen R, Saretzki T (eds) Politik und Technik. Analysen zum Verhältnis von technologischem, politischem und staatlichen Wandel am Anfang des 21.Jahrhunderts. PVS special edition 31/2000. Westdeutscher Verlag, Wiesbaden, pp 457-475

Coutard O (1999) The Governance of Large Technical Systems. Routledge, London

CSTM (2001) Electricity in Flux: Sociotechnical Change in the Dutch Electricity System, 1970-2000. Project report. Centre for Clean Technology and Environmental Policy at the University of Twente, Bilthoven

Dierkes M, Hoffmann U, Marz L (1992) Leitbild und Technik: zur Entstehung und Steuerung technischer Innovationen. edn Sigma, Berlin

Diekmann J, Hopf R, Ziesing HJ, Kleemann M, Krey V, Markewitz P, Martinsen D, Vögele S, Eichhammer W, Jochem E, Mannsbart W, Schlomann B, Schön M, Wietschel M, Matthes FC, Cames M, Harthan R (2003) Politikszenarien für den Klimaschutz – Langfristszenarien und Handlungsempfehlungen ab 2012 (Politikszenarien III). Deutsches Institut für Wirtschaftsforschung (DIW), Forschungszentrum Jülich, Fraunhofer Institut für Systemtechnik und Innovationsforschung, Öko-Institut, Berlin Jülich Karlsruhe

Dosi G (1988) The Nature of the Innovative Process. In: Dosi G, Freeman C, Nelson R, Silverberg G, Soete L (eds) Technical Change and Economic Theory. Pinter, London New York, pp 221-238

Eising R (2000) Liberalisierung und Europäisierung. Die regulative Reform der Elektrizitätsversorgung in Großbritannien, der Europäischen Gemeinschaft und der Bundesrepublik Deutschland. Leske+Budrich, Opladen

Enquete-Kommission (2002) Nachhaltige Energieversorgung unter den Bedingungen der Globalisierung und der Liberalisierung. Deutscher Bundestag, Berlin. http://www.bundestag.de/parlament/kommissionen/archiv/ener/index.html. Last accessed: Mar 9, 2005

Fischedick M, Hanke T, Hennicke P, Lechtenböhmer S, Merten F, Viefhues D (1999/2001) Bewertung eines Ausstiegs aus der Kernenergie aus klimapolitischer und volkswirtschaftlicher Sicht. Mit überarbeiteter Zusammenfassung vom Januar 2001. Wissenschaftszentrum Nordrhein-Westfalen, Wuppertal Institut für Klima, Umwelt, Energie, Öko-Institut, Wuppertal Freiburg Bremen Berlin

Fischedick M, Nitsch J, Lechtenböhmer S, Hanke T, Barthel C, Jungbluth C, Assmann D, vor der Brüggen T, Trieb F, Nast M, Langniß O, Brischke LA (2002) Langfristszenarien für eine nachhaltige Energienutzung in Deutschland. Umweltbundesamt, Berlin

Geels F (2001) Technological Transitions as Evolutionary Reconfiguration Processes: A Multi-Level Perspective and a Case-Study. Presented at the conference: The Future of Innovation Studies, organized by Eindhoven Centre for Innovation Studies (ECIS), Eindhoven

Geels F (2002a) Understanding Technological Transitions: A Critical Literature Review and a Pragmatic Conceptual Synthesis. Presented at the conference: Twente workshop on Transitions and System Innovations, Twente

Geels F (2002b) Understanding the dynamics of technological transitions. Twente University Press, Enschede

Grin J, Grunwald A (2000) Vision Assessment: Shaping Technology in 21st Century Society. Towards a Repertoire for Technology Assessment. Springer, Berlin Heidelberg New York

Holst Jørgensen B, Nielsen O, Reuss T, Wehnert T (2004) EurEnDel. Technology and Social Visions for Europe's Energy Future. A Europe-wide Delphi Study. Institut für Zukunftsstudien und Technologiebewertung, Berlin

Hughes TP (1983) Networks of Power: Electrification 1880-1930. Johns Hopkins University Press, Baltimore

IEA, OECD (2003) Energy to 2050. Scenarios for a Sustainable Future. International Energy Agency, Organization for Economic Cooperation and Development, Paris

Jäger T, Mertens J, Karger C (2004) Integrierte Mikrosysteme der Versorgung: Szenariobeschreibungen. http://www.mikrosysteme.org/documents/IMV_Szenarien.pdf. Last accessed: Mar 15, 2005

Kemp R (1994) Technology and the Transition to Environmental Sustainability. The Problem of Technological Regime Shifts. Futures 26:1023-1046

Konrad K (2004) Prägende Erwartungen. Szenarien als Schrittmacher der Technikentwicklung. edn Sigma, Berlin

La Porte TR (1991) Responding to Large Technical Systems: Control or Anticipation. Kluwer, Dordrecht

Larsen H, Sønderberg Petersen L (2002) Risø Energy Report 1. New and Emerging Technologies. Options for the Future. Risø National Laboratory, Roskilde

Matthes FC, Cames M (2000) Energiewende 2020: Der Weg in eine zukunftsfähige Energiewirtschaft. Heinrich-Böll-Stiftung, Berlin

Mayntz R, Hughes TP (1988) The Development of Large Technical Systems. Campus, Frankfurt New York

Nakicenovic N, Riahi, K (2002) An Assessment of Technological Change Across Selected Energy Scenarios. International Institute for Applied Systems Analysis (IIASA), Laxenburg

Nelson RR (2000) Recent Evolutionary Theorizing About Economic Change. In: Ortmann G, Sydow J, Türk K (eds) Theorien der Organisation. Die Rückkehr der Gesellschaft. Westdeutscher Verlag, Opladen, pp 81-123

Nelson RR, Winter SG (1982) An Evolutionary Theory of Economic Change. Bellknap, Cambridge Massachussets

Nitsch J, Krewitt W, Nast M, Viebahn P, Gärtner S, Pehnt M, Reinhardt G, Schmidt R, Uihlein A, Barthel C, Fischedick M, Merten F (2004) Ökologisch optimierter Ausbau der Nutzung erneuerbarer Energien in Deutschland. Deutsches Zentrum für Luft- und Raumfahrt, Institut für Energie- und Umweltforschung, Wuppertal Institut für Klima, Umwelt, Energie, Stuttgart Heidelberg Wuppertal

Norgaard RB (1994) Development Betrayed. The End of Progress and a Coevolutionary Revisioning of the Future. Routledge, London

Patterson W (1999) Transforming Electricity. The Coming Generation of Change. Earthscan, London

Pehnt M, Fischer C, Sauter R, Cames M, Schneider L, Voß, JP, Grashof K, Praetorius B, Schumacher K (2004) MicroCHP - a sustainable innovation? Presented at the conference: Erfolgreiche Energieinnovationsprozesse, February 2004, Graz

Pfaffenberger W, Hille M (2004) Investitionen im liberalisierten Energiemarkt: Optionen, Marktmechanismen, Rahmenbedingungen. bremer energie institut, Bremen

Pierson P (2000) Increasing Returns, Path Dependence, and the Study of Politics. American Political Science Review 94(2):251-267

Prognos AG (2000) Energiereport III. Die längerfristige Entwicklung der Energiemärkte im Zeichen von Wettbewerb und Umwelt. Schaeffer-Poeschel, Stuttgart

Rip A (1995) Introduction of New Technology: Making Use of Recent Insights from Sociology and Economics of Technology. Technology Analysis & Strategic Management, 7(4):417-431

Rip A (2002) Co-Evolution of Science, Technology and Society. An Expert Review, Berlin-Brandenburgische Akademie der Wissenschaften. Enschede.. www.sciencepolicystudies.de/dok/expertise-rip.pdf. Last accessed: Mar 9, 2005

Rip A, Kemp R (1998) Technological Change. In: Rayner S, Malone EL (eds) Human Choice and Climate Change. Batelle Press, Ohio, pp 327-399
Robinson JB (1982) Energy Backcasting. A Proposed Method of Policy Analysis. Energy Policy, 10:337-344
Robinson JB (2003) Future Subjunctive: Backcasting as Social Learning. Futures 35:839-856
Shell International (2001) Energy Choices, Needs and Possibilities. Scenarios to 2050. Global Business Environment. Shell International Ltd, London
Schneider V, Werle R (1998) Co-Evolution and Development of Large Technical Systems in Evolutionary Perspective. In: García CE, Sanz-Menédez L (eds) Management and Technology, Vol. 5. European Commission, Luxemburg, pp 12-29
Smil V (2003) Energy at the Crossroads: Global Perspectives and Uncertainties. Massachusetts Institute of Technology Press, Cambridge Massachusetts
Summerton J (1992) Changing Large Technical Systems. Westview, Boulder Colorado
van Lente H (1993) Promising Technologies: The Dynamics of Expectations in Technological Development. Twente University Press, Enschede
van Lente H, Rip A (1998) Expectations in Technological Developments: An Example of Prospective Structures to be Filled in by Agency. In: Disco C, van der Meulen, BJR (eds) Getting New Things Together. Walter de Gruyter, Berlin New York, pp 195-220
Velte D (2004) The EurEnDel Scenarios. Europe's Energy System by 2030. Institut für Zukunftsforschung und Technologiebewertung, Berlin
Voß JP (2000) Institutionelle Arrangements zwischen Zukunfts- und Gegenwartsfähigkeit: Verfahren der Netzregelung im liberalisierten deutschen Stromsektor. In: Prittwitz V (ed) Institutionelle Arrangements in der Umweltpolitik. Zukunftsfähigkeit durch innovative Verfahrenskombination? Leske+Budrich, Opladen, pp 227-254
Voß JP, Kemp R (2005) Reflexive Governance and Sustainable Development. In: Voß JP, Bauknecht D, Kemp R (eds) Reflexive Governance for Sustainable Development. Edward Elgar, Cheltenham (manuscript submitted)
Voß JP, Konrad K, Truffer B (2005) Sustainability Foresight. Reflexive Governance in the Transformation of Utility Systems. In: Voß JP, Bauknecht D, Kemp R (eds) Reflexive Governance for Sustainable Development. Edward Elgar, Cheltenham (manuscript submitted)
Weyer J, Kirchner U, Riedl L, Schmidt JFK (1997) Technik, die Gesellschaft schafft. edn Sigma, Berlin

Chapter 3: The Future Heat Market and the Potential of Micro Cogeneration

BGW (2003) Bundesverband der deutschen Gas- und Wasserwirtschaft. www.bgw.de. Last accessed: Mar 15, 2005

Destatis (2003) Bevölkerung 2050. 10. koordinierte Bevölkerungsvorausberechnung. Statistisches Bundesamt, Wiesbaden

Enquete-Kommission (2002) Nachhaltige Energieversorgung unter den Bedingungen der Globalisierung und der Liberalisierung. Deutscher Bundestag, Berlin http://www.bundestag.de/parlament/kommissionen/archiv/ener/index.html. Last accessed: March 15, 2005

Fischedick M, Nitsch J (2002) Langfristszenarien für eine nachhaltige Energieversorgung. Umweltbundesamt, Research Report 20097104, Wuppertal Stuttgart

IEU (2004) Initiativkreis Erdgas und Umwelt. www.ieu.de. Last accessed: Oct 21, 2004

Krammer T (2001) Brennstoffzellenanlagen in der Hausenergieversorgung. Instrumentarien zur Potenzialanalyse. Dissertation at the Technische Universität München

Krewitt W, Pehnt M, Temming H, Fischedick M (2004) Brennstoffzellen in der Kraft-Wärme-Kopplung. Ökobilanzen, Szenarien, Marktpotenziale. Erich Schmidt Verlag, Berlin

Nast M (2004) Chancen und Perspektiven der Nahwärme im zukünftigen Energiemarkt. Conference: Nahwärme 2004. VDI-Verlag, Düsseldorf

Statisches Bundesamt (2003a) Statistisches Jahrbuch 2003. Statistisches Bundesamt, Wiesbaden

Statistisches Bundesamt (2003b) Mikrozensus 2002. Wiesbaden

Chapter 4: Economics of Micro Cogeneration

ASUE (2003) Infodienst neue Produkte. August 2003. Transferstelle für neue Produkte der Arbeitsgemeinschaft für sparsamen und umweltfreundlichen Energieverbrauch, Essen. www.transferstelle.info. Last accessed: March 15, 2005

ASUE (2004) Infodienst neue Produkte. August 2004. Transferstelle für neue Produkte der Arbeitsgemeinschaft für sparsamen und umweltfreundlichen Energieverbrauch, Essen. www.transferstelle.info. Last accessed: March 15, 2005

BMWA (2005) Monatliche Erdgasbilanz und Entwicklung der Grenzübergangspreise ab 1991. Bundesministerium für Wirtschaft und Arbeit, Berlin

FHG-ISI, DIW, GfK, IEU, TUM (2004) Energieverbrauch der privaten Haushalte und des Sektors Gewerbe, Handel, Dienstleistungen (GHD). Fraunhofer-Institut für Systemtechnik und Innovationsforschung, Deutsches Institut für Wirtschaftsforschung, GfK Marketing Services, GfK Panel Services Consumer Research, Institut für Energetik und Umwelt, Technische Universität München, Karlsruhe Berlin Nürnberg Leipzig München

European Commission (2003) Second Benchmarking Report on the Implementation of the Internal Electricity and Gas Market. Commission staff working paper SEC(2003)448. Brussels

European Commission (2005a) Electricity Prices. Data 1990-2004. Office for Official Publications of the European Communities, Luxembourg
European Commission (2005b) Gas Prices. Data 1990-2004. Office for Official Publications of the European Communities, Luxembourg
Enquete-Kommission (2002) Nachhaltige Energieversorgung unter den Bedingungen der Globalisierung und der Liberalisierung. Deutscher Bundestag, Berlin. www.bundestag.de/parlament/kommissionen/archiv/ener/index.html. Last accessed: March 15, 2005
Krewitt W, Pehnt M, Temming H, Fischedick M (eds) (2004) Stationäre Brennstoffzellen – Umweltauswirkungen, Rahmenbedingungen und Marktpotenziale. Study for the Ministerium für Umwelt, Naturschutz und Reaktorsicherheit (BMU). Erich Schmidt Verlag, Berlin
Krzikalla N, Schrader K (2002) Untersuchung von Einflussgrößen auf die Höhe der Belastungen der Endkunden aus dem EEG. Kurzgutachten. Büro für Energiewirtschaft und Technische Planung (BET), Aachen
Schlesinger M, Eckerle K, Haker K, Hobohm J, Hofer P, Scheelhaase JD (2000) Energiereport III. Die längerfristige Entwicklung der Energiemärkte im Zeichen von Wettbewerb und Umwelt. Schäffer Poeschel, Stuttgart
ZSW (2000) BHKW-Plan. Version 1.05.00. Excel Based Design Software for Small CHP Plants. Zentrum für Sonnenenergie- und Wasserstoff-Forschung, Stuttgart

Chapter 5: Environmental Impacts of Micro Cogeneration

Arndt U, Duschl A, Köhler D, Schwaegerl P (2004) Energiewirtschaftliche Bewertung dezentraler KWK-Systeme für die Hausenergieversorgung. Koordinationsstelle der Wasserstoff-Initiative Bayern (WIBA), Forschungsstelle für Energiewirtschaft, München
AUSTAL2000 (2003) Programme version 1.0. Ing.-Büro Janicke for the Environmental Protection Agency
Bilharz M (2003) Individuelle Ökobilanzen für einen nachhaltigen Konsum: eine Bewertung. Diskussionsbeitrag Nr. 109, Universität St. Gallen. www.iwoe.unisg.ch/org/iwo/web.nsf/SysWebRessources/db109/$File/db_109.pdf. Last accessed: June 22, 2005
CML (1992) Heijungs R, Guinée JB, Huppes G, Lankreijer RM, Haes HAU, Sleeswijk AW, Environmental Life Cycle Assessment of Products. Guide and Backgrounds. Institute of Environmental Sciences CML, Leiden
EEA (2001) Indicator Fact Sheet Signals 2001 – Chapter Air Pollution. European Environment Agency, Copenhagen themes.eea.eu.int/Specific_media/air/indicators/acidification/AP_1.pdf. Last accessed: March 9, 2005
EEA (2004) Technical report 2/2004. Annual European Community Greenhouse Gas Inventory 1990-2002 and Inventory Report 2004. European Environment Agency, Copenhagen
Enquete-Kommission (2002) Nachhaltige Energieversorgung unter den Bedingungen der Globalisierung und der Liberalisierung. Deutscher

Bundestag, Berlin. www.bundestag.de/parlament/kommissionen/archiv/ener/index.html. Last accessed: March 15, 2005

ETC-ACC (2004) An Initial Assessment of Member States' National Programmes and Projections under the National Emission Ceiling Directive (2001/81/EC). Summary paper, ETC-ACC Technical Paper 2003-8. European Topic Centre on Air and Climate Change, Bilthoven

EU (2004) Directive 2004/8/EC of the European Parliament and of the Council of 11 February 2004 on the promotion of cogeneration based on a useful heat demand in the internal energy market and amending directive (92/42/EE). European Union, Bruxelles

Franke B, Vogt R (2000) MBA-Machbarkeitsstudie Dortmund. Report for the Bezirksregierung Arnsberg (unpublished). Institut für Energie- und Umweltforschung (IFEU), Heidelberg

Frischknecht R, Emmenegger MF (2003) Strommix und Stromnetz. Ecoinvent: Sachbilanzen von Energiesystemen: Grundlagen für den ökologischen Vergleich von Energiesystemen. Swiss Centre for Life Cycle Inventories, Dübendorf

Gärtner SO, Reinhardt GA (2003) Erweiterung der Ökobilanz für RME. Report for the Union zur Förderung von Öl- und Proteinpflanzen, Berlin

Genennig B, Hoffmann VU (1996) Sozialwissenschaftliche Begleituntersuchung zum Bund-Länder-1000 Dächer-Photovoltaik-Programm. Umweltinstitut Leipzig, Leipzig

Giegrich J, Detzel A (2000) Gesamtökologischer Vergleich graphischer Papiere. UBA-Texte 22/00. Institut für Energie- und Umweltforschung (IFEU), Heidelberg

Gottron F (2001) Energy Efficiency and the Rebound Effect: Does Increasing Efficiency Decrease Demand? Congressional Research Service, Washington

Haas R, Ornetzeder M, Hametner K, Wroblewski A, Hübner M (1999) Socio-Economic Aspects of the Austrian 200 kWp-Photovoltaic Rooftop Programme. Solar Energy 66:183-191

Haas R, Biermayr P, Baumann B, Schriefl E, Skopetz H (2001) Erneuerbare Energieträger und Energieverbrauchsverhalten. Bundesministerium für Verkehr, Innovation und Technologie, Wien

ISO (1997) Environmental Management – Life Cycle Assessment – Principles and framework. ISO 14040

Jensen OM (2003) Visualisation Turns Down Energy Demand. In: ECEEE summer study 2003 proceedings. European Council for an Energy-Efficient Economy, Stockholm, pp 451-454

Karsten A (1998) In der Praxis gewonnene Erfahrungen mit Photovoltaikanlagen in Hamburger Privathaushalten, Unternehmen und öffentlichen Einrichtungen. Hamburger Universität für Wirtschaft und Politik, Hamburg

Kleemann M, Heckler R, Kolb G, Hille M (2000) Die Entwicklung des Energiebedarfs zur Wärmebereitstellung in Gebäuden. bremer energie institut, Bremen

Krewitt W, Pehnt M, Temming H, Fischedick M (2004) Brennstoffzellen in der Kraft-Wärme-Kopplung. Ökobilanzen, Szenarien, Marktpotenziale. Erich Schmidt Verlag, Berlin

Kristensen PG, Jensen JK, Nielsen M, Illerup JB (2004) Emission factors for gas fired CHP units < 25 MW. Danish Gas Technology Centrem, Horsholm

Lehmann K, Vogel M, Hiller W (2003) Dezentrale Einspeiser im NS-Netz: Einfluss auf Planung, Betrieb und Netznutzung. Envia Mitteldeutsche Energie AG, Halle

Lovins A, Datta EK, Feiler T, Rábago KR, Swisher JN, Lehmann A, Wicker K (2002) Small is Profitable: The Hidden Benefits of Making Electrical Resources the Right Size. Rocky Mountain Institute, Boulder

McCalley LT (2003) From Motivation and Cognition Theories to Everyday Applications and Back Again: the Case of Product-Integrated Information and Feedback. ECEEE summer study 2003 proceedings. European Council for an Energy-Efficient Economy, Stockholm, pp1151-1157

Menges R (2003) Individual Demand for Energy Services. In: ECEEE summer study 2003 proceedings. European Council for an Energy-Efficient Economy, Stockholm, pp 1123-1134

Pehnt M (2002) Ganzheitliche Bilanzierung von Brennstoffzellen in der Energie- und Verkehrstechnik. Fortschritt-Berichte Reihe 6. VDI Verlag, Düsseldorf

Pehnt M (2003) Assessing Future Energy and Transport Systems: The Case of Fuel Cells. Part 1: Methodological Aspects. International Journal of Life Cycle Analysis 6:283-289

Pehnt M (2005) Dynamic Assessment of Renewable Energy Technologies, Renewable Energy, in print

Senertec (1992-2003) Emission Measurements of Herzogtum Lauenburg 1992, Schleswig-Holstein 1992 and 1994, update 2003. Schweinfurt

UBA (1995) Methodik der produktbezogenen Ökobilanz. Wirkungsbilanz und Bewertung. UBA-Texte 23/95. Umweltbundesamt, Berlin

van Elburg M (2001) Adapting People to Policies or the Other Way Round? (unpublished) Contribution to the Seminar: Energy Efficiency and Consumer Preferences - Two Sides of the Same Coin?, Aukrug, Oct 24 2001

Wilhite H, Norgard JS (2003) A Case for Self-Deception in Energy Policy. In: ECEEE summer study 2003 proceedings. European Council for an Energy-Efficient Economy, Stockholm, pp 249-257

Chapter 6: From Consumers to Operators: The Role of Micro Cogeneration Users

ALLBUS (2002) Allgemeine Bevölkerungsumfrage der Sozialwissenschaften. www.gesis.org/Datenservice/ALLBUS/index.htm. Last accessed: March 15, 2005

Beier G (1999) Kontrollüberzeugungen im Umgang mit Technik. Report Psychologie, 24(9): 684-693

DENA (2003) Initiative Solarwärme Plus. Marktforschungsergebnisse – Kurzfassung (unpublished). Deutsche Energie Agentur, Berlin

Genennig B, Hoffmann VU (1996) Sozialwissenschaftliche Begleituntersuchung zum Bund-Länder-1000 Dächer-Photovoltaik-Programm. Umweltinstitut Leipzig, Leipzig

Greenpeace (1996) Marktanalyse Photovoltaik. Der deutsche Solarmarkt nach der Greenpeace-Cyrus-Kampagne. archiv.greenpeace.de/GP_DOK_3P/HINTERGR/C04HI13.htm. Last accessed: March 15, 2005

Haas R, Ornetzeder M, Hametner K, Wroblewski A, Hübner M (1999) Socio-Economic Aspects of the Austrian 200 kWp-Photovoltaic-Rooftop Programme. Solar energy 66(3):183-199

Haas R, Biermayr P, Baumann B, Schriefl E, Skopetz H (2001) Erneuerbare Energieträger und Energieverbrauchsverhalten. Bundesministerium für Verkehr, Innovation und Technologie, Wien

Hackstock R, Könighofer K, Ornetzeder M, Schramm W (1992) Übertragbarkeit der Solaranlagen - Selbstbautechnologie. Kurzfassung. Bundesministerium für Wissenschaft und Forschung, Wien

Hocke-Bergler P, Stolle M (2003) Attachment to the ITAS-report: Ergebnisse der Bevölkerungsumfragen und der Medienanalyse zum Thema Endlagerung radioaktiver Abfälle. www.akend.de/projekte/pdf/anlagen1-7.pdf, www.akend.de/projekte/pdf/anlagen8-15.pdf, www.akend.de/projekte/pdf/anlage16.pdf. Last accessed: March 15, 2005

Hübner G, Felser G (2001) Für Solarenergie. Konsumenten- und Umweltpsychologie strategisch anwenden. Asanger, Heidelberg

Karsten A (1998) In der Praxis gewonnene Erfahrungen mit Photovoltaikanlagen in Hamburger Privathaushalten, Unternehmen und öffentlichen Einrichtungen. Hamburger Universität für Wirtschaft und Politik, Hamburg

Katzbeck G (1997) Der Einfluss von Photovoltaikanlagen auf das Energieverbrauchsverhalten österreichischer Haushalte. Thesis (Diplomarbeit) at the Technische Universität Wien

Kuckartz U, Grunenberg H (2002) Umweltbewusstsein in Deutschland 2002. Ergebnisse einer bundesweiten Repräsentativstudie. Bundesministerium für Umwelt, Naturschutz und Reaktorsicherheit (BMU), Berlin

Polzer G (2003) Spaichingen. Von der Sonne verwöhnt. Lokale Agenda 21 – Arbeitskreis Energie, Spaichingen

Reif R (2000) Photovoltaik-Siedlung: Aufbau von 23 dachintegrierten PV-Anlagen auf 22 Einfamilienhäusern und dem dazugehörigen Kindergarten im Zuge eines gemeinschaftlichen, räumlich zusammengehörenden Siedlungsbauprojektes. Abschlussbericht. edok01.tib.uni-hannover.de/edoks/e001/313318409l.pdf. Last accessed: March 15, 2005

Rogers EM (1995) Diffusion of Innovations. The Free Press, New York London Toronto Sydney Tokyo Singapore

Rohracher H, Suschek-Berger J, Schwärzler G (1997) Verbreitung von Biomasse-Kleinanlagen. Situationsanalyse und Handlungsempfehlungen. Bundesministerium für Verkehr, Innovation und Technologie, Wien

Villiger A, Wüstenhagen R, Meyer A (2000) Jenseits der Öko-Nische. Birkhäuser, Basel Boston

Chapter 7: Micro Cogeneration – Setting of an Emerging Market

BGW (2004) Erdgas – Wunschenergie Nr. 1. Bundesverband der deutschen Gas- und Wasserwirtschaft, Berlin

BMWA (2003) Bericht des Bundesministeriums für Wirtschaft und Arbeit an den Deutschen Bundestag über die energiewirtschaftlichen und wettbewerblichen Wirkungen der Verbändevereinbarungen (Monitoring-Bericht). Bundesministerium für Wirtschaft und Arbeit, Berlin

BNE (2004) Energiemarkt heute. Bundesverband Neuer Energieanbieter, Berlin

Brunekreeft G, Twelemann S (2005) Regulation, Competition and Investment in the German Electricity Market: RegTP or REGTP. Energy Journal, forthcoming.

Bundeskartellamt (2003) Bericht des Bundeskartellamtes über seine Tätigkeit in den Jahren 2001/2002 sowie über die Lage und Entwicklung auf seinem Aufgabengebiet. Bonn

COGEN Europe (2004) Micro-CHP Fact Sheet Germany. Brussels

Connor P, Mitchell C (2002) A Review of Four European Regulatory Systems and their Impact on the Deployment of Distributed Generation. Report for SUSTELNET - Policy and Regulatory Roadmaps for the Integration of Distributed Generation and the Development of Sustainable Electricity Networks. University of Warwick, Coventry

Ernst&Young (2003) Stadtwerkestudie 2003 - Erfolgreiche Geschäftsstrategien für Stadtwerke und regionale Energieversorgungsunternehmen. Düsseldorf Frankfurt

Jörß W, Jorgensen BH, Löffler P, Morthorst PE, Uyterlinde M, Sambeek E, Wehnert T (2003) Decentralized Power Generation in the Liberalized EU Energy Markets. Springer, Berlin Heidelberg New York

Leprich U (2004) Aktive Stromnetzbetreiber als Stützpfeiler eines stärker dezentralisierten Stromsystems: Rahmenbedingungen und Anreize. Presentation at the conference: KWK als Perspektive - Perspektiven der KWK, B.KWK, Berlin

Leprich U, Bauknecht D (2003) Review of Current Electricity Policy and Regulation - German Study Case. Report for SUSTELNET - Policy and Regulatory Roadmaps for the Integration of Distributed Generation and the Development of Sustainable Electricity Networks. Institut für Zukunfts-EnergieSysteme (IZES), Öko-Institut, Saarbrücken Freiburg Berlin

Leprich U, Thiele A (2004) Rahmen- und Erfolgsbedingungen für die weitere Verbreitung von Brennstoffzellen und anderen Klein-KWK-Anlagen in Deutschland. Survey for the Umweltbundesamt. Institut für ZukunftsEnergie-Systeme (IZES), Saarbrücken

Madlener R, Schmid C (2003) Combined Heat and Power Generation in Liberalized Markets and a Carbon-Constrained World. GAIA 12(2):114-120
Monopolkommission (2004) Wettbewerbspolitik im Schatten "nationaler Champions". Fünfzehntes Hauptgutachten der Monopolkommission 2002/2003. Berlin Bonn
Preissl B, Solimene L (2003) The Dynamics of Clusters and Innovation - Beyond Systems and Networks. Physica-Verlag, Heidelberg
Schmidt H (2004) Wettbewerb auf dem deutschen Strommarkt steht fast still. Frankfurter Allgemeine Zeitung, Frankfurt, Sept 24, 2004
VDEW (2004) Jahresbericht 2003. Verband der Elektrizitätswirtschaft, Frankfurt
VKU (2003) Survey of Cogeneration Investment Activities among VKU Members. Verband kommunaler Unternehmen, Berlin
WhisperTech (2004) Whisper Tech Signs $300 Million Agreement. Whisper Tech Media Release, Christchurch, Aug 12, 2004
Williamson OE (1985) The Economic Institutions of Capitalism. The Free Press, New York

Chapter 8: Institutional Framework and Innovation Policy for Micro Cogeneration in Germany

BDI (2001) Vereinbarung zwischen der Regierung der Bundesrepublik Deutschland und der deutschen Wirtschaft zur Minderung der CO_2-Emissionen und der Förderung der Kraft-Wärme-Kopplung in Ergänzung zur Klimavereinbarung vom 9.11.2000. Bundesverband der deutschen Industrie, Berlin
BDI et al. (2001) Verbändevereinbarung über Kriterien zur Bestimmung von Netznutzungsentgelten für elektrische Energie und über Prinzipien der Netznutzung (VVII+). Bundesverband der deutschen Industrie, Berlin www.strom.de/wysstr/stromwys.nsf/01db8e410336942bc1256b020042d7d1/30270521fe1f711cc1256b210060f08c/$FILE/_uapin4ok4dpi6atj5e9imirj2c5p7arj7954im_.pdf. Last accessed: March 20, 2005
BKWK (2004): Aktualisierung der Position des B.KWK zum EnWG. Bundesverband Kraft-Wärme-Kopplung, Berlin. www.bkwk.de/bkwk/download/politik/Anmerkungen. Last accessed: March 20, 2005
BKWK (2005) Änderungsvorschläge zur StromNEV, Mar 1, 2005. Bundesverband Kraft-Wärme-Kopplung, Berlin
BMU (2004) The Ecological Tax Reform: Introduction, Continuation and Further Development to an Ecological Financial Tax Reform. Bundesministerium für Umwelt, Naturschutz und Reaktorsicherheit, Berlin www.bmu.de/publikationen/doc/5406.php. Last accessed: March 15, 2005
BMWA (2003) Energieforschung. Bundesministerium für Wirtschaft und Arbeit, Berlin. www.bmwa.bund.de/Navigation/Technologie-und-Energie/Energiepolitik/energieforschung.html. Last accessed: March 20, 2005

BNE (2004) Vermiedene Netznutzungsentgelte durch dezentrale Stromeinspeisung. Bundesverband Neuer Energieanbieter, Berlin www2.neue-energieanbieter.de/uploads/04_09_10%20vermiedene%20NNE_bne%20Position.pdf. Last accessed: Mar 20, 2005

COGEN europe (2003) Micro-CHP Needs Specific Treatment in the European Directive on Cogeneration. Brussels

Corsten (2003) Interview with Mr Corsten, Ruhrgas AG, responsible for the Initiative Brennstoffzelle, Oct 26, 2003

DBU (2003) Einzelergebnisse DBU-Projektdatenbank. Feldversuch eines Kleinst-BHKWs auf Basis eines Stirlingmotors. Aktenzeichen 03873/01. Deutsche Bundesstiftung Umwelt, Osnabrück. datenbanken.wiminno.com/dbu/pdf/A-03873.pdf. Last accessed: March 20, 2005

Deutscher Bundestag (1998) Gesetz über die Elektrizitäts- und Gasversorgung (Energiewirtschaftsgesetz EnWG). In: Bundesgesetzblatt (Teil I). Berlin, p 730

Deutscher Bundestag (2000) Gesetz zum Schutz der Stromerzeugung aus Kraft-Wärme-Kopplung (Kraft-Wärme-Kopplungsgesetz). In: Bundesgesetzblatt. Berlin, pp 703-704

Deutscher Bundestag (2002a) Gesetz für die Erhaltung, die Modernisierung und den Ausbau der Kraft-Wärme-Kopplung (Kraft-Wärme-Kopplungsgesetz KWK-G). In: Bundesgesetzblatt. Berlin, pp 1092-1096

Deutscher Bundestag (2002b) Verordnung über energiesparenden Wärmeschutz und energiesparende Anlagentechnik bei Gebäuden (Energieeinsparverordnung EnEV). Berlin

Deutscher Bundestag (2004) Gesetzentwurf der Bundesregierung. Entwurf eines Zweiten Gesetzes zur Neuregelung des Energiewirtschaftsrechts (EnWG). Bundestags-Drucksache 15/3917, Berlin. dip.bundestag.de/btd/15/039/1503917.pdf. Last accessed: March 20, 2005

ECON (2004) EU Emission Trading Scheme and the effect on the price of electricity, Stockholm www.econ.no/oslo/econreports.nsf/0/A24A2458697EF109C1256F08003D20EE?OpenDocument. Last accessed: March 19, 2005

Esser H (2000) Soziologie, Spezielle Grundlagen, vol 5. Institutionen. Campus, Frankfurt New York

European Commission (2003) Directive 2003/87/EC of the European Parliament and of the Council of Oct 13, 2003 Establishing a Scheme for Greenhouse Gas Emission Allowance Trading within the Community and Amending Council Directive 96/61/EC. In: Official Journal of the European Union (L 275), pp 32-46

Gilles V, Ocran GW, Gray R, Woodbridge D (2003) German Electricity Wholesale Market – Three Shocks for the Price of one. UBS Investment Research, London

Glante N (2003) Einigung über KWK-Richtlinie ohne Vermittlung. Potsdam. www.glante.de/index2.php?id=61. Last accessed: Jan 9, 2004

Götze (2003) Interview with Mr Götze, Kreditanstalt für Wiederaufbau, responsible for the KfW-Programm zur CO_2-Minderung and the KfW-CO_2-Gebäudesanierungsprogramm on Oct 25, 2003

Harrison J, Redford S (2001) Domestic CHP – What are the Potential Benefits? EA Technology Limited, Capenhurst www.est.org.uk/uploads/documents/benefits.pdf. Last accessed: March 15, 2005

Hegner HD (2002) Die Energieeinsparverordnung – neue Möglichkeiten für Planung und Ausführung im Neubau und bei der Modernisierung. Bundesministerium für Verkehr, Bauen und Wohnen, Berlin

Hohmann, W (2005) Interview with Mr Hohmann, hessenENERGIE GmbH, Small-sized CHP department on Mar 4, 2005

Horn M, Harthan R, Matthes FC and Ziesing HJ (2004) Ermittlung der Potenziale für die Anwendung der Kraft-Wärme-Kopplung und der erzielbaren Minderung der CO_2-Emissionen einschließlich Bewertung der Kosten (Verstärkte Nutzung der Kraft-Wärme-Kopplung). 3. und 4. Zwischenbericht. Berlin

Kafke (2003) Interview with Mr Kafke, Bundesverband Verbraucherzentralen, responsible for coordinating energy advice services on Nov 26, 2003

Kalkutschki (2003) Interview with Mr Kalkutschki, responsible for the Research and Development activities of the Bundesministerium für Wirtschaft und Arbeit on Nov 25, 2003

Kemp R (1996) The Transition from Hydrcarbons: The Issues for Policy. In: Faucheux S, Pearce D, Proops JLR (eds): Models of Sustainable Development. Edward Elgar, Aldershot, pp 151-175

Kohlhaas M (2003) Energy Taxation and Competitiveness – Special Provisions for Business in Germany's environmental tax reform. DIW Discussion Paper 349. Deutsches Institut für Wirtschaftsforschung, Berlin www.diw.de/deutsch/produkte/publikationen/diskussionspapiere/docs/papers/dp349.pdf. Last accessed: March 15, 2005

Kohlhaas M, Mayer B (2004) Ecological Tax Reform in Germany: Economic and political analysis of an evolving policy. In: Hatch MT (ed): Environmental Policymaking: Assessing the Use of Alternative Policy Instruments. State University of New York Press, Albany New York (anticipated)

Leprich U, Bauknecht D (2003a) Regulatory Road Map for Germany – Creating a Level Playing Field for Centralized and Decentralized Power Plants. Institut für Zukunftsenergiesysteme and Öko-Institut, Saarbrücken and Freiburg

Leprich U, Bauknecht D (2003b) Review of Current Electricity Policy and Regulation – German Study Case. Institut für Zukunftsenergiesysteme and Öko-Institut, Saarbrücken and Freiburg. www.sustelnet.net/docs/wp2/wp2-germany-sc.pdf. Last accessed: March 20, 2005

Malinowski (2003) Interview with Dr Malinowski, Projektträger Jülich at the Forschungszentrum Jülich GmbH, responsible for fuel cell federal research funding on Oct 22, 2003

Matthes FC, Cames M, Deuber O, Repenning J, Koch M, Harinsch J, Kohlhaas M, Schumacher K, Ziesing HJ (2003) Auswirkungen des europäischen

Emissionshandelssystems auf die deutsche Industrie. Öko-Institut, Deutsches Institut für Wirtschaftsforschung (DIW), Ecofys, Berlin Köln

Meixner H (2003) KWK im Spannungsfeld von Politik und Markt. Zubauchancen von BHKW unter dem KWK-G. In: Energie-Agenturen Deutschland e.V. and Bundesverband Kraft-Wärme-Kopplung e.V. (eds) Berliner Energietage 2003, Berlin

Meixner H, Stein R (2002) Blockheizkraftwerke – Ein Leitfaden für Anwender. TÜV-Verlag, Köln

Meyer-Krahmer F, Kuntze U and Walz R (1998) Innovation and Sustainable Development. Lessons for Innovation Policies? Introduction and Overview. In: Meyer-Krahmer F (ed) Innovation and Sustainable Development. Lessons for Innovation Policies. Physica, Heidelberg, pp 3-33

Mez L, Matthes FC (2002) Private Planwirtschaft. Die Übernahme von Ruhrgas durch E.on brächte keinen volkswirtschaftlichen Nutzen. In: Die Zeit 14/2002

Minett S (2004) The Role of Micro CHP in the European CHP Directive. Personal communication. Brussels

Monopolkommission (2004) 15. Hauptgutachten der Monopolkommission 2002/2003. Bundestags-Drucksache 15/3610, Berlin. dip.bundestag.de/btd/15/036/1503610.pdf. Last accessed: March 20, 2005

Mühlstein J (2003) Vermiedene Netznutzungsentgelte der dezentralen Einspeisung. Kurzgutachten. Energie & Management, Herrsching

Peisker (2003) Interview with Mr Peisker, Projektträger Jülich at the Forschungszentrum Jülich GmbH, Communication Department on Nov 25, 2003

Rip A, Misa TJ, Schot, JP (1995) Managing Technology in Society. The Approach of Constructive Technology Assessment. Pinter, London

Schlösser (2003) Interview with Ms Schlösser, Deutsche Energie Agentur, responsible for the European RTD Framework Programme on Dec 10, 2003

Simonis G (2001) Die TA-Landschaft in Deutschland – Potenziale reflexiver Techniksteuerung. In: Simonis G, Martinsen R, Saretzki T (eds) Analysen zum Verhältnis von technologischem, politischem und staatlichen Wandel am Anfang des 21.Jahrhunderts. PVS special edition 31/2000. Westdeutscher Verlag, Wiesbaden, pp 425-456

Stronzik M, Cames M (2002) Final Report for the Preparation of an Opinion on the Proposed Directive on the Implementation of EU-Wide Emissions Trading COM (2001) 581. Berlin, Mannheim www.oeko.de/oekodoc/44/2002-005-en.pdf. Last accessed: Dec 11, 2004

Telges K (2003) Interview with Dr. Telges, Transferstelle neue Produkte der Arbeitsgemeinschaft für sparsamen und umweltfreundlichen Energieverbrauch on Dec 10, 2003

Traube K (2002) Vorschlag der EU-Kommission für eine KWK-Richtlinie. Bundesverband Kraft-Wärme-Kopplung, Berlin

Traube K (2003) Bisherige Auswirkungen des KWK- Gesetzes auf den Zubau kleiner KWK-Anlagen. Bundesverband Kraft-Wärme-Kopplung, Berlin

VDEW (1996) Technische Richtlinie "Parallelbetrieb von Eigenerzeugungsanlagen mit dem Niederspannungsnetz des EVU", Frankfurt

VDEW (2004) Günstiger Strom für Industrie. Vereinigung Deutscher Elektrizitätswerke, Frankfurt. www.strom.de/wysstr/stromwys.nsf/WYSInfoDokumentePunktmk22Lookup/5DE6A3503E22956FC1256F2C00437418?OpenDocument&WYSEbene0N=Fakten&WYSEbene1N=Daten&WYSEbene2N=nachDatum&WYSEbene3N=&&WYSEbene4N=& Last accessed: Dec 11, 2004

VDN (2003) Distribution Code 2003. Regeln für den Zugang zu Verteilungsnetzen. Berlin. http://www.vdn-berlin.de/global/downloads/Publikationen/DistributionCode 2003.pdf. Last accessed: March 20, 2005

Vergragt P, Weaver P, Jansen L, van Grootveld G, van Spiegel E (2000) Sustainable Technology Development. Greenleaf Publishing, Sheffield

VKU (2004) Entwurf eines Gesetzes zur Neuregelung des Rechts der Erneuerbaren Energien im Strombereich. Verband kommunaler Unternehmen, Köln

http://www.vku.de/vku/themen/eeg/eeg_15.pdf Last accessed: Mar 15, 2005

Voß JP, Barth R, Ebinger F (2001) Institutionelle Innovationen im Bereich Energie- und Stoffströme. In: Öko-Institut (ed) Abschlussbericht zu einer Sondierungsstudie im BMBF-Förderschwerpunkt sozial-ökologische Forschung. Freiburg Darmstadt Berlin

Chapter 9: Embedding Micro Cogeneration in the Energy Supply System

Arndt U, Duschl A, Köhler D, Schwaegerl P (2004) Energiewirtschaftliche Bewertung dezentraler KWK-Systeme für die Hausenergieversorgung. Perspektiven einer Wasserstoff-Energiewirtschaft – Teil 5. Wasserstoff-Initiative Bayern (wiba), München

BCG (2003) Keeping the lights on. Navigating Choices in European Power Generation. Boston Consulting Group, Boston

Boehlert T, Bodendieck M (2004) Marktchancen in der Stromerzeugung. Die Sicht des Gasversorgers. Presented at the conference: Marktchancen in der Stromerzeugung. Handlungsoptionen für Stadtwerke und Händler, organized by Enervis, Berlin

Casten T (2003) Preventing Blackouts. Whether to Spend or Save our Way out of the Problem. Cogeneration & On-Site Power Production 4(6):24-29

CEIDS (2002) Self-Healing Grid Element program plan, March 2002. Consortium for Electric Infrastructure to Support a Digital Society. Electric Power Research Institute, Palo Alto

DER (2003) US DoE Office of Distributed Energy Resources, www.eere.energy.gov/de. Last accessed: March 15, 2005

EU (2003) Hydrogen and Fuel cells - a Vision of our Future. High level group Draft report v 4.8. European Union, Brussels

European Commission (ed) (2001) Green Paper. Towards a European Strategy for the Security of Energy Supply. Luxembourg

Fischedick M, Nitsch J (2002) Langfristszenarien für eine nachhaltige Energieversorgung. Umweltbundesamt, Research Report 20097 ,104, Wuppertal Stuttgart

Gilles V, Ocran GW, Gray R, Woodbridge D (eds) (2003) German Electricity Wholesale Market. London

Hilger C (2002) Impact of Decentralized Generation on System Operation. VGB Power Tech 6:53-56

IEA (2002) Distributed Generation in Liberalized Electricity Markets. International Energy Agency, Paris

Jarret K, Hedgecock J, Gregory R, Warham T (2004) Technical Guide to the Connexion of Generation to the Distribution Network. Department of Trade and Industry, London

Jenkins N, Allan R, Kirschen D, Strbac G (2000) Embedded Generation. Institution of Electrical Engineers (IEE), London

Jensen J (2002) Integrating CHP and Wind Power. How Western Denmark is leading the way. Cogeneration and On-Site Power Production 3(6):55-62

Kempton W, Tomic J, Letendre S, Brooks A, Lipman T (2001) Vehicle-to-Grid Power: Battery, Hybrid, and Fuel Cell Vehicles as Resources for Distributed Electric Power in California. University of Delaware, Delaware

Lauersen B (2001) District Heating in Denmark – Where Are We Now? Euroheat & Power 4:16-20

Mendez VH, Rivier J, de la Fuente JI, Gomez T, Arceluz J, Marin J (2002): Impact of Distributed Generation on Distribution Losses. Universidad Pontificia Comillas, Madrid

Mertens F (2004) Integration von Brennstoffzellen in Versorgungsstrukturen. Brennstoffzellen in der Kraft-Wärme-Kopplung. In: Krewitt W, Pehnt M, Fischedick M, Temming H (eds)Erich Schmidt Verlag, Berlin

Nast M (2003) Nahwärme - ein unverzichtbares Strukturelement. In: Fischedick M, Nitsch J (eds) Langfristszenarien für eine nachhaltige Energienutzung in Deutschland. Umweltbundesamt, Research Report 20097104, Stuttgart Wuppertal

Öko-Institut (2002) Security of Supply – A Challenge for Energy Policy in Europe. Freiburg Berlin Darmstadt

Pitz V, Brandl M, Hauptmeier M, Horenkamp W, Lehmer D, Schwan M (2003) Systemtechnische Anforderungen an elektrische Verteilnetze bei flächendeckendem Einsatz dezentraler Energieversorgungsanlagen. Presented at the ETG-Conference

Rempel H (2000) Geht die Kohlenwasserstoff-Ära zu Ende? Presentation at the DGMK/BGR event: Geowissenschaften für die Exploration und Produktion: Informationsbörse für Forschung und Industrie, Hannover. www.bgr.de/b123/kw_aera/kw_aera.htm. Last accessed: March 15, 2005

Schlemmermeier B (2003) Langfristige Erdgasverstromung. Bestehen Versorgungsrisiken? Symposium of the Gesellschaft für Energiewissenschaft und Energiepolitik (GEE): Langfristige Struktur der Stromversorgung, Berlin

Van Overbeeke F, Roberts V (2002) Active Networks as Facilitators for Embedded Generation. Cogeneration and On-Site Power Production 3(2): 37-42

Varming S, Nielson JE (2004) Review of Technical Options and Constraints for Integration of Distributed Generation in Electricity Networks. Result of the project Sustelnet (Policy and Regulatory Roadmaps for the Integration of Distributed Generation and the Development of Sustainable Electricity Networks). www.sustelnet.net/results.html Last accessed: March 15, 2005

VDEW (2001) Eigenerzeugungsanlagen am Niederspannungsnetz. Richtlinie für Anschluss und Parallelbetrieb von Eigenerzeugungsanlagen am Niederspannungsnetz. Verband der Elektrizitätswirtschaft, Frankfurt Heidelberg

VDN (2003) Transmission Code 2003. Verband der Netzbetreiber, Berlin

Chapter 11: Micro Cogeneration in North America

EIA(2000) Energy Information Administration www.eia.doe.gov/cneaf/electricity/public/t01p01p1.html. Last accessed: March 15, 2005

Flynn WM (2004) PSC Votes to Approve New Natural Gas Delivery Rates to Provide Opportunities for Self-Generation by Residential Electricity Customers. Public Service Commission, New York www3.dps.state.ny.us/pscweb/WebFileRoom.nsf/0/7BD36EA9409427AD85256ECB005AEE98/$File/pr04051.pdf?OpenElement. Last accessed: March 15, 2005

IES Industrial Extension Service. www.ies.ncsu.edu. Last accessed: March 15, 2005

Chapter 12: Micro Cogeneration in Britain

BRE (2003) The Potential for Community Heating in the UK. BRE for Carbon Trust. November 2003.

Cambridge Econometrics (2003) Modelling Good Quality CHP capacity to 2010. Revised Projections. Cambridge

DEFRA (2004) The Government's Strategy for Combined Heat and Power to 2010. Department of Environment, Food and Rural Affairs, London

Dresner S, Ekins P (2004) Climate Change and Fuel Poverty. Policy Studies Institute, London

DTI Department of Trade and Industry. www.dti.gov.uk/energy/inform/energy_prices/tables/table_221.xls. Last accessed: March 15, 2005

DTI (1999) UK energy sector indicators. Department of Trade and Industry, London

ECI (2004) The practicalities of Developing Renewable Energy Standby Capacity and Intermittency. Environmental Change Institute, Oxford

EEA (2004) News Release, Dec 21, 2004. European Environment Agency, Copenhagen

ENA (2003) Recommendations for the Connection of Small-Scale Embedded Generators (up to 16A per phase) in Parallel with Public Low-Voltage Distribution Networks. Engineering Recommendation G83/1. Electricity Networks Association, Wellington

Energy Act (2004). www.hmso.gov.uk/acts/acts2004/20040020.htm. Last accessed: March 15, 2005

Energy White Paper (2003) Our Energy Future-Creating a Low Carbon Economy. www.dti.gov.uk/energy/whitepaper/ourenergyfuture.pdf. Last accessed: March 15, 2005

Harrison J (2002) Options for Upgrading Residential CHP. Paper presented at the Conference COGEN Europe

Ilex (2002) The Carbon Saving Potential of Micro-CHP. Ilex Energy Consulting, Oxford

ILEX (2004) Carbon Displacement by Micro CHP Generation. Ilex Energy Consulting, Oxford

PIU (2002) The Energy Review. Performance and Innovation Unit, London

RCEP (2000) Energy – The changing climate. 22nd Report, June 2000. Royal Commission on Environmental Pollution, London

SIAM (2005) System Integration of Additional Microgeneration, Department of Trade and Industry, London

Chapter 13: Micro Cogeneration in Japan

Cogeneration Center (2004) Cogeneration Center of Japan. www.cgc-japan.com/english/e_top.html. Last accessed: March 15, 2005

16.2 Internet Links

This appendix offers a sample of internet links of micro cogeneration developers and vendors worldwide. Further links may be found on the information internet pages listed below.

Reciprocating Engines

Aisin	www.aisin.com
BTB	www.btb-energietechnik.de
EAW Energieanlagenbau GmbH	www.EAW-Energieanlagenbau.de
Giese Energie- und Regeltechnik	www.Energiese.de
GLIZIE	www.glizie.de

Honda	www.honda.com
Hubert Tippkötter GmbH	www.tippkoetter.de
KraftWerK GmbH	www.kwk.info
KW Energie Technik	www.kw-energietechnik.de
Öko-Energiesysteme	www.energie-as.de
Power Plus Technologies	www.ecopower.de
RK Energietechnik	www.rkenergie.de
Sanyo	www.sanyo.co.jp/cmg/e/
SenerTec	www.senertec.de
Spilling Energie Systeme	www.powertherm.de
VectorCoGen	www.vectorcogen.com
Wilhelm Schmitt GmbH	www.schmitt-mayen.de
Yanmar	www.yanmar.co.jp/english/index.htm

Stirling Engines

ENATEC Micro-cogen B.V.	www.enatec.com
Microgen	www.microgen.com
powerbloc GmbH	www.powerbloc.biz
SIG Swiss Industrial	www.sig-group.com
Sigma Elektroteknisk A.S.	www.sigma-el.com
Solo Kleinmotoren GmbH	www.stirling-engine.de
Stirling Technology Company	www.stirlingtech.com
Stirling Technology, Inc.	www.stirling-tech.com
Sunmachine	www.sunmachine.de
Sunpower, Inc.	www.sunpower.com
Tamin Enterprises	www.tamin.com
Whisper Tech Ltd.	www.whispertech.co.nz www.whispergen.com

Fuel Cells

Avista Labs	www.avistalabs.com
Ballard Power Systems	www.ballard.com
Caterpillar	www.cat.com

Dais Analytic	www.daisanalytic.com
DCH Technology	www.dcht.com
European fuel cell GmbH	www.europeanfuelcell.de
FuelCell Energy, Inc.	www.fce.com
GE MicroGeneration	www.gemicrogen.com
H Power Corp.	www.hpower.com
IdaTech Northwest Power Systems	www.idatech.com
International Fuel Cells United Technologies	www.internationalfuelcells.com
Kyocera	global.kyocera.com
Matsushita Electric Industry	www.mei.co.jp
Mitsubishi Materials	www.mmc.co.jp/english
Plug Power	www.plugpower.com
Proton Energy Systems	www.protonenergy.com
Sanyo	www.sanyo.co.jp
Siemens Westinghouse	www.spcf.siemens.com
Sulzer Hexis	www.hexis.com
Sunmachine	www.sunmachine.de
Sure Power	www.hi-availability.com
Toshiba	www.toshiba.co.jp
TOTO	www.toto.co.jp
Vaillant	www.vaillant.com

Further Technologies

Cogen Micro	www.cogenmicro.com.au
Energetix	www.energetixgroup.co.uk
Enginion	www.enginion.com
OTAG	www.otag.de

Information Links to International Websites

Association for the Promotion of Cogeneration Association Technique	www.cogen.org

Energie Environnement	www.atee.fr
Baxi Group	www.baxi.com
CHP Club	www.chpclub.com
COGEN Canada	www.cogencanada.ca
Cogen España	www.cogenspain.org
COGEN European Distributed Generation	www.distributed-generation.com
COGEN Nederland	www.cogen.nl
COGEN Vlaandern	www.cogenvlaanderen.be
Combined Heat and Power Association	www.chpa.co.uk
Delta Energy and Environment	www.delta-ee.com
Energetix	www.energetixgroup.com
Euroheat & Power	www.euroheat.org
Japan Cogeneration Center	www.cgc-japan.com
MicroCHAP	www.microchap.info
Micropower	www.micropower.co.uk
Midwest Cogeneration Association	www.cogeneration.org
Österreichischer Energiekonsumenten Verband	www.oekv-energy.at
SBGI	www.sbgi.org.uk
Schweizerischer Verband für Wärmekraftkopplung	www.waermekraftkopplung.ch
Swedish District Heating Association	www.fjarrvarme.org
US CHP Association	uschpa.admgt.com
World Alliance for Decentralized Energy	www.localpower.org

Information Links to German-Based Websites

ASUE Arbeitsgemeinschaft für sparsamen und umweltfreundlichen Energieverbrauch	www.asue.de

BHKW - Information on small cogeneration units	www.bhkw.de www.bhkw-info.de www.minibhkw.de
BHKW Infozentrum	www.bhkw-infozentrum.de
Bine Informationsdienst	www.bine.info
B.KWK Bundesverband Kraft-Wärme-Kopplung	www.bkwk.de
HessenEnergie	www.hessenenergie.net

16.3 The Authors

Martin Cames

Martin Cames, born in 1959, is a senior research fellow at the Öko-Institut – Institute for applied ecology at the Berlin office. After a vocational training and working as locksmith and later as master craftsman he was trained as an economist. In 1993 he finished his studies at the Free University Berlin with a diploma in economics (Diplom-Volkswirt). He has more than 10 years experience in research and policy consultancy relating to energy & climate change issues, such as energy policy and employment, environmental instruments (emissions trading, green tradable certificates, eco tax etc.), and national & international climate change policy. Since 2003 he is working on his PhD which is addressing innovation incentives to the electricity industry induced by the introduction of the European emissions trading scheme.

Martin Cames
Öko-Institut – Institute for Applied Ecology
Novalisstr. 10
10115 Berlin
Germany
Phone ++49-(0)30-280 486-83
m.cames@oeko.de www.oeko.de

Michael Colijn

Michael Colijn, born in 1973, is Director of Michael Colijn Ltd. The company is specialized in project development in the area of energy, economics and environment. Michael is both lobbyist in European regulation and liberalization for micro generation, and business developer on behalf of several micro-generation companies. He also sits on the national electro-standards committee in the Netherlands. Prior to starting his own company in 2005, Michael worked for BG Group as European Regulations Manager for its microCHP subsidiary Microgen, and for Shell as global Marketing Manager in solar energy. Michael has an MSc in Environmental Technology, specialized in Energy Policy, from Imperial College London, and a BSc Physics with Environmental Studies from the University of London, Royal Holloway College. He is also writing his PhD in financial engineering for decentralized power at the Technical University of Delft.

Michael Colijn
MICHAEL COLIJN LTD (UK)
Energy - Economics - Environment
Vrijheidslaan 17-I
1079 KB Amsterdam
Netherlands
Phone ++31-(0)20-644 4077
michael@michaelcolijn.com www.michaelcolijn.com

Dr. Corinna Fischer

Dr. Corinna Fischer, born in 1971, works as a researcher at the Environmental Policy Research Center, Free University of Berlin. She studied political science and psychology at the Philipps-Universität Marburg and Free University of Berlin. Her PhD thesis, completed at the Institute for Socialization Research and Empirical Social Sciences, Technical University of Chemnitz, in spring 2001, deals with the motivation of East German youth to participate in an environmental organization. Corinna worked for the Forum Environment and Health, Deutsches Hygiene-Museum Dresden, and for the Enquete Commission "Sustainable Berlin" at the Berlin city parliament. Her main research interests are environmental behavior, sustainable consumption, environmental movements and political participation. Corinna also works in political education and as a trainer and counselor for environmental organizations.

Dr. Corinna Fischer
Forschungsstelle für Umweltpolitik – Environmental Policy Research Center
Freie Universität Berlin
Ihnestr. 22

14195 Berlin
Germany
Phone ++49-(0)30-838 54494
cofiffu@zedat.fu-berlin.de www.fu-berlin.de/ffu

Jeremy Harrison

Jeremy Harrison, born 1955, graduated BSC from Edinburgh University in 1979. His early career was spent designing low energy housing construction systems. He later spent three years as Export Manager for ABB Fläkt in Enköping, Sweden, responsible for development of low energy, domestic indoor climate systems. From 1993-2003, he worked with EA Technology, the UK energy consultancy, leading a number of micro CHP evaluation and trial programmes, as well as work on the impact of DG on distribution networks. During this period he also spent two years with the micro CHP developer, Sigma in Norway.
He is now Technology Development Manager with Powergen, the UK energy supply subsidiary of E.ON plc, where he is responsible for a number of novel domestic energy technologies including micro CHP. He also maintains a personal interest in micro CHP, running a public access information site: www.microchap.info

Jeremy Harrison
Powergen Retail Limited
Wyvern House
Colliers Way
Nottingham NG6 8AT
United Kingdom
Phone ++44 870 419 1673
jeremy.harrison@powergen.co.uk

Dr. Martin Pehnt

Dr. Martin Pehnt, born 1970, studied Physics and Energy Management in Tübingen, Stuttgart, Berlin and Boulder (USA) and obtained a Masters Degree in 1996 and a PhD in Energy Technology in 2002. Research period at the National Renewable Energy Laboratory NREL (Golden, Colorado). 1997-2001 Scientist at the German Aerospace Center, Institute for Technical Thermodynamics, Department for Systems Analysis and Technology Assessment. Since 2001 Senior Scientist at the Institut für Energie- und Umweltforschung Heidelberg IFEU. His research focuses on technology and environmental assessment of innovative energy and transport systems, such as fuel cells, cogeneration, renewable energies, bio fuels; environmental communication.

Dr. Martin Pehnt
Institute for Energy and Environmental Research IFEU
Wilckensstr. 3
69120 Heidelberg
Germany
Phone ++49-(0)6221-4767-36
martin.pehnt@ifeu.de www.ifeu.de

Dr. Barbara Praetorius

Barbara Praetorius, born 1964 in Berlin, is senior research associate at DIW Berlin, Department of Energy, Transportation and Environment and leads the interdisciplinary research project "Transformation and Innovation in Power Systems" (TIPS). She studied Economics and Political Science in Frankfurt (Main) and Aix/Marseille, France and obtained her Masters degree (Diplom) in Economics in 1990 and a Licence és des Sciences Economiques in 1987. Her Ph.D on the reform of the electricity sector in South Africa was completed in February 2000 at Free University Berlin. 1990-1992 research assistant at Future Energies Forum, Bonn. Visiting scholar at Lawrence Berkeley National Laboratory (California, USA) in 1995. Several research periods at the Energy and Development Research Centre (Cape Town, South Africa) 1996-1998. Her research focuses on the assessment and concepts of energy and climate policy instruments, electricity markets and regulation, innovation, distributed generation, and energy sector reform in South Africa.

Dr. Barbara Praetorius
DIW Berlin (German Institute for Economic Research)
Königin-Luise-Str. 5
14195 Berlin
Germany
Phone ++49-(0)30-89789-676
bpraetorius@diw.de www.diw.de

Yasushi Santo

Yasushi Santo is Visiting Professor of the Kwansei Gakuin University, Graduate School of Policy Study, and Board Director of Green Energy.com. He entered the Osaka Gas Co., Ltd. after graduating from the

University of Tokyo, Law Department, in 1961 and worked there until 2001. During the period, he studied at the University of Michigan, Graduate School of Business Administration from 1966 to 67 as a Fulbright Scholar. He worked as project manager for the field-test of a 40 kW on-site Phosphoric Acid fuel cell unit early 1980's. His last job at Osaka Gas was to lead the Research Institute for Culture, Energy and Life. He was born in Osaka in 1938 and now lives in Nara, Japan.

Yasushi Santo
Nishi-tomigaoka 8-8-4
Nara 631-0006
Japan
santo@kcn.ne.jp
Phone ++81 (0)742-44 9423

Lambert Schneider

Lambert Schneider, born in 1971, studied energetic, chemical and environmental engineering and environmental economics at the Technical University of Berlin. After a masters degree in 1999 he worked for one year at Berliner Energieagentur, an energy service company. Since 2000 he is researcher in the energy & climate division at Öko-Institut (Institute for Applied Ecology) and focuses on international climate policy and power technologies. Ongoing PhD on the role of CHP in the future electricity system.

Lambert Schneider
Öko-Institut – Institute for Applied Ecology
Novalisstr. 10
10115 Berlin
Germany
Phone ++49-(0)30-280486 74
l.schneider@oeko.de www.oeko.de

Katja Schumacher

Katja Schumacher, born 1968, studied at the University of Bonn and the University of California at Berkeley and received her graduate degree in economics from the University of Bonn in 1996. From 1996 to 2000, Katja Schumacher worked as a senior research associate in the Energy Analysis Department at Lawrence Berkeley National Laboratory. In October 2000 Katja Schumacher joined the Energy, Transport and Environment

Department at DIW. Katja Schumacher's main fields of research include national and global climate change policy, in particular the design and implementation of flexible mechanisms, as well as the estimation of costs and benefits of climate change mitigation using energy-economy models. She currently works on her dissertation on modelling the effects of innovation and intervention in the German electric Power system within the interdisciplinary research project "Transformation and Innovation in Power Systems" (TIPS).

Katja Schumacher
DIW Berlin (German Institute for Economic Research)
Königin-Luise Str. 5
14195 Berlin
Germany
Phone: +49-(0)30-89789675
kschumacher@diw.de www.diw.de

Jon Slowe

Jon Slowe, born in 1969, is a director of Delta Energy and Environment, a specialized decentralized energy research and consultancy company based in the UK and Belgium and serving clients around the world. At Delta, Jon operates the E Source Micro-CHP Service on behalf of Platts. Previously, Jon managed Platts research into distributed energy, carrying out European and global research on distributed energy (DE) topics such as cogeneration markets, utility DE strategies, and emerging DE technologies. Before joining Platts, Jon worked for two and a half years with the UK Energy Saving Trust, managing their energy services and cogeneration programs. Jon has an MSc with distinction in Environmental Technology from Imperial College, London, and a First Class BSc (Hons) in Physics from Warwick University.

Jon Slowe
Delta Energy and Environment
4th Floor, 111 Union Street
Glasgow G1 3TA
United Kingdom
Phone ++44-(0)141-227-3982
jon.slowe@delta-ee.com www.delta-ee.com

Jan-Peter Voß

Jan-Peter Voß, born 1972. He studied Political Science and Economics at Freie Universität Berlin and London School of Economics and Political Science (LSE) and obtained a Masters Degree in 1999 in Political Science. Research internships with the government of Peru, 1996, and the UN Climate Change Secretariat, 1997. Since 1999 Research fellow and Project Leader at the Öko-Institut - Institute for Applied Ecology in Freiburg, since 2003 in Berlin. Ongoing PhD on governance innovation in energy systems at University of Twente (NL). His research focuses on energy policy, network industries, socio-ecological transformation and sustainability assessment, governance and innovation theory, methods of transdisciplinary knowledge production.

Jan-Peter Voß
Öko-Institut – Institute for Applied Ecology
Novalisstr. 10
10115 Berlin
Germany
Phone ++49-(0)30-280486-62
j.voss@oeko.de www.oeko.de www.sustainable-transformation.net

Sylvia Westermann

Sylvia Westermann, born 1961 in Lutherstadt Wittenberg, studied Civil Engineering at the Technical University of Weimar and Energy Management in Berlin. From 1993 to 1997 she worked as a project engineer for a micro CHP manufacturer and as a freelance energy consultant. She is currently working at the engineering office ITG Planungs- und Energieberatungs GmbH in Schönebeck/Elbe as well as teaching at the Handwerkskammer in Magdeburg.

Sylvia Westermann
Goethestraße 8
39167 Niederndodeleben
Germany
Phone ++49-(0)39204-5794
s.westermann@t-online.de
www.enerwest.de

16.4 Abbreviations

AC	Alternating Current
ASUE	Arbeitsgemeinschaft für sparsamen und umweltfreundlichen Energieverbrauch (Working Group for the Economic and Environmentally Friendly Use of Energy), Germany
BAFA	Bundesanstalt für Außenhandel (Federal Agency for Economy and Export Control), Germany
BCG	Boston Consulting Group
BDI	Bundesverband der deutschen Industrie (Federation of German Industries), Germany
BEB	BEB Erdgas GmbH (a natural gas transport and storage company), Germany
BET	Büro für Energiewirtschaft und Technische Planung (a consulting firm in the energy industry), Germany
BGR	Bundesanstalt für Geowissenschaften und Rohstoffe, Germany
BGW	Bundesverband der deutschen Gas- und Wasserwirtschaft (Federal Association of the German Gas and Water Business), Germany
BHKW	Blockheizkraftwerk (small combined heat and power plant)
BKWK	Bundesverband Kraft-Wärme-Kopplung (Federal Association for Combined Heat and Power), Germany
BMBF	Bundesministerium für Bildung und Forschung (German Federal Ministry for Education and Research)
BMU	Bundesministerium für Umwelt, Naturschutz und Reaktorsicherheit (German Federal Ministry for the Environment, Nature Conservation and Nuclear Safety)
BMWA	Bundesministerium für Wirtschaft und Arbeit (German Federal Ministry for Economics and Labour)

BNE	Bundesverband Neuer Energieanbieter (Federal Association of New Energy Suppliers), Germany
Btu	British Thermal Units
CC	Combined Cycle
CEIDS	Consortium for Electric Infrastructure to Support a Digital Society, U.S.A.
CH_4	Methane
CHP	Combined Heat and Power
CHPA	Combined Heat & Power Association
CHPQA	Quality Assurance for Combined Heat and Power, UK
CML	Centrum voor Milieuwetenschappen Leiden (Institute for Environmental Sciences Leiden), Netherlands
CO	Carbon monoxide
CO_2	Carbon dioxide
COSPP	Cogeneration and On-Site Power Production
COGEN Europe	European Association for the Promotion of Cogeneration
COP	Conference of the Parties (to the United Nations Framework Convention on Climate Change)
CSTM	Centrum voor Schone Technologie en Mileubeleid (Centre for Clean Technology and Environmental Policy), Netherlands
DBU	Deutsche Bundesstiftung Umwelt (Federal German Environmental Foundation)
DC	Direct Current
DEFRA	Department of Environment, Food and Rural Affairs
DENA	Deutsche Energie Agentur (German Energy Agency)
DER	Distributed Energy Resources
DG	Distributed Generation
DGMK	Deutsche Wissenschaftliche Gesellschaft für Erdöl, Erdgas und Kohle (German Society for Petroleum and Coal Science and Technology)
DH	District Heating
DNO	Distribution Network Operator
DIW	Deutsches Institut für Wirtschaftsforschung (German Institute for Economic Research)

DLR	Deutsches Zentrum für Luft- und Raumfahrt (German Aerospace Center)
DoE	Department of Energy, USA
DSIRE	Database of State Incentives for Renewable Energy
DTI	Department of Trade and Industry, UK
EC	European Commission
ECEEE	European Council for an Energy-Efficient Economy
ECN	Energieonderzoek Centrum Nederland
ECI	Environmental Change Institute
ECIS	Eindhoven Centre for Innovation Studies, Netherlands
ECO	Energy Co-Opportunity
EEA	European Environment Agency, Denmark
EEC	Energy Efficiency Commitment
EEW	Energieverband Elbe-Weser Beteiligungsholding, Germany
EEX	European Energy Exchange
EIA	Energy Information Administration
EnBW	Energie Baden-Württemberg (Energy Baden Württemberg), Germany
EnEV	Energieeinsparverordnung (Energy Conservation Decree)
EnWG	Energiewirtschaftsgesetz (Energy Industry Law)
EPRI	Electric Power Research Institute
ESCO	Energy Contracting and Service Companies
EST	Energy Saving Trust, UK
ETC-ACC	European Topic Centre on Air and Climate Change
ETR	Ecological Tax Reform
EU	European Union
EurEnDel	European Energy Delphi
EWE	Energieversorgung Weser-Ems AG (Energy Supply Weser-Ems), Germany
FHG-ISI	Fraunhofer-Institut für Systemtechnik und Innovationsforschung, Germany
FLOX	Flameless oxidation
FY	Fiscal Year
GbR	Gesellschaft des bürgerlichen Rechtes (non-trading partnership)

GEE	Gesellschaft für Energiewissenschaft und Energiepolitik (Society for Energy Science and Energy Policy), Germany
GHD	Gewerbe, Handel, Dienstleistungen (Industry, Trade, Services)
GHG	Greenhouse Gas
GmbH	Gesellschaft mit beschränkter Haftung (Limited Liability Company)
GUS	Gemeinschaft unabhängiger Staaten (CIS, Commonwealth of Independent States)
GW	Gigawatt
H_2	Hydrogen
HH	Half-Hourly
HHI	Herfindahl-Hirschman Index
HP	Heat pump
HV	High voltage
ICT	Information and communication technology
IEA	International Energy Agency
IEE	Institution of Electrical Engineers
IE	Institut für Energetik und Umwelt (Institute for Energy and Environment), Germany
IEU	Initiativkreis Erdgas und Umwelt
IFEU	Institut für Energie- und Umweltforschung (Institute for Energy and Environmental Research), Germany
IGU	International Gas Union
IIASA	International Institute for Applied Systems Analysis, Austria
IMV	Integrierte Mikrosysteme der Versorgung (Integrated Micro-Systems of Supply, a German research project)
ISDN	Integrated Services Digital Network
ISO	International Organization for Standardization
ITAS	Institut für Technikfolgenabschätzung und Systemanalyse (Institute for Technology Assessment and Systems Analysis), Germany
IZES	Institut für ZukunftsEnergieSysteme (Institute for Future Energy Systems), Germany
KfW	Kreditanstalt für Wiederaufbau (KfW banking group), Germany
KUT test	Kompetenz im Umgang mit Technik-Test

	(test measuring subjective technical competence)
KWK-G	Kraft-Wärme-Kopplungs-Gesetz (Combined Heat and Power Law)
kVA	kilo Volt Ampère
LCA	Life Cycle Assessment
LNG	Liquefied natural gas
LV	Low voltage
LPG	Liquefied petroleum gas
MV	Medium voltage
MOP	Meter operator
NEC	National emission ceilings
NETA	New Electricity Trading Arrangements
NH_3	Ammonia
NHH	Non-half-hourly
NMVOC	Non methane volatile organic compounds
NO_x	Nitrogen oxides
N_2O	Nitrous oxide
NPV	Net Present Value
NREL	National Renewable Energy Laboratory
NRW	Nordrhein-Westfalen (North Rhine Westphalia)
nTPA	Negotiated third party access
O_2	Oxygen
ODPM	Office of the Deputy Prime Minister, UK
OECD	Organization for Economic Cooperation and Development
OFGEM	Office of Gas and Electricity Markets, UK
O&M	Operation and maintenance
PAFC	Phosphoric acid fuel cell
PEFC	Polymer electrolyte fuel cell
PEMFC	Proton exchange membrane fuel cell
PIU	Performance and Innovation Unit
PSC	Public Service Commission
PV	Photovoltaics
QUANGO	Quasi non-governmental organization
RCEP	Royal Commission on Environmental Pollution
R&D	Research and Developement
RegTP	Regulierungsbehörde für Telekommunikation und Post (German Regulatory Authority for Telecommunications and Posts, future regulation unit for electricity)
RME	Rapeseed methyl ester

RPI-X%	Retail price index
RWE AG	Rheinisch-Westfälische Elektrizitätswerks Aktiengesellschaft (Rhenanian-Westphalian Electricity Untility AG), Germany
SBGI	Society of British Gas Industries
SECA	Solid State Energy Conversion Alliance
SIAM	System integration of additional microgeneration
SO_2	Sulphur dioxide
SOFC	Solid Oxide Fuel Cells
SPRU	Science and Technology Policy Research, UK
ST	Solarthermal collector
TPV	Thermophotovoltaics
TSO	Transmission system operator
TUM	Technische Universität München (Technical University of Munich), Germany
TUoS	Transmission use of system
UBA	Umweltbundesamt (Federal Environmental Agency of Germany)
UCTE	Union for the Coordination of Transmission of Electricity
UK	United Kingdom
UMTS	Universal Mobile Telecommunications System
USA	United States of America
UBS	Swiss Bank
VAT	Value added tax
VDEW	Verband der Elektizitätswirtschaft (German Electricity Association)
VDI	Verein Deutscher Ingenieure (Association of German Engineers)
VDN	Verband der Netzbetreiber (Association of German Network Operators)
VGB	Verband der Großkessel-Besitzer (Federation of Large Boiler Owners), Germany
VKU	Verband kommunaler Unternehmen (Association of Municipal Suppliers), Germany
VNG	Verbundnetz Gas
VVII+	Verbändevereinbarung II plus (Industry Association's Agreement on grid use charges and grid use), Germany
WADE	World Alliance for Decentralized Energy

WBZU	Weiterbildungszentrum Brennstoffzelle Ulm (Fuel cell education and training center Ulm)
WEE	Weser-Ems-Energiebeteiligungen (Weser-Ems Energy Holdings), Germany
WI	Wuppertal Institut für Klima, Umwelt, Energie (Wuppertal Institute for Climate, Environment, Energy)
wiba	Wasserstoff-Initiative Bayern
ZSW	Zentrum für Sonnenenergie- und Wasserstoff-Forschung (Centre for Solar Energy and Hydrogen Research), Germany

Printing: Krips bv, Meppel
Binding: Stürtz, Würzburg